Femtosecond Laser Pulses

Claude Rullière (Ed.)

Femtosecond Laser Pulses

Principles and Experiments

Second Edition

With 296 Figures, Including 3 Color Plates,
and Numerous Experiments

 Springer

Professor Dr. Claude Rullière
Centre de Physique Moleéculaire Optique et Hertzienne (CPMOH)
Université Bordeaux 1
351, cours de la Libération
33405 TALENCE CEDEX, France

and

Commissariat á l'Energie Aromique (CEA)
Centre d'Etudes Scientifiques et Techniques d'Aquitarine
BP 2
33114 LE BARP, France

Library of Congress Cataloging-in-Publication Data
Femtosecond laser pulses: principles and experiments / Claude Rullière, (ed.). – [2nd ed.].
    p. cm. – (Advanced texts in physics, ISSN 1439-2674)
    Includes bibliographical references and index.
    ISBN 3-387-01769-0 (acid-free paper)
    1. Laser pulses, Ultrashort. 2. Nonlinear optics. I. Rullière, Claude, 1947–   . II. Series.
QC689.5L37F46   2003
621.36'6—dc22   2003062207

ISBN 0-387-01769-0   Printed on acid-free paper.

9 8 7 6 5 4 3 2 1   SPIN 10925751

springeronline.com

Preface

This is the second edition of this advanced textbook written for scientists who require further training in femtosecond science. Four years after publication of the first edition, femtosecond science has overcome new challenges and new application fields have become mature. It is necessary to take into account these new developments. Two main topics merged during this period that support important scientific activities: attosecond pulses are now generated in the X-UV spectral domain, and coherent control of chemical events is now possible by tailoring the shape of femtosecond pulses. To update this advanced textbook, it was necessary to introduce these fields; two new chapters are in this second edition: "Coherent Control in Atoms, Molecules, and Solids" (Chap. 11) and "Attosecond Pulses" (Chap. 12) with well-documented references.

Some changes, addenda, and new references are introduced in the first edition's ten original chapters to take into account new developments and update this advanced textbook which is the result of a scientific adventure that started in 1991. At that time, the French Ministry of Education decided that, in view of the growing importance of ultrashort laser pulses for the national scientific community, a Femtosecond Centre should be created in France and devoted to the further education of scientists who use femtosecond pulses as a research tool and who are not specialists in lasers or even in optics.

After proposals from different institutions, Université Bordeaux I and our laboratory were finally selected to ensure the success of this new centre. Since the scientists involved were located throughout France, it was decided that the training courses should be concentrated into a short period of at least 5 days. It is certainly a challenge to give a good grounding in the science of femtosecond pulses in such a short period to scientists who do not necessarily have the required scientific background and are in some cases involved only as users of these pulses as a tool. To start, we contacted well-known specialists from the French femtosecond community; we are very thankful that they showed enthusiasm and immediately started work on this fascinating project.

Our adventure began in 1992 and each year since, generally in spring, we have organized a one-week femtosecond training course at the Bordeaux University. Each morning of the course is devoted to theoretical lectures concerning different aspects of femtosecond pulses; the afternoons are spent in the laboratory, where a very simple experimental demonstration illustrates each point developed in the morning lectures. At the end of the afternoon, the saturation threshold of the attendees is generally reached, so the evenings are devoted to discovering Bordeaux wines and vineyards, which helps the otherwise shy attendees enter into discussions concerning femtosecond science.

A document including all the lectures is always distributed to the participants. Step by step this document has been improved as a result of feedback from the attendees and lecturers, who were forced to find pedagogic answers to the many questions arising during the courses. The result is a very comprehensive textbook that we decided to make available to the wider scientific community; i.e., the result is this book.

The people who will gain the most from this book are the scientists (graduate students, engineers, researchers) who are not necessarily trained as laser scientists but who want to use femtosecond pulses and/or gain a real understanding of this tool. Laser specialists will also find the book useful, particularly if they have to teach the subject to graduate or PhD students. For every reader, this book provides a simple progressive and pedagogic approach to this field. It is particularly enhanced by the descriptions of basic experiments or exercises that can be used for further study or practice.

The first chapter simply recalls the basic laser principles necessary to understand the generation process of ultrashort pulses. The second chapter is a brief introduction to the basics behind the experimental problems generated by ultrashort laser pulses when they travel through different optical devices or samples. Chapter 3 describes how ultrashort pulses are generated independently of the laser medium. In Chaps. 4 and 5 the main laser sources used to generate ultrashort laser pulses and their characteristics are described. Chapter 6 presents the different methods currently used to characterize these pulses, and Chap. 7 describes how to change these characteristics (pulse duration, amplification, wavelength tuning, etc.). The rest of the book is devoted to applications, essentially the different experimental methods based on the use of ultrashort laser pulses. Chapter 8 describes the principal spectroscopic methods, presenting some typical results, and Chap. 9 addresses mainly the problems that may arise when the pulse duration is as short as the coherence time of the sample being studied. Chapter 10 describes typical applications of ultrashort laser pulses for the characterisation of electronic devices and the electromagnetic pulses generated at low frequency. Chapter 11 is an overview of the coherent control physical processes making it possible to control evolution channels in atoms, molecules and solids. Several examples of oriented reactions in this chapter illustrate the possible applications of such a technique. Chapter 12 introduces the attosecond pulse generation by femtosecond pulse-matter interaction. It is designed for a best understanding of the physics principles sustaining attosecond pulse creation as well as the encountered difficulties in such processes.

I would like to acknowledge all persons and companies whose names do not directly appear in this book but whose participation has been essential to the final goal of this adventure. My colleague Gediminas Jonusauskas was greatly involved in the design of the experiments presented during the courses and at the end of the chapters in this book. Danièle Hulin, Jean-René Lalanne and Arnold Migus gave much time during the initial stages, particularly in writing the first version of the course document. The publication of this book would not have been possible without their important support and contribution. My colleagues Eric Freysz, François Dupuy, Frederic Adamietz and Patricia Segonds also participated in the organization of the courses, as did the post-doc and PhD students Anatoli Ivanov, Corinne Rajchenbach, Emmanuel Abraham, Bruno Chassagne and Benoit Lourdelet.

Essential financial support and participation in the courses, particularly by the loan of equipment, came from the following laser or optics companies: B.M. Industries, Coherent France, Hamamatsu France, A.R.P. Photonetics, Spectra-Physics France, Optilas, Continuum France, Princeton Instruments SA and Quantel France.

I hope that every reader will enjoy reading this book. The best result would be if they conclude that femtosecond pulses are wonderful tools for scientific investigation and want to use them and know more.

Bordeaux, April 2004   *Claude Rullière*

Contents

9 Coherent Effects in Femtosecond Spectroscopy: A Simple Picture Using the Bloch Equation

10 Terahertz Femtosecond Pulses

Contributors

T. Amand
Laboratoire de Physique de la Matière
Condensée, INSA/CNRS,
Complexe Scientifique de Rangueil,
F-31077 Toulouse Cedex 4, France

V. Blanchet
Laboratoire Collisions Agrégats Réactivité
CNRS UMR 5589
Université Paul Sabatier
118 Route de Narbonne
31062 Toulouse Cedex, France

A. Bonvalet
Laboratoire d'Optique et Biosciences (LOB)
CNRS UMR 7645 – INSERM U541-X-
ENSTA
Ecole Polytechnique
91128 Palaiseau Cedex (France)

E. Constant
Centre Lasers Intenses at Applications
(CELIA)
UMR 5107 (Université Bordeaux I-CNRS-
CEA)
351 cours de la Libération
33405 Talence Cedex, France

B. Couillaud
Coherent, 5100 Patrick Henry Drive,
Santa Clara, CA 95054, USA

A. Ducasse
Centre de Physique Moléculaire
Optique et Hertzienne, Université
Bordeaux I, 351 cours de la Libération,
F-33405 Talence Cedex, France

B. Girard
Laboratoire Collisions Agrégats Réactivité
CNRS UMR 5589
Université Paul Sabatier
118 Route de Narbonne
31062 Toulouse Cedex, France

C. Hirlimann
Institut de Physique et Chimie
des Matériaux de Strasbourg (IPCMS),
UMR7504 CNRS-ULP-ECPM,
23 rue du Loess, BP 43
F-67034 Strasbourg Cedex2, France
ch@valholl.u-strasbg.fr

M. Joffre
Laboratoire d'Optique et Biosciences (LOB)
CNRS UMR7645 – INSERM U541-X-
ENSTA
Ecole Polytechnique
91128 Palaiseau Cedex (France)
manuel.joffre@polytechnique.fr

X. Marie
Laboratoire de Physique de la
Matière Condensée, INSA/CNRS,
Complexe Scientifique de Rangueil
F-31077 Toulouse Cedex 4, France

E. Mével
Centre Lasers Intenses at Applications
(CELIA)
UMR 5107 (Université Bordeaux I-CNRS-
CEA)
351 cours de la Libération
33405 Talence Cedex, France

J. Oberlé
Centre de Physique Moléculaire
Optique et Hertzienne (CPMOH),
UMR5798 (CNRS-Université Bordeaux I)
351 Cours de la Libération,
F-33405 Talence, France
oberle@cpmoh.u-bordeaux.fr

C. Rullière
Centre de Physique Moléculaire
Optique et Hertzienne (CPMOH)
UMR5798 (CNRS-Université Bordeaux I)
351 cours de la Libération,
F-33405 Talence Cedex, France
rulliere@cribx1.u-bordeaux.fr
and
Commissariat á l'Energie Atomique (CEA)
CESTA BPNo2
33114-Le Barp (FRANCE)
claude.rulliere@cea.fr

F. Salin
Centre Lasers Intenses et Applications
(CELIA)
UMR 5107 (Université Bordeaux I-CNRS-
CEA)
351 cours de la Libération
33405 Talence Cedex, France
salin@celia.u-bordeaux.fr

L. Sarger
Centre de Physique Moléculaire
Optique et Hertzienne (CPMOH),
UMR5798 (CNRS-Université Bordeaux I)
351 Cours de la Libération,
F-33405 Talence, France
sarger@cpmoh.u-bordeaux.fr

1

Laser Basics

C. Hirlimann

With 18 Figures

1.1 Introduction

Lasers are the basic building block of the technologies for the generation of short light pulses. Only two decades after the laser had been invented, the duration of the shortest produced pulse had shrunk down six orders of magnitude, going from the nanosecond regime to the femtosecond regime. "Light amplification by stimulated emission of radiation" is the misleading meaning of the word "laser". The real instrument is not only an amplifier but also a resonant optical cavity implementing a positive feedback between the emitted light and the amplifying medium. A laser also needs to be fed with energy of some sort.

1.2 Stimulated Emission

Max Planck, in 1900, found a theoretical derivation for the experimentally observed frequency distribution of black-body radiation. In a very simplified view, a black body is the thermal equilibrium between matter and light at a given temperature. For this purpose Planck had to divide the phase space associated with the black body into small, finite volumes. Quanta were born. The distribution law he found can be written as

$$I(\omega)\,\mathrm{d}\omega = \frac{\hbar\omega^3\,\mathrm{d}\omega}{\pi^2 c^2(\mathrm{e}^{\hbar\omega/kT}-1)},\tag{1.1}$$

where $I(\omega)$ stands for the intensity of the angular frequency distribution in the small interval $\mathrm{d}\omega$, $\hbar = h/2\pi$, h is a constant factor which was later named after Planck, k is Boltzmann's constant, T is the equilibrium temperature and c the velocity of light in vacuum. Planck first considered his findings as a heretical mathematical trick giving the right answer; it took him sometime to realize that quantization has a physical meaning.

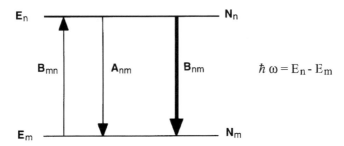

Fig. 1.1. Energy diagram of an atomic two-level system. Energies E_m and E_n are measured with reference to some lowest level

In 1905, Albert Einstein, though, had to postulate the quantization of electromagnetic energy in order to give the first interpretation of the photo-electric effect. This step had him wondering for a long time about the compatibility of this quantization and Planck's black-body theory. Things started to clarify in 1913 when Bohr published his atomic model, in which electrons are constrained to stay on fixed energy levels and may exchange only energy quanta with the outside world. Let us consider (see Figure 1.1) two electronic levels n and m in an atom, with energies E_m and E_n referenced to some fundamental level; one quantum of light, called a photon, with energy $\hbar\omega = E_n - E_m$, is absorbed with a probability B_{mn} and its energy is transferred to an electron jumping from level m to level n. There is a probability A_{nm} that an electron on level n steps down to level m, emitting a photon with the same energy. This spontaneous light emission is analogous to the general spontaneous energy decay found in classical mechanical systems. In the year 1917, ending his thinking on black-body radiation, Einstein came out with the postulate that, for an excited state, there should be another de-excitation channel with probability B_{nm}: the "induced" or "stimulated" emission. This new emission process only occurs when an electromagnetic field $\hbar\omega$ is present in the vicinity of the atom and it is proportional to the intensity of the field. The quantities A_{nm}, B_{nm}, B_{mn} are called Einstein's coefficients.

Let us now consider a set of N atoms, of which N_m are in state m and N_n in state n, and assume that this set is illuminated with a light wave of angular frequency ω such that $\hbar\omega = E_n - E_m$, with intensity $I(\omega)$. At a given temperature T, in a steady-state regime, the number of absorbed photons equals the number of emitted photons (equilibrium situation of a black body). The number of absorbed photons per unit time is proportional to the transition probability B_{mn} for an electron to jump from state m to state n, to the incident intensity $I(\omega)$ and to the number of atoms in the set N_m. A simple inversion of the role played by the indices m and n gives the number of electrons per unit time relaxing from state n to state m by emitting a photon under the influence of the electromagnetic field. The last contribution to the interaction, spontaneous emission, does not depend on the intensity but only on the number of electrons in state n and on the transition probability A_{mn}. This can be simply formalized in a simple energy conservation equation

$$N_m B_{mn} I(\omega) = N_n B_{nm} I(\omega) + N_n A_{nm}. \tag{1.2}$$

Boltzmann's law, deduced from the statistical analysis of gases, gives the relative populations on two levels separated by an energy $\hbar\omega$ at temperature T, $N_n/N_m = \exp(-\hbar\omega/kT)$. When applied to (1.2) one gets

$$B_{mn} I(\omega)\, e^{\hbar\omega/kT} = A_{nm} + B_{nm} I(\omega) \tag{1.3}$$

and

$$I(\omega) = \frac{A_{nm}}{B_{mn}\, e^{\hbar\omega/kT} - B_{nm}}. \tag{1.4}$$

This black-body frequency distribution function is exactly equivalent to Planck's distribution (1.1). At this point it is important to notice that Einstein wouldn't have succeeded without introducing the stimulated emission. Comparison of expressions (1.1) and (1.4) shows that $B_{mn} = B_{nm}$: for a photon the probability to be absorbed equals the probability to be emitted by stimulation. These two effects are perfectly symmetrical; they both take place when an electromagnetic field is present around an atom.

Strangely enough, by giving a physical interpretation to Planck's law based on photons interacting with an energy-quantized matter, Einstein has made the spontaneous emission appear mysterious. Why is an excited atom not stable? If light is not the cause of the spontaneous emission, then what is the hidden cause? This point still gives rise to a passionate debate today about the role played by the fluctuations of the field present in the vacuum. Comparison of expressions (1.1) and (1.4) also leads to $A_{nm}/B_{nm} = \hbar\omega^3/\pi^2 c^2$, so that when the light absorption probability is known then the spontaneous and stimulated emission probabilities are also known.

According to Einstein's theory, three different processes can take place during the interaction of light with matter, as described below.

1.2.1 Absorption

In this process one photon from the radiation field disappears and the energy is transferred to an electron as potential energy when it changes state from E_m to E_n. The probability for an electron to undergo the absorption transition is B_{mn}.

1.2.2 Spontaneous Emission

When being in an excited state E_n, an electron in an atom has a probability A_{nm} to spontaneously fall to the lower state E_m. The loss of potential energy gives rise to the simultaneous emission of a photon with energy $\hbar\omega = E_n - E_m$. The direction, phase and polarization of the photon are random quantities.

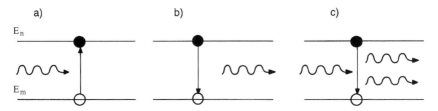

Fig. 1.2. The three elementary electron–photon interaction processes in atoms: (**a**) absorption, (**b**) spontaneous emission, (**c**) stimulated emission

1.2.3 Stimulated Emission

This contribution to light emission only occurs under the influence of an electromagnetic wave. When a photon with energy $\hbar\omega$ passes by an excited atom it may stimulate the emission by this atom of a twin photon, with a probability B_{nm} strictly equal to the absorption probability B_{mn}. The emitted twin photon has the same energy, the same direction of propagation, the same polarization state and its associated wave has the same phase as the original inducing photon. In an elementary stimulated emission process the net optical gain is two.

1.3 Light Amplification by Stimulated Emission

In what follows we will discuss the conditions that have to be fulfilled for the stimulated emission to be used for the amplification of electromagnetic waves.

What we need now is a set of N atoms, which will simulate a two-level material. The levels are called E_1 and E_2 (Fig. 1.3).

Their respective populations are N_1 and N_2 per unit volume; the system is illuminated by a light beam of n photons per second per unit volume with individual energy $\hbar\omega = E_2 - E_1$. The absorption of light in this medium is proportional to the electronic transition probability, to the number of photons at position z in the medium and to the number of available atoms in state 1 per unit volume.

To model the variation of the number of photons n as a function of the distance z inside the medium, the use of energy conservation leads to the

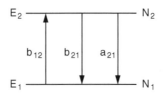

Fig. 1.3. Energy diagram for a set of atoms with two electronic levels

following differential equation:

$$\frac{\mathrm{d}n}{\mathrm{d}z} = (N_2 - N_1)b_{12}n + a_{21}N_2, \tag{1.5}$$

where $b_{12} = b_{21}$ and a_{21} are related to the Einstein coefficients by constant quantities. For the sake of simplicity we will neglect the spontaneous emission process and thus the number of photons as a function of the propagation distance is given as

$$n(z) = n_0\, \mathrm{e}^{(N_2 - N_1)b_{12}z}, \tag{1.6}$$

with $n_0 = n(0)$ being the number of photons impinging on the medium.

When $N_2 < N_1$, expression (1.6) simply reduces to the usual Beer–Lambert law for absorption, $n(z) = n_0\, \mathrm{e}^{-\alpha z}$, where $\alpha = (N_1 - N_2)b_{12} > 0$ is the linear absorption coefficient. This limit is found with any absorbing material at room temperature: there are more atoms in the ground state ready to absorb photons than atoms in the excited state able to emit a photon.

When $N_1 = N_2$, expression (1.6) shows that the number of photons remains constant along the propagation distance. In this case the full symmetry between absorption and stimulated emission plays a central role: the elementary absorption and stimulated emission processes are balanced. If spontaneous emission had been kept in expression (1.6), a slow increase of the number of photons with distance would have been found due to the spontaneous creation of photons.

When $N_2 > N_1$, there are more excited atoms than atoms in the ground state. The population is said to be "inverted". Expression (1.6) can be written $n(z) = n_0\, \mathrm{e}^{gz}$, $g = (N_1 - N_2)b_{12}$ being the low-signal gain coefficient. This process is very similar to a chain reaction: in an inverted medium each incoming photon stimulates the emission of a twin photon and its descendants too. The net growth of the number of photons is exponential but does not exactly correspond to the fast doubling every generation mentioned at the end of Sect. 1.2.3. Because the emitted photons are resonant with the two-level system, some of them are reabsorbed; also, some of the electrons available in the excited state are lost for stimulated emission because of their spontaneous decay. The elementary growth factor is therefore less than 2.

1.4 Population Inversion

To build an optical oscillator, the first step is to find how to amplify light waves, and we have just seen that amplification is possible under the condition that there exist some way to create an inverted population in some material medium.

1.4.1 Two-Level System

Let us first consider, again, the two-electronic-level system (Fig. 1.3). Electrons, because they have wave functions that are antisymmetric under inter-

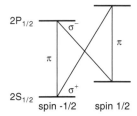

Fig. 1.4. Sketch of the Zeeman structure of rubidium atoms in the vapour phase

change of particles, obey the Fermi–Dirac statistical distribution. With ΔE being the energy separation between the two levels, the population ratio for a two-electronic-level system is given by

$$\frac{N_2}{N_1} = \frac{1}{e^{\Delta E/kT} + 1}. \tag{1.7}$$

When the temperature T goes to $0\,\mathrm{K}$ the population ratio also goes to 0. At $0\,\mathrm{K}$ the energy of the system is zero: all the electrons are in the ground state, $N_2 = 0$, $N_1 = N_0$ the total number of electrons. In contrast, when the temperature goes towards infinity $(T \to \infty)$, the population ratio goes to one-half, $N_2 = N_1/2$. In the high-temperature limit the electrons are equally distributed between the ground and excited states: an inverted population regime cannot be reached by just heating a material. It is not possible, either, to create an inverted population in a two-level system by optically exciting the electrons: at best there can be as many absorbed as emitted photons.

1.4.2 Optical Pumping

Optical pumping was proposed, in 1958, by Alfred Kastler as a way to produce inverted populations of electrons. Kastler studied the spectroscopic properties of rubidium atoms in the vapour phase, under the influence of a weak magnetic field. The Zeeman structure of the gas is shown in Fig. 1.4 and the splitting of the substates of the ground state is small enough that they can be considered as equally occupied. Selection rules imply that π transitions are not sensitive to the polarization state of light. With circularly polarized light, σ^+ or σ^- transitions are possible, depending on the right $(+)$ or left $(-)$ handed character of the polarization state. When a σ^+ polarization is chosen to excite the system the $2P_{1/2}$ (spin 1/2) sublevel is enriched. From this state the atoms can return to the ground state through a σ transition with probability 2/3 or a π transition with probability 1/3, and thus the $2S_{1/2}$ (spin 1/2) sublevel is enriched compared to the other sublevel of the ground state. A population inversion is realized.

Two-level systems are seldom found in natural systems, so that the difficulty pointed out in Sect. 1.4.1 is basically of an academic nature. Real electronic structures are rather complicated series of states and for the sake of

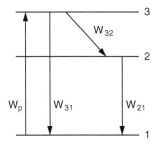

Fig. 1.5. Three-level system used to model the population inversion in optical pumping

simplicity we will consider a three-level system illuminated with photons of energy $\hbar\omega = E_3 - E_1$ (Fig. 1.5). Electrons are promoted from state 1 to state 3 with a probability per unit time W_p, which accounts for both the absorption cross-section of the material and the intensity of the incoming light.

From state 3, electrons can decay either to state 2 with probability W_{32} or to the ground state with probability W_{31}. We assume now that the transition probability from state 3 to state 2 is much larger than to state 1 ($W_{32} \gg W_{31}$). The electronic transition $2 \rightarrow 1$ is supposed to be the radiative transition of interest and we suppose that state 2 has a large lifetime compared to state 3 ($W_{21} \ll W_{32}$). State 2 is called a metastable state.

Rate equations can be easily derived that model the dynamical behavior of such a three-level system:

$$\frac{\mathrm{d}N_3}{\mathrm{d}t} = W_p N_1 - W_{32} N_3 - W_{31} N_3, \qquad (1.8a)$$

$$\frac{\mathrm{d}N_2}{\mathrm{d}t} = W_{32} N_3 - W_{21} N_2. \qquad (1.8b)$$

State 3 is populated from state 1 with probability W_p and in proportion to the population N_1 of state 1 (first right term in (1.8a)); it decays with a larger probability to state 2 than to state 1 and in proportion to its population N_3 (two decay terms in (1.8a)). State 2 is populated from state 3 according to W_{32} and N_3 (source term in (1.8b)) and decays to state 1 through spontaneous emission of light and in proportion to its population N_2 (decay term in (1.8b)).

A steady dynamical behavior is a solution of (1.8) and corresponds to constant state populations with time; in this regime time derivatives vanish and (1.8) simply gives on

$$\frac{N_2}{N_1} \approx \frac{W_p}{W_{21}} \left(1 - \frac{W_{31}}{W_{32}}\right). \qquad (1.9)$$

When the pumping rate is large enough to overcome the spontaneous emission between states 2 and 1 ($W_p \gg W_{21}$), then (1.9) shows that the average number

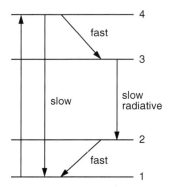

Fig. 1.6. Sketch of the four-level system found in most laser gain media

of atoms in state 2 can be larger that the average number of atoms in state 1: the population is inverted between states 2 and 1. This population inversion is reached when (i) the pumping rate is large enough to overcome the natural decay of the metastable level, (ii) the electronic decay from the pumping state to the radiative state is faster than any other decay, and (iii) the radiative decay time is long enough to ensure that the intermediate metastable state is substantially overoccupied. Because of all these stringent physical conditions the three-level model might seem to be unrealistic; this is actually not the case, for it mimics quite accurately the electronic structure and dynamics found in chromium ions dissolved in alumina (ruby)!

A large variety of materials have shown their ability to sustain a population inversion when, some way or another, energy is fed to their electronic system. Optical pumping still remains a common way of producing population inversion but many other ways have been developed to reach that goal, e.g. electrical excitation, collisional energy transfer and chemical reaction. Most of the efficient media used in lasers have proved to be four-level structures as far as population inversion is concerned (Fig. 1.6).

In these systems, state 3 is populated and the various transition probabilities are as follows: $W_{43} \gg W_{41}$, $W_{21} \gg W_{32}$, $W_{21} \approx W_{43}$. Therefore in the steady-state regime the population of states 4 and 2 can be kept close to zero and the population inversion contrast can be made larger than in a three level-system ($N_3 - N_2 \gg N_2$). This favorable situation is relevant to argon and krypton ion lasers, dye lasers and the neodymium ion in solid matrices, for example.

1.4.3 Light Amplification

Once a population inversion is established in a medium it can be used to amplify light. In order to simplify calculations let us consider a medium in which a population inversion is realized ($\Delta N = N_1 - N_2$, $\Delta N < 0$) in a three-level system (Fig. 1.5). In such a medium, the intensity of a low-intensity electromagnetic wave propagating along z is proportional to the population inversion ΔN and to the wave intensity. This can be formally written as

$$\frac{dI(z)}{dz} = -I(z)\,\Delta N \sigma_{21}, \tag{1.10}$$

where σ_{21}, the proportionality factor, is called the stimulated-emission cross-section of the transition. It depends on both the medium and the wavelength of the light. Equation (1.10) describes light amplification in an inverted-population medium.

During the amplification process a depletion of the inverted level is to be expected due to the stimulated emission process itself: ΔN must depend on the intensity of the light. This dependence can be qualitatively explored using (1.9) as a starting point. If an efficient gain medium is assumed, then $W_{31}/W_{32} \ll 1$ can be neglected and one gets

$$\Delta N \approx N_2 \left(\frac{W_{21}}{W_{\mathrm{p}}} - 1 \right). \tag{1.11}$$

One can also make the assumption that $N_2 \approx N$, all the available atoms (N) being in the inverted state 2. This regime corresponds to a situation where the pumping is strong so that $W_{21} \ll W_{\mathrm{p}}$. In the framework of these approximations, an inverted expansion leads to

$$\Delta N \approx -N \Big/ \left(1 + \frac{W_{21}}{W_{\mathrm{p}}} \right). \tag{1.12}$$

Replacing the probabilities by intensities because they are only involved in ratios gives the following expression, in which I_{s} is a constant intensity depending on the gain medium:

$$\Delta N \approx -N \Big/ \left(1 + \frac{I(z)}{I_{\mathrm{s}}} \right). \tag{1.13}$$

When expression (1.13) is introduced into (1.10) we obtain

$$\frac{1}{I(z)} \frac{dI(z)}{dz} = g_0 \Big/ \left(1 + \frac{I(z)}{I_{\mathrm{s}}} \right), \tag{1.14}$$

where $g_0 = N\sigma_{21} > 0$ is the low-intensity gain. The saturation intensity I_{s} allows one to distinguish between the low-and high-intensity regimes for light amplification in a gain medium.

1.4.3.1 Low-Intensity Regime, $I(z) \ll I_{\mathrm{s}}$. In this regime the evolution of the light intensity as a function of the distance z in the medium simplifies to $I(z) = I(0)e^{g_0 z}$. Starting from its incoming value $I(0)$, the intensity grows as an exponential function along the propagation direction. This behavior could be intuitively predicted from the previous discussion on the stimulated emission process and the chain reaction.

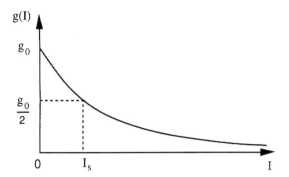

Fig. 1.7. Simple hyperbolic intensity dependence of the gain in a light amplifier

1.4.3.2 High-Intensity Regime, $I(z) \gg I_s$. When, in the gain medium, the intensity becomes larger than the saturation intensity, (1.14) reduces to $I(z) = I(0) + I_s g_0 z$; the intensity only grows as a linear function of the distance z. In the high-intensity regime the amplification process is much less efficient than in the low-intensity regime. The gain is said to saturate in the high-intensity regime.

It is far beyond the scope of this introduction to laser physics to rigorously discuss gain saturation; we will only focus on the simple hyperbolic model

$$g(I) = \frac{g_0}{1 + I/I_s}. \tag{1.15}$$

In this very approximate framework, the saturation intensity I_s is the intensity for which the gain value is reduced by a factor of two. This is very similar to the saturation of the absorption in saturable absorbers, and this again arises from the fact that light absorption and stimulated emission are symmetrical effects. The simplest gain model simply mimics the Beer–Lambert law for absorption: $I(z) = I(0)e^{gz}$, where g is given by (1.15).

Gain saturation is of prime importance in the field of ultrashort light pulse generation; this mechanism is a key ingredient for pulse shortening. But it also becomes a limiting factor in the process of amplifying ultrashort pulses: the intensity in these pulses rapidly reaches the saturation value. Beam broadening and pulse stretching are ways used to overcome this difficulty, as will be described in the following chapters.

1.5 Amplified Spontaneous Emission (ASE)

Spontaneous light emission from an excited medium is isotropic: photons are randomly emitted in every possible space direction with equal probability; the polarization states are also randomly distributed when the emission takes place from an isotropic medium. Stimulated emission, on the contrary, retains

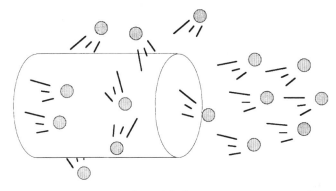

Fig. 1.8. Schematic illustration of amplified spontaneous emission (ASE). Spontaneously emitted photons are amplified when propagating along the major dimension of the gain medium

the characteristics of the inducing waves. This memory effect is responsible for the unwanted amplified spontaneous emission which takes place in laser amplifiers.

Most of the gain media in which a population inversion is created have a geometrical shape such that one of their dimensions is larger than the others (Fig. 1.8). At the beginning of the population inversion, when net gain becomes available, there are always spontaneously emitted photons propagating in directions close to the major dimension of the medium which trigger stimulated emission. In both space and phase, ASE does not have good coherence properties, because it is seeded by many incoherently, spontaneously emitted photons.

Amplified spontaneous emission is a problem when using light amplifiers in series to amplify light pulses: the ASE emitted by one amplifying stage is further amplified in the next stage and competes for gain with the useful signal. ASE returns in oscillators are also undesirable; they may damage the solid-state gain medium or induce temporal instabilities. To overcome these difficulties it is necessary to use amplifier decoupling.

1.5.1 Amplifier Decoupling

1.5.1.1 Static Decoupling. For light pulses that are not too short ($>$ 100 fs), a Faraday polarizer (Fig. 1.9) can be used to stop any return of linearly polarized light. Depending on the wavelength, properly chosen materials exhibit a strong rotatory power when a static magnetic field is applied. Adjustment of the magnetic field intensity and of the length of the material allows one to rotate the linear polarization of a light beam by a 45° angle.

Owing to the pseudovector nature of a magnetic field, the polarization rotation direction is reversed for a beam propagating in the reverse direction.

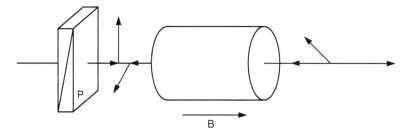

Fig. 1.9. Schematic of a Faraday polarizer. P is a linear polarizer; the cylinder is a material exhibiting strong rotatory power under the influence of a magnetic field B

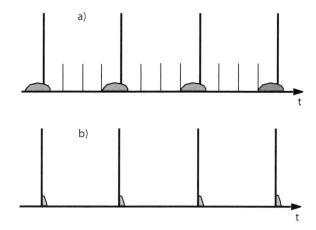

Fig. 1.10. (a) Train of pulses from a laser amplifier consisting of weak, unamplified pulses with high repetition rate and low-repetition-rate, amplified, short pulses, associated with long-lasting ASE. (b) Cleaning-up produced by a saturable absorber

Therefore the linear polarization of a reflected beam is rotated 90° and can be stopped by an analyzer.

1.5.1.2 Dynamic Decoupling. Amplifier stages are often decoupled using saturable absorbers. Absorption saturation (see Chap. 2) is a nonlinear optical effect that is symmetrical to gain saturation. It can be described by replacing the gain g by the absorption α in expression (1.15) and changing the sign in the evolution of the intensity with propagation. For a low-intensity signal, the intensity decreases as an exponential function with distance, while it only decreases linearly at high intensity. Only short, intense pulses can cross the saturable absorber. As an example we will consider a light output of an amplifier stage consisting of a superposition of light pulses: a high-repetition-rate train of weak, short pulses and a low-repetition-rate train of intense, short pulses superimposed on low-intensity, long-lasting ASE pulses (Fig. 1.10a).

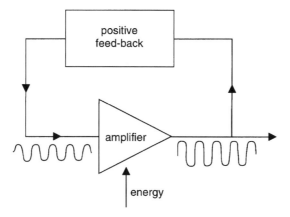

Fig. 1.11. General sketch of an oscillator. An oscillator is basically made of an amplifier and a positive feedback. The feedback must ensure a constructive interference between the input and amplified waves

The optical density of the saturable absorber is adjusted in such a way that it only saturates when crossed by the amplified pulses. When the pulse train crosses the saturable absorber the unamplified pulses and the leading edge of the ASE are absorbed. In order to improve the energy ratio between the amplified pulses and the remaining ASE the saturable absorber must be chosen to have a short recovery time. That way the long-lasting trailing part of the ASE can be partly absorbed. Malachite green, for example, with a 3 ps recovery time, has been widely used as a stage separator in dye amplifiers.

1.6 The Optical Cavity

We now know how to create and use stimulated emission to amplify light. From a very general point of view an oscillator is the association of an amplifier with a positive feedback (Fig. 1.11). The net gain of the amplifier must be larger than one in order to overcome the losses, including the external coupling. The phase change created by the feedback loop must be an integer multiple of 2π in order to maintain a constructive interference between the input and amplified waves. What is the way to use an amplifier in the building of an optical oscillator?

1.6.1 The Fabry–Pérot Interferometer

In the year 1955, *Gordon* et al. [1.1] developed the ammonia maser, clearly proving, in the microwave wavelength range, the possibility to amplify weak signals using stimulated emission. A metallic box, with suitably chosen size and shape, surrounding a gain medium was proved to create efficient positive feedback, allowing the device to run as an oscillator. In order for the oscillation to operate in a single mode the size of the box had to be of the order of a few wavelengths, i.e. a few centimeters. A large number of research groups tried to transpose the technique to the visible range using appropriate gain media. They were stopped by the necessity to design an optical box having a volume of the order of λ^3, which in the visible means 0.1 μm^3. A tractable much larger box would have had a large number of modes fitting the gain bandwidth, and this was expected to create mode beating which in turn would destroy the build-up of a coherent oscillation.

In the following years, *Schawlow* and *Townes* [1.2], as well as *Basov* and *Prokhorov* [1.3], came up with calculations showing that the number of modes in an optical cavity could be greatly reduced by confining light in only one dimension to create a feedback. The Fabry–Pérot resonator then came into the picture.

A gain medium would be put between the two high-reflectivity ($\approx 100\%$) mirrors of a Fabry–Pérot interferometer so that a coherent wave could be constructed after several round trips of the light through the amplifier. At the starting time of the device, when the inverted population was established in the gain medium, a unique spontaneously emitted photon propagating along the cavity axis would start stimulated emission, increasing the number of coherent photons. If, after a round trip between the mirrors, the gain was larger than the losses then the intensity of the visible electromagnetic wave would increase as an exponential function after each round trip, and a self-sustained oscillation would start. But in a cavity, owing for example to diffraction at the edge of the mirrors or spurious reflections and absorption, photons are lost. The value of the gain which overcomes the losses is called the laser threshold.

1.6.2 Geometric Point of View

We will now focus on the properties of an optical cavity, and specifically look for the necessary conditions that must be fulfilled so that the cavity can accommodate an infinite number of round trips of the light. The cavity under consideration consists simply of two concave, spherical mirrors with radii of curvature R_1 and R_2, spaced by a length L (Fig. 1.12). From a geometric point of view, Fig. 1.12 shows that the light ray coincident with the mirrors' axis will repeat itself after an arbitrary number of back-and-forth reflections from the mirrors. Other rays may or may not escape the volume defined by the two mirrors. A cavity is said to be stable when there exists at least one family of rays which never escape. When there is no such ray the cavity is said to be unstable; any ray will eventually escape the volume defined by the mirrors.

A very simple geometric method allows one to predict whether a given cavity is stable or not [1.4]. Consider now the cavity defined in Fig. 1.13. Two circles having their centers on the axis of the cavity are drawn with their diameters equal to the radii of curvature of the mirrors so as to be tangent to the reflecting face of the mirrors. The center of each is coincident with the

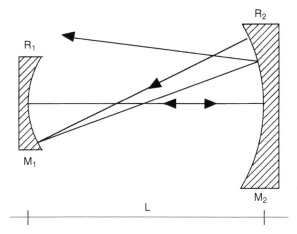

Fig. 1.12. Optical cavity consisting of two concave mirrors. Two ray paths are shown. The axial ray is stable, the other is not

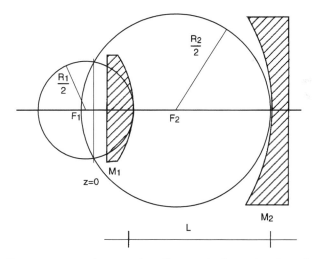

Fig. 1.13. Concave–convex laser cavity. Geometric determination of the stability of the cavity [1.4]

real or virtual focus point of the mirror. The straight line joining the points of intersection of the two circles crosses the axis of the cavity at a point which defines the position of the beam waist of the first-order transverse mode. If the two circles do not cross, the cavity is unstable.

1.6.3 Diffractive-Optics Point of View

Geometrical optics is not enough to estimate quantitatively the properties of the modes which may be established in a Fabry–Pérot cavity. The full calculation of these modes is rather tedious and we will concentrate only on some of the most important features. In a Fabry–Pérot interferometer light bounces back and forth from one mirror to the other with a constant time of flight, so that its dynamics is periodic by nature. A wave propagating inside a cavity remains unchanged after one period in the simple case in which the polarization direction does not change. The propagation is governed by diffraction laws because of the finite diameter of the mirrors, but also because of the presence of apertures inside the cavity; the most common aperture is simply the finite diameter of the gain volume. From a mathematical point of view, the radial distribution of the electric field in a given mode inside the cavity is described through a two-dimensional spatial Fourier transformation. For the periodicity condition to be respected it is therefore necessary that the function describing the radial distribution is its own Fourier transform. The Gaussian function is its own Fourier transform and is therefore the very basis of the transverse electromagnetic structure for spherical-mirror cavities.

It can be shown that the electric field distribution of the fundamental transverse mode in a spherical-mirror cavity can be written as

$$E(x, y, z) = E_0 \frac{W_0}{W(z)} \exp\left\{ -i\left[k\left(z + \frac{x^2 + y^2}{2R(z)} \right) - \Phi(z) \right] \right\}$$

$$\times \exp - \frac{x^2 + y^2}{W^2(z)}, \tag{1.16}$$

where

$$R(z) = z\left(1 + \left(\frac{\pi W_0^2}{\lambda z} \right)^2 \right), \qquad \Phi(z) = \tan^{-1}\left(\frac{\lambda z}{\pi W_0^2} \right). \tag{1.17}$$

Here $R(z)$ is the radius of curvature of the wave surface (Fig. 1.14), $\Phi(z)$ is the phase as a function of the distance z,

$$W^2(z) = W_0^2 \left(1 + \left(\frac{\lambda z}{\pi W_0^2} \right)^2 \right) \tag{1.18}$$

is the radius of the beam and

$$k = \frac{2\pi}{\lambda} \tag{1.19}$$

is the propagation factor in vacuum.

The propagation origin $z = 0$ is chosen to be coincident with the position of the minimum radius of the beam, the beam waist W_0. When $z = 0$ then $R(0) = \Phi(0) = \infty$, the wave surface is plane and the electric field amplitude

$$E(x, y, 0) = E_0 \exp - \left(x^2 + y^2 / W_0^2 \right) \tag{1.20}$$

decays as a Gaussian function along the radius of the beam.

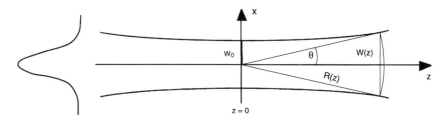

Fig. 1.14. Schematic structure of a Gaussian light beam in the vicinity of a focal volume. The y axis is perpendicular to the paper sheet. The $z = 0$ origin is at the minimum radius W_0 of the beam

Along the axis, the radius of curvature of the wave surface $R(z)$ varies as a hyperbola and its asymptotes make an angle θ with the axis such that $\tan \theta = \lambda/\pi W_0$. This angle is a good definition of the beam divergence.

For large z the hyperbola may be replaced by its asymptotes and the radius of curvature varies linearly with the distance z; $R(z) \approx z$ when z goes to infinity in (1.17). In this long-distance regime the amplitude of the electric field varies as the inverse of the beam radius, i.e. $E(z) \approx W^{-1}(z)$, along z, and as a Gaussian function along the radius. Putting aside the Gaussian attenuation, the fundamental mode behaves like a spherical wave. Such a wavefront has the right shape to fit nicely the surface of spherical mirrors.

This structure of the fundamental transverse mode is referred to as TEM$_{00}$ (fundamental transverse electric and magnetic mode). Many other transverse modes can propagate inside the cavity; they can be expressed as the super-position of higher-order fundamental modes TEM$_{nm}$. These modes can be calculated by multiplying the fundamental lowest-order mode (1.16) by the Hermite polynomials of integer orders n and m,

$$ H_n \left(\frac{\sqrt{2}\, x}{W(z)} \right), \qquad H_m \left(\frac{\sqrt{2}\, y}{W(z)} \right), \tag{1.21} $$

and multiplying the phase term $\Phi(z)$ by $(1 + n + m)$.

1.6.4 Stability of a Two-Mirror Cavity

The problem is now to find which Gaussian mode with a far-field spherical behavior can fit a given pair of spherical mirrors with radii R_1 and R_2, spaced by a distance L. In Fig. 1.15, the mirror positions z_1 and z_2 are measured from the yet unknown position of the beam waist. For a cavity to be stable it must be able to accommodate a mode in which spherical wavefronts will fit the reflecting surfaces of the two spherical mirrors. From a formal point of view, one simply has to make the radii of curvature of the wavefront, given by (1.17), equal to the radii of curvature of the mirrors; adding the conservation of length leads to the three equations

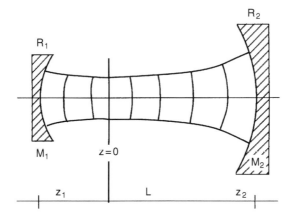

Fig. 1.15. Schematic diagram of a simple transverse Gaussian beam fitting a two-spherical-mirror cavity

$$R_1 = -z_1 - \frac{z_R^2}{z_1}, \tag{1.22}$$

$$R_2 = +z_2 + \frac{z_R^2}{z_2}, \tag{1.23}$$

$$L = z_2 - z_1, \tag{1.24}$$

where $z_R = \pi W_0/\lambda$, called the Rayleigh range, is the distance, measured from the beam waist position, where the radius of the beam is equal to $\sqrt{2}\,W_0$. This length defines the focal volume, which, to first order in z, is almost cylindrical.

These simple equations have been generally solved using the special cavity parameters

$$g_1 = 1 - \frac{L}{R_1} \quad \text{and} \quad g_2 = 1 - \frac{L}{R_2}, \tag{1.25}$$

tying the distances z_1, z_2, z_R to the geometric cavity parameters R_1, R_2, L. Solving the equations in this new notation leads to the following results:

– the beam waist position measured from the mirror position

$$z_1 = \frac{g_2(1 - g_1)}{g_1 + g_2 - 2g_1g_2}L, \qquad z_2 = \frac{g_1(1 - g_2)}{g_1 + g_2 - 2g_1g_2}L; \tag{1.26}$$

– the beam waist radius,

$$W_0^2 = \frac{L\lambda}{\pi}\sqrt{\frac{g_1g_2(1 - g_1g_2)}{(g_1 + g_2 - 2g_1g_2)^2}}; \tag{1.27}$$

– the beam radius at the surface of the mirrors,

$$W_1^2 = \frac{L\lambda}{\pi}\sqrt{\frac{g_2}{g_1(1 - g_1g_2)}}, \qquad W_2^2 = \frac{L\lambda}{\pi}\sqrt{\frac{g_1}{g_2(1 - g_1g_2)}}. \tag{1.28}$$

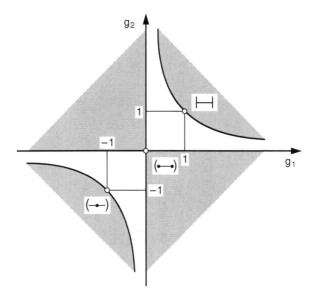

Fig. 1.16. Stability diagram for a laser cavity. The shaded areas define the set of values of the cavity parameters g_1 and g_2 for which the cavity is unstable [1.5]

From (1.27) one notices that the radius of the beam may only be a real number if the argument of the square root function is a positive and finite number. This leads to the following inequalities:

$$0 \le g_1 g_2 \le 1, \tag{1.29}$$

which put stringent conditions on the mirrors' radii of curvature and on their spacing.

The stability conditions (1.29), define a hyperbola $g_1 g_2 = 1$ in the g_1, g_2 plane. The two white regions in Fig. 1.16 correspond to stable cavities when g_1 and g_2 are both positive or negative. The shaded regions correspond to unstable resonators. Three specific, commonly used cavities are shown in the figure: (i) the symmetrical concentric resonator ($R_1 = R_2 = -L/2$, $g_1 = g_2 = -1$), (ii) the symmetrical confocal resonator ($R_1 = R_2 = -L$, $g_1 = g_2 = 0$) and (iii) the planar resonator ($R_1 = R_2 = \infty$, $g_1 = g_2 = 1$). The fact that a cavity is optically unstable does not mean that it cannot produce any laser oscillation, nor does it mean that its emitted intensity is necessarily unstable. It only means that the number of round trips of the light it allows is limited.

Some gain media are short-lived (a few nanoseconds) compared to the cavity period; it is of no use in this situation to pile up round trips. On the contrary, it might be of great help to use an unstable cavity accommodating the right number of round trips. But as the number of passes of the light in the gain medium is limited, such unstable cavities do not show good transverse modal qualities. This situation is encountered, for example, in high gain lasers like exciplex lasers or copper-vapor lasers.

1.6.5 Longitudinal Modes

When it comes to the use of lasers as short-pulse generators, the most important property of optical resonators is the existence of longitudinal modes. Transverse modes, as we saw, are a geometric consequence of light propagation, while longitudinal modes are a time–frequency property. In other words, we now know how to apply a feedback to a gain medium; we need to explore the conditions under which this feedback can constructively interfere with the main signal. Fabry–Pérot interferometers were originally developed as high-resolution bandpass filters. The interferential treatment of optical resonators can be found in most textbooks on optics; here, as a remainder of the time–frequency duality, we will consider a time-domain analysis of the Fabry–Pérot interferometer. This way of looking will prove useful in the understanding of mode-locking.

An electromagnetic field can be established between two parallel mirrors only when a wave propagating in one direction adds constructively with the wave propagating in the reverse direction. The result of this superposition is a standing wave, which is established if the distance L between the two mirrors is an integer multiple of the half-wavelength of the light. Writing τ for the period of the wave and c for its velocity in vacuum, 3×10^8 m/s, and remembering that $\lambda = c\tau$, the standing wave condition is

$$\frac{mc\tau}{2} = L, \qquad m \in N^+, \qquad (1.30)$$

which fixes the value of the positive integer m. The cavity has a specific period $T = m\tau$, which is also a round-trip time of flight $T = 2L/c$.

For a typical laser cavity with length $L = 1.5$ m, the period is $T = 10$ ns and the characteristic frequency $\nu = 100$ MHz. These numbers do fix the repetition rate of mode-locked lasers and also the period of the pulse train.

In the continuous-wave (CW) regime the amplification process in a laser cavity is basically coherent and linear; the gain balances the losses. At any point inside the cavity the signal which can be observed at some instant will be repeated unchanged after the time T has elapsed. In the time domain the electromagnetic field in a laser cavity can be seen as a periodic repetition of the same distribution such that $\varepsilon(t) = \varepsilon(t + nT)$, n being an integer (see Fig. 1.17).

$E(\omega)$ is the Fourier transform of one period of the electric field; its spectral extent is determined by the spectral bandwidth of the various active and passive elements present in the cavity. The Fourier transformation is a linear operation, and therefore the Fourier transform of the total electric field, from 0 to N periods, is the simple sum of the delayed partial Fourier transforms,

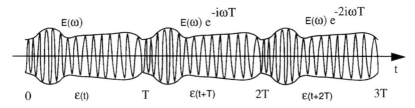

Fig. 1.17. Schematic representation, in the time domain, of the electric field at some point in a laser cavity. The field is repeated unchanged so that $\varepsilon(t) = \varepsilon(t+nT)$ after each period $T = 2L/c$

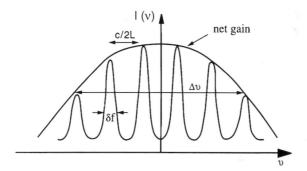

Fig. 1.18. Schematic emission spectrum of a laser

$$E^N(\omega) = \sum_{n=0}^{N-1} e^{-i\omega nT} E(\omega) = \frac{1 - e^{-iN\omega T}}{1 - e^{-i\omega T}} E(\omega). \qquad (1.31)$$

The resulting sum is a geometric series which gives the power spectrum for N periods,

$$I^N(\omega) = |E^N(\omega)|^2 = \frac{1 - \cos N\omega T}{1 - \cos \omega T} I(\omega) = \frac{\sin^2 (N\omega T/2)}{\sin^2 (\omega T/2)} I(\omega). \qquad (1.32)$$

When N goes to infinity, the intensity response of a laser cavity tends towards an infinite periodic series of Dirac δ distributions spaced by the quantity $\delta\omega = 2\pi/T = \pi c/L$, or $\delta\nu = c/2L$. As is well known from the usual frequency analysis, the Fabry–Pérot cavity only allows specific frequencies to pass through; the energy is quantized.

Real laser spectra are more likely to look like Fig. 1.18, where the bandwidth δf of the axial modes is finite and governed by the resonator finesse, depending on the reflectivity of the mirrors. Moreover, the number of active modes is also finite, depending on the bandwidth of the net gain $\Delta\nu$.

1.7 Here Comes the Laser!

The first operating laser was set up by *Maiman* [1.6], at this time working with the Hughes aircraft company, in the middle of the year 1960, and the first gas laser by A. Javan at MIT at the end of that same year.

Ruby was used as the gain medium in Maiman's laser. Ruby is alumina, Al_2O_3, also known as corundum or sapphire, in which a small fraction of the Al^{3+} ions are replaced by Cr^{3+} ions. The electronic structure of $Cr^{3+}: Al_2O_3$ consists of bands and discrete states. The absorption takes place in green and violet bands in the spectrum, giving the material its pink color, and the emission at 694.3 nm takes place between a discrete state and the ground state. The overall structure is that of a three-level system. Inversion of the electronic population was produced by a broadband, helical flash-tube surrounding the ruby rod and the resonator was simply made from the parallel, polished ends of the rod, which were silver coated for high reflectivity.

Very rapidly following Maiman's achievement, A. Javan came out with the first gas laser, in which a mixture of helium and neon was continuously excited by an electric discharge [1.7].

1.8 Conclusion

In this chapter some laser basics were introduced as a background to the understanding of ultrashort laser pulse generation. It was not the aim to present details of the physics of lasers, but rather to point out specific key points necessary to understand the following chapters. For more information on laser physics we recommend the references contained in the "Further Reading" section of this chapter.

1.9 Problems

1. Remove the stimulated emission in (1.2) and show that the intensity distribution does not vanish to zero when the frequency does. In the pre-Planck era this was called the 'infrared catastrophe'.
2. Solve the photon population evolution equation

$$\frac{dn}{dz} = (N_2 - N_1)b_{12}n + a_{21}N_2$$

taking into account the spontaneous emission term $a_{21}N_2$.
(a) Show that in the low-and negative-temperature limits ($T \rightarrow 0$, $N_2 \ll N_1$ and $T \rightarrow -\infty$, $N_2 \gg N_1$) there is no qualitative change in the response of the medium.
(b) In the high temperature-limit ($T \rightarrow \infty$, $N_2 = N_1$), show that the number of photons grows linearly with the distance.

3. Assume a 1.5 m long linear laser cavity. One of the mirrors is flat($R_1 = \infty$).
   (a) Using the diagram technique, what is the minimum radius of curvature R_2 of the other mirror for the cavity to be optically stable? (Answer: $R_2/2 = 1.5\,\mathrm{m}$). What is the position of the beam waist? (Answer: $z_1 = 0$, $z_2 = 1.5\,\mathrm{m}$).
   Turning now to the analytical expressions in Sect. 1.6.4, answer the following questions.
   (b) Check the position of the beam waist by directly calculating z_1 and z_2. What is the diameter of the beam waist for an operating wavelength $\lambda = 514.5\,\mathrm{nm}$? (Answer: $2W_0 = 832\,\mu\mathrm{m}$).
   (c) What are the radii of the first fundamental transverse mode at the cavity mirrors? (Answer: $2W_1 = 2W_0$, $2W_2 = 1.44\,\mathrm{mm}$).
   (d) Calculate the Rayleigh range for which the mode can be considered as cylindrical. (Answer: $z_\mathrm{R} = 1.05\,\mathrm{m}$).

Further Reading

W. Koechner: *Solid-State Laser Engineering*, 4th edn., Springer Ser. Opt. Sci., Vol. 1 (Springer, Berlin, Heidelberg 1996)

J.R. Lalanne, S. Kielich, A. Ducasse: *Laser–Molecule Interaction and Molecular Nonlinear Optics* (Wiley, New York 1996)

B. Saleh, M. Teich: *Fundamentals of Photonics* (Wiley, New York 1991)

K. Shimoda: *Introduction to Laser Physics*, 2nd edn., Springer Ser. Opt. Sci., Vol. 44 (Springer, Berlin, Heidelberg 1991)

A.E. Siegman: *Lasers* (University Science Books, Mill Valley, CA 1986)

O. Svelto: *Principles of Lasers*, 3rd edn. (Plenum Press, New York 1989)

J.T. Verdeyen: *Laser Electronics* (Prentice Hall, New Jersey 1989)

A. Yariv: *Quantum Electronics*, 3rd edn. (Wiley, New York 1989)

Historial References

[1.1] J.P. Gordon, H.J. Zeiger, C.H. Townes: Phys. Rev. **99**, 1264 (1955)
[1.2] A.L. Schawlow, C.H. Townes: Phys. Rev. **112**, 1940 (1958)
[1.3] N.G. Basov, A.M. Prokhorov: Sov. Phys.-JETP **1**, 184 (1955)
[1.4] G.A. Deschamps, P.E. Mast: Proc. Symposium on Quasi-Optics, ed. J. Fox (Brooklyn Polytechnic Press, New York 1964); P. Laures: Appl. Optics **6**, 747 (1967)
[1.5] H. Kogelnik, T. Li: Appl. Optics **5**, 1550 (1966)
[1.6] T.H. Maiman: Nature **187**, 493 (1960)
[1.7] A. Javan, W.R. Bennet, D.R. Herriot: Phys. Rev. Lett. **6**, 106 (1961)

2

Pulsed Optics

C. Hirlimann

With 23 Figures

2.1 Introduction

Optics is the field of physics which comprises knowledge on the interaction between light and matter. When the superposition principle can be applied to electromagnetic waves or when the properties of matter do not depend on the intensity of light, one speaks of linear optics. This situation occurs with regular light sources such as light bulbs, low-intensity light-emitting diodes and the sun. With such low-intensity sources the reaction of matter to light can be characterized by a set of parameters such as the index of refraction, the absorption and reflection coefficients and the orientation of the medium with respect to the polarization of the light. These parameters depend only on the nature of the medium. The situation changed dramatically after the development of lasers in the early sixties, which allowed the generation of light intensities larger than a kilowatt per square centimeter. Actual large-scale short-pulse lasers can generate peak powers in the petawatt regime. In that large-intensity regime the optical parameters of a material become functions of the intensity of the impinging light. In 1818 Fresnel wrote a letter to the French Academy of Sciences in which he noted that the proportionality between the vibration of the light and the subsequent vibration of matter was only true because no high intensities were available. The intensity dependence of the material response is what usually defines nonlinear optics. This distinction between the linear and nonlinear regimes clearly shows up in the polynomial expansion of the macroscopic polarization of a medium when it is illuminated with an electric field $\boldsymbol{E}$:

$$
\begin{aligned}
\frac{\boldsymbol{P}}{\varepsilon_0} = {} & \chi^{(1)} \cdot \boldsymbol{E} && \textit{Linear optics}, \text{ index, absorption} \\
& + \chi^{(2)} : \boldsymbol{E}\boldsymbol{E} && \textit{Nonlinear optics}, \text{ second-harmonic} \\
& && \text{generation, parametric effects} \\
& + \chi^{(3)} : \boldsymbol{E}\boldsymbol{E}\boldsymbol{E} && \text{third-harmonic generation, nonlinear} \\
& && \text{index} \\
& + \cdots
\end{aligned}
\tag{2.1}
$$

In this expansion, the linear first-order term in the electric field describes linear optics, while the nonlinear higher-order terms account for nonlinear optical effects (ε_0 is the electric permittivity).

The development of ultrashort light pulses has led to the emergence of a new class of phase effects taking place during the propagation of these pulses through a material medium or an optical device. These effects are mostly related to the wide spectral bandwidth of short light pulses, which are affected by the wavelength dispersion of the linear index of refraction. Their analytical description requires the Taylor expansion of the light propagation factor k as a function of the angular frequency ω,

$$k(\omega) = k(\omega_0) + k'(\omega - \omega_0) + \tfrac{1}{2}k''(\omega - \omega_0)^2 + \ldots \tag{2.2}$$

Contrary to what happens in "classical" nonlinear optics, these nonlinear effects occur for an arbitrarily low light intensity, provided one is dealing with short (< 100 fs) light pulses. Both classes of optical effects can be classified under the more general title of "pulsed optics."

2.2 Linear Optics

2.2.1 Light

If one varies either a magnetic or an electric field at some point in space, an electromagnetic wave propagates from that point, which can be completely determined by Maxwell's equations.

If the magnetic field has an amplitude which is negligible when compared to the electric field, the propagation equation for light can be written as

$$\nabla^2 \boldsymbol{E} = \frac{1}{c^2}\frac{\partial^2 \boldsymbol{E}}{\partial t}, \qquad \frac{1}{c^2} = \mu_0 \varepsilon_0, \tag{2.3}$$

where c is a parameter, usually called the velocity of light, depending on the electric and the magnetic permittivity, ε_0 and μ_0 respectively, of the material medium in which the wave is propagating. Equation 2.3 is a second-order differential equation, for which retarded plane waves are the simplest propagating solutions,

$$E_y = \mathrm{Re}\left(E_0\, \mathrm{e}^{\mathrm{i}\omega(t - x/c)}\right). \tag{2.4}$$

This particular solution 2.4 describes the propagation of a transverse electric field E_y along the positive x axis. The amplitude of the electric field varies periodically as a cosine function in time with angular frequency ω and in space with wavelength $\lambda = 2\pi c/\omega$. At any given point x along the propagation axis the amplitude has the same value as it had at the earlier time $t - x/c$, when it was at the origin ($x = 0$).

A rewriting of 2.4 as

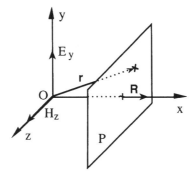

Fig. 2.1. Plane-wave propagation

$$E_y = \mathrm{Re}\left(E_0\, \mathrm{e}^{\mathrm{i}(\omega t - \boldsymbol{k}\cdot\boldsymbol{r})}\right), \qquad |\boldsymbol{k}| = \frac{\omega}{c} = \frac{2\pi}{\lambda}, \qquad (2.5)$$

allows the introduction of the wave vector $\boldsymbol{k}$ of the light. Figure 2.1 shows the geometry of the propagation of a plane wave. Let us consider the plane P orthogonal to the propagation vector $\boldsymbol{k}$, at distance x from the origin O. For any point in this plane, at distance $\boldsymbol{r}$ from the origin O, the scalar product $\boldsymbol{k}\cdot\boldsymbol{r} = kx = \omega x/c$ is constant; therefore, at a given time t, plane P is a plane of equal phase or equal time delay for a plane wave. A plane wave with a wavelength in the visible range ($400 > \lambda > 800\,\mathrm{nm}$) has an angular frequency ω equal to a few petahertz.

It is to be noticed that a plane wave, being a simple sine or cosine function, has an infinite duration and its spectrum, which contains only one angular frequency ω, is a δ distribution. A plane wave is the absolute opposite of a light pulse!

Up to this point light has been considered as a wave, but its particle behavior cannot be ignored. The wave–particle duality has the following meaning. The wave description of light is continuous in both time and space, while the particle description is discrete. In our classical culture these two visions are antithetical and one cannot be reduced to the other. Nowadays physics says and experiments show that light is both continuous and discrete. In the particle description light is made of photons, energy packets equal to the product of the frequency and Planck's constant $h\nu = \hbar\omega$. Photons have never been observed at rest and as they travel at the speed of light, relativity implies that their mass would be zero.

Fermat's principle states that the path of a ray propagating between two fixed point, A and B must be a stationary path, which in a mathematical point of view translates as

$$\delta \int \boldsymbol{k} \cdot \mathrm{d}\boldsymbol{l} = 0, \qquad (2.6)$$

δ standing for a variation of the path integral and $\mathrm{d}\boldsymbol{l}$ for an elementary path element anywhere between A and B.

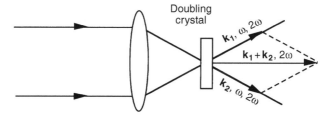

Fig. 2.2. Energy and momentum conservation in the doubling geometry used to record background-free autocorrelation traces of short optical pulses. Both the energy and the momentum are conserved during the mixing process

The path of the light is either the fastest one or the slowest one (in the latter some cases of reflection geometry with concave mirrors).

For a material particle with momentum p, Maupertuis's principle states that the integrated action of the particle between two fixed points A and B must be a minimum, which translates as

$$\delta \int p \cdot \mathrm{d}l = 0. \tag{2.7}$$

By analogy, the wave vector k for the light can be seen as the momentum of a zero-mass particle travelling along the light rays.

When light interacts with matter, both the energy and the momentum are conserved quantities. As an example, let us consider the dual-beam frequency-doubling process used in the background-free autocorrelation technique (Fig. 2.2).

In this geometry, two beams are incident, at an angle α, on a doubling crystal that can mix two photons with angular frequency ω and momenta k_1 and k_2, and produce one photon with angular frequency ω_2 and momentum k. Energy conservation implies that $\omega_2 = 2\omega$, while momentum conservation yields $k = k_1 + k_2$, $k = 2k_1$ or $k = 2k_2$. Five rays therefore exit the doubling material: two rays are at the fundamental frequency ω in the directions k_1 and k_2 of the incident rays, and superimposed on them are two rays with doubled frequency 2ω and momentum vectors $2k_1$ and $2k_2$. The fifth ray, with angular frequency 2ω, is oriented along the bisector of the angle α, corresponding to the geometrical sum $k_1 + k_2$.

2.2.2 Light Pulses

It is quite easy to produce "gedanken" light pulses. Let us start with a monochromatic plane wave, previously defined as (Fig. 2.3)

$$E_y = \mathrm{Re}\left(E_0\, e^{i\omega_0 t}\right). \tag{2.8}$$

The time representation of the field is an unlimited cosine function. Constructing a light pulse implies multiplying 2.8 by a bell-shaped function. To

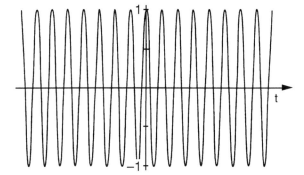

Fig. 2.3. Schematic time evolution of the electric field of a monochromatic plane wave. Notice that because of the finite width of the figure this picture is already representative of a light pulse having a rectangular envelope

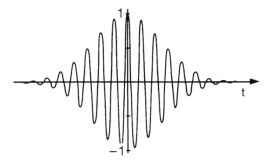

Fig. 2.4. Time evolution of the electric field in a Gaussian-shaped pulse. This pulse is built up by multiplying a cosine function by a Gaussian envelope function

simplify further calculation, we choose to multiply by a Gaussian function. A Gaussian pulse can be written

$$E_y = \mathrm{Re}\left(E_0\,\mathrm{e}^{(-\Gamma t^2 + \mathrm{i}\omega_0 t)}\right) \tag{2.9}$$

and its time evolution is shown in Fig. 2.4.

Γ is the shape factor of the Gaussian envelope; it is proportional to the inverse of the squared duration t_0, i.e. $\Gamma \propto t_0^{-2}$.

Let us now turn to the spectral content of a light pulse. This can be obtained by calculating the modulus of the Fourier transform of the time evolution function of the pulse. As said earlier, a plane wave oscillates with the unique angular frequency ω_0 and its Fourier transform is a Dirac distribution $\delta(\omega_0)$ (Fig. 2.5).

The Fourier transform of a Gaussian pulse is also a Gaussian function (Fig. 2.6). Therefore the frequency content of a light pulse is larger than the

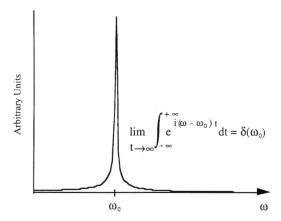

Fig. 2.5. Numerical Fourier transform of the truncated cosine function shown in Fig. 2.3. As the width of the cosine function grows larger and larger ($t_0 \rightarrow \infty$), the Fourier transform tends toward a Dirac distribution $\delta(\omega_0)$ with zero width

$$\lim_{t \rightarrow \infty} \int_{-\infty}^{+\infty} e^{i(\omega - \omega_0)t} \, dt = \delta(\omega_0)$$

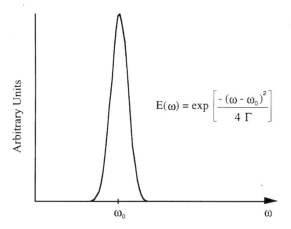

$$E(\omega) = \exp\left[\frac{-(\omega - \omega_0)^2}{4\,\Gamma}\right]$$

Fig. 2.6. Numerical Fourier transform of the Gaussian pulse shown in Fig. 2.4. The transformed function is also a Gaussian function with a frequency width proportional to Γ

unique frequency of a plane wave. The mathematical expression for the spectrum is given in Fig. 2.6 and the width of the spectrum is proportional to Γ.

2.2.3 Relationship Between Duration and Spectral Width

We have just empirically observed that the spectral width and the duration of a pulse are related quantities. What is the exact relationship? We start from the general time and frequency Fourier transforms of a pulse:

$$\varepsilon(t) = \frac{1}{2\pi} \int_{-\infty}^{+\infty} E(\omega)\, e^{-i\omega t}\, d\omega, \qquad E(\omega) = \int_{-\infty}^{+\infty} \varepsilon(t)\, e^{i\omega t}\, dt. \qquad (2.10)$$

When the duration and the spectral width of the pulse are calculated using the standard statistical definitions

$$\langle \Delta t \rangle = \frac{\int_{-\infty}^{+\infty} t |\varepsilon(t)|^2\, dt}{\int_{-\infty}^{+\infty} |\varepsilon(t)|^2\, dt},$$

$$\langle \Delta \omega^2 \rangle = \frac{\int_{-\infty}^{+\infty} \omega^2 |E(\omega)|^2\, d\omega}{\int_{-\infty}^{+\infty} |E(\omega)|^2\, d\omega}, \qquad (2.11)$$

it can be shown that these quantities are related through the following universal inequality:

$$\Delta t\, \Delta \omega \geq \tfrac{1}{2}. \qquad (2.12)$$

This classical-physics relationship, which leads to the quantum-mechanical time–energy uncertainty principle, has several important consequences in the field of ultrashort light pulses.

- In order to produce a light pulse with a given duration it is necessary to use a broad enough spectral bandwidth. A Gaussian-shaped pulse lasting for one picosecond (10^{-12} s) has a minimum spectral bandwidth of 441 MHz ($\Delta \omega = 4.41 \times 10^{11}$ Hz). If the central frequency ν_0 of the pulse lies in the visible part of the electromagnetic spectrum, say $\nu_0 = 4.84 \times 10^{14}$ Hz (wavelength $\lambda_0 = 620$ nm), then the relative frequency bandwidth is $\Delta \nu / \nu_0 \approx 10^{-3}$. But for a 100 times shorter pulse ($\Delta t = 10$ fs), $\Delta \nu / \nu_0 \sim 0.1$. As $|\Delta \lambda / \lambda_0| = \Delta \nu / \nu_0$, the wavelength extension of this pulse is 62 nm, covering 15 % of the visible window of the electromagnetic spectrum. Taking into account the wings of the spectrum, a 10 fs pulse actually covers most of the visible window!
- Equality to 1/2 in 2.12 can only be reached with Gaussian time and spectral envelopes. The Gaussian pulse shape "consumes" a minimum amount of spectral components. When the equality is reached in (2.12), the pulse is called a Fourier-transform-limited pulse. The phase variation of such a pulse has a linear time dependence as described by (2.9); in other words, the instantaneous frequency is time-independent.
- For a given spectrum, one pulse envelope can be constructed that has the shortest possible duration.
- The shortest constructed pulse can only be transform-limited if its spectrum is symmetrical.
- If a transform-limited pulse is not Gaussian-shaped, then the equality in an expression similar to 2.12 applies, but for a constant quantity larger than 1/2 which depends on the shape of the pulse.

From the experimental point of view, half-maximum quantities are easier to measure; the Fourier inequality is then usually written as $\Delta \nu \Delta t = K$, where

$\Delta\nu$ is the frequency full width at half-maximum and Δt the half maximum duration. K is a number which depends on the shape of the pulse. Table 2.1 gives values of K for some symmetrical pulse shapes.

Shape	$\varepsilon(t)$	K
Gaussian function	$\exp[-(t/t_0)^2/2]$	0.441
Exponential function	$\exp[-(t/t_0)/2]$	0.140
Hyperbolic secant	$1/\cosh(t/t_0)$	0.315
Rectangle	$-$	0.892
Cardinal sine	$\sin^2(t/t_0)/(t/t_0)^2$	0.336
Lorentzian function	$[1+(t/t_0)^2]^{-1}$	0.142

Table 2.1. Values of K for various pulse shapes, in the inequality $\Delta\nu \, \Delta t \geq K$, when $\Delta\nu$ and Δt are half-maximum quantities

Let us now consider a simple Gaussian light pulse

$$E_y = \mathrm{Re}\left(E_0 \, e^{(-\Gamma t^2 + i\omega_0 t)}\right). \tag{2.13}$$

The instantaneous frequency is obtained by calculating the time derivative of the phase,

$$\omega(t) = \partial\Phi/\partial t = \omega_0. \tag{2.14}$$

In this situation the angular frequency is constant and equals the central angular frequency ω_0. The light pulse is transform-limited; $\Delta\nu \, \Delta t = 0.441$. Let us now suppose that the phase of the pulse obeys a quadratic law in time,

$$E_y = \mathrm{Re}\left(E_0 \, e^{[-\Gamma t^2 + i(\omega_0 t - at^2)]}\right), \qquad \Gamma \in \Re, \tag{2.15}$$

then the instantaneous angular frequency varies linearly with time (Fig. 2.7):

$$\omega(t) = \partial\Phi/\partial t = \omega_0 + at, \qquad \alpha > 0. \tag{2.16}$$

When a quadratic time dependence is added to the original phase term, the instantaneous frequency is more red in the leading part of the pulse and more blue in the trailing part (compare Fig. 2.4). The pulse is said to be "chirped." This result describes the self-phase modulation process in the limit when $t \ll \Gamma$ (see (2.54) and (2.57)).

2.2.4 Propagation of a Light Pulse in a Transparent Medium

What happens to a short optical pulse propagating in a transparent medium? Because of its wide spectral width and because of group velocity dispersion in transparent media, it undergoes a phase distortion inducing an increase of its duration. This happens with any optical element and needs to be properly corrected for in the course of experiments.

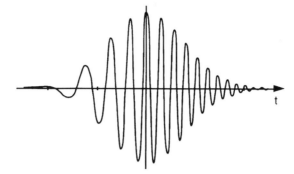

Fig. 2.7. Time evolution of the electric field in a Gaussian pulse having a quadratic time dependence of the phase

The frequency Fourier transform of a Gaussian pulse has already been given as

$$E_0(\omega) = \exp\left(\frac{-(\omega - \omega_0)^2}{4\Gamma}\right). \tag{2.17}$$

After the pulse has propagated a distance x, its spectrum is modified to

$$E(\omega, x) = E_0(\omega)\exp[-ik(\omega)x], \qquad k(\omega) = n\omega/c, \tag{2.18}$$

where $k(\omega)$ is now a frequency-dependent propagation factor. In order to allow for a partial analytical calculation of the propagation effects, the propagation factor is rewritten using a Taylor expansion as a function of the angular frequency, assuming that $\Delta\omega \ll \omega_0$ (this condition is only weakly true for the shortest pulses). Applying the Taylor's expansion

$$k(\omega) = k(\omega_0) + k'(\omega - \omega_0) + \tfrac{1}{2}k''(\omega - \omega_2)^2 + \ldots, \tag{2.19}$$

where

$$k' = \left(\frac{dk(\omega)}{d\omega}\right)_{\omega_0} \tag{2.20}$$

and

$$k'' = \left(\frac{d^2k(\omega)}{d\omega^2}\right)_{\omega_0}, \tag{2.21}$$

to 2.18, the pulse spectrum becomes

$$E(\omega, x) = \exp\left(-ik(\omega_0)x - ik'x(\omega - \omega_0)\right.$$
$$\left. - \left(\frac{1}{4\Gamma} + \frac{i}{2}k''\right)(\omega - \omega_0)^2\right]. \tag{2.22}$$

The time evolution of the electric field in the pulse is then derived from the calculation of the inverse Fourier transform of 2.22,

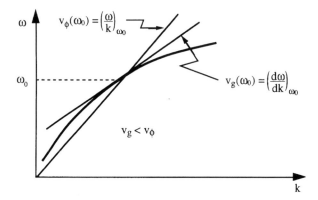

Fig. 2.8. Schematic relationship between the phase and group velocities in an ordinary transparent medium. Note that this relationship does not depend on the sign of the curvature of the dispersion curve $\omega(k)$

$$\varepsilon(t,x) = \frac{1}{2\pi} \int_{-\infty}^{+\infty} E(\omega, x)\, e^{i\omega t}\, d\omega, \qquad (2.23)$$

so that

$$\varepsilon(t,x) = \sqrt{\frac{\Gamma(x)}{\pi}}\, \exp\left[i\omega_0\left(t - \frac{x}{v_\phi(\omega_0)}\right)\right] \times \exp\left[-\Gamma(x)\left(t - \frac{x}{v_g(\omega_0)}\right)^2\right], \qquad (2.24)$$

where

$$v_\phi(\omega_0) = \left(\frac{\omega}{k}\right)_{\omega_0}, \quad v_g(\omega_0) = \left(\frac{d\omega}{dk}\right)_{\omega_0}, \quad \frac{1}{\Gamma(x)} = \frac{1}{\Gamma} + 2ik''x. \qquad (2.25)$$

In the first exponential term of 2.24, it can be observed that the phase of the central frequency ω_0 is delayed by an amount x/v_ϕ after propagation over a distance x. Because the phase is not a measurable quantity, this effect has no observable consequence. The phase velocity $v_\phi(\omega)$ measures the propagation speed of the plane-wave components of the pulse in the medium. These plane waves do not carry any information, because of their infinite duration.

The second term in 2.24 shows that, after propagation over a distance x, the pulse keeps a Gaussian envelope. This envelope is delayed by an amount x/v_g, v_g being the group velocity. Figure 2.8 shows schematically the relationship between the phase and the group velocities.

The phase velocity $v_\phi(\omega_0) = (\omega_0/k)$ is the slope of a straight line starting at the origin and crossing the dispersion curve $\omega(k)$ where the angular frequency equals ω_0. The group velocity $v_g(\omega_0) = (d\omega/dk)_{\omega_0}$ is the slope of the line tangential to the dispersion curve at the same point. In ordinary matter, $v_g < v_\phi$.

From the expression for the propagation factor $k = 2\pi/\lambda$ and the expression for the wavelength in a medium $\lambda = 2\pi c/\omega n(\omega)$, $n(\omega)$ being the index of refraction, one gets

$$v_\phi = \frac{c}{n(\omega)}, \tag{2.26}$$

$$v_g = \frac{d\omega}{dk} = \frac{1}{(dk/d\omega)}, \qquad \frac{dk}{d\omega} = \frac{1}{c}\left(n(\omega) + \omega\frac{dn(\omega)}{d\omega}\right), \tag{2.27}$$

$$v_g \approx v_\phi\left(1 - \frac{\omega}{n(\omega)}\frac{dn(\omega)}{d\omega}\right). \tag{2.28}$$

The second term in 2.24 also shows that the pulse envelope is distorted during its propagation because its form factor $\Gamma(x)$, defined as

$$\frac{1}{\Gamma(x)} = \frac{1}{\Gamma} + 2ik''x, \tag{2.29}$$

depends on the angular frequency ω through $k''(\omega)$,

$$k'' = \left(\frac{d^2k}{d\omega^2}\right)_{\omega_0} = \frac{d}{d\omega}\left(\frac{1}{v_g(\omega)}\right)_{\omega_0}. \tag{2.30}$$

This term is called the "group velocity dispersion".

Rewriting $\Gamma(x)$ as

$$\Gamma(x) = \frac{\Gamma}{1 + \xi^2 x^2} - i\frac{\Gamma\xi x}{1 + \xi^2 x^2}, \qquad \xi = 2\Gamma k'' \tag{2.31}$$

and substituting 2.31 into the second term of the right-hand side of 2.24 yields the following expression:

$$\exp\left[-\frac{\Gamma}{1 + \xi^2 x^2}\left(t - \frac{x}{v_g}\right)^2 + i\frac{\Gamma\xi x}{1 + \xi^2 x^2}\left(t - \frac{x}{v_g}\right)^2\right]. \tag{2.32}$$

The real part of 2.32 is still a delayed Gaussian function. Its form factor

$$\frac{\Gamma}{1 + \xi^2 x^2} \tag{2.33}$$

is always smaller than the original one Γ, which means that the pulse undergoes a duration broadening. Figure 2.9 shows a sketch of the pulse envelope broadening during its propagation through a transparent medium.

The phase, i.e. imaginary part in 2.32, contains a quadratic time term and we have already seen that this creates a linear frequency chirp in the pulse (Sect. 2.2.3).

In summary, the propagation of a short optical pulse through a transparent medium results in a delay of the pulse, a duration broadening and a frequency chirp.

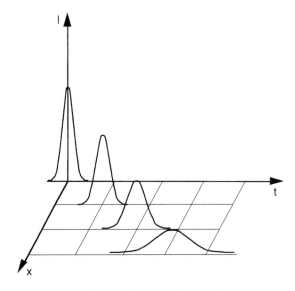

Fig. 2.9. Numerical calculation of the intensity envelope of a pulse propagating along x, in a lossless, transparent medium. The pulse broadens with time but, from energy conservation, its time-integrated intensity remains constant

2.2.4.1 Dispersion Parameter of a Transparent Medium. The dispersion of an index of refraction is usually tabulated as a function of the wavelength of light in vacuum. We therefore need to recalculate the dispersion as a function of the wavelength. From 2.30 we obtain

$$k'' = -\frac{\lambda^2}{2\pi c}D, \qquad D = \frac{1}{L}\frac{dt_g}{d\lambda}, \tag{2.34}$$

with t_g being the group delay induced by propagation over length L. D is called the dispersion parameter. The group delay t_g is calculated using 2.28 for the group velocity and the simple definition $t_g = L/v_g$, which leads to

$$k'' = \frac{\lambda^3}{2\pi c^2}\frac{d^2 n}{d\lambda^2}. \tag{2.35}$$

The sign of k'' depends on the curvature of the dispersion of the index $d^2 n/d\lambda^2$.

A one-resonance Drude model is the simplest way to describe the electronic properties of matter. In this very simple model the variation of the index of refraction in the vicinity of the resonance looks as shown in Fig. 2.10.

For wavelengths larger than the resonance wavelength the curvature of the index dispersion curve is positive (upward concavity) and the group velocity dispersion is positive ($k'' > 0$). This situation is the most usual one, encountered in ordinary optical glasses in the visible range. The index of refraction diminishes as the wavelength increases, which in turn implies an increase of

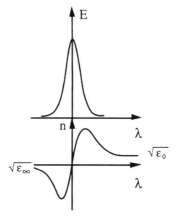

Fig. 2.10. (*Top*) sketch of an electronic resonance, from which the index of refraction (*bottom*) can be calculated

the group velocity: in a light pulse propagating through a transparent medium, the instantaneous frequency varies from its lowest value in the leading edge to its highest value in the trailing edge. Notice from 2.34 that a positive group velocity dispersion corresponds to a negative dispersion parameter D.

For wavelengths below the electronic resonance the situation is reversed and the group velocity dispersion is negative. This situation can be found in silica optical fibers, where a single resonance takes place, due to OH vibrations. For wavelengths around $1.55\,\mu$m, the group velocity is negative, allowing the propagation of optical solitons.

2.2.4.2 Time Compression with a Pair of Gratings. In order to correct for group-velocity-dispersion distortions, several optical devices have been designed that have an overall negative group velocity dispersion. As an example we consider a pair of transmission gratings R_1 and R_2. These gratings have a groove spacing d and their separation is l (Fig. 2.11).

A light ray, with wavelength λ, impinges on grating R_1 with an angle of incidence γ and is scattered with an angle θ. The gratings are set in such a way that their wavelength dispersions are reversed, which implies that the exiting ray at point B is parallel to the incident ray. P_1 and P_2 are wave planes at the entrance A and exit B of the system. P_2 crosses the emerging ray at point C. Between points A and B the light travels a distance $b = l/\cos\theta$. The diffraction due to a grating can be written as

$$d(\sin\gamma + \sin\theta) = \lambda. \tag{2.36}$$

In order to calculate the dispersion from 2.34, the group delay experienced by the light must first be evaluated. In this specific case, where propagation takes place only in air, the group delay is simply equal to the travel time of

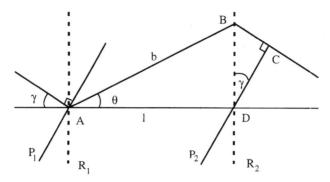

Fig. 2.11. Optical path through a pair of transmission gratings

light along ABC,

$$t = L/c = (AB + BC)/c, \qquad BC = DB \sin \gamma = b \sin \theta \sin \gamma, \qquad (2.37)$$

$$t = \frac{b}{c}(1 + \sin \theta \sin \gamma), \qquad (2.38)$$

and the dispersion parameter is expressed as

$$D = \frac{1}{b}\frac{dt}{d\lambda}, \qquad (2.39)$$

which from 2.36, applying a small-angle approximation, yields

$$D = \frac{\lambda}{cd^2}\left[1 - \left(\frac{\lambda}{d} - \sin \gamma\right)^2\right]^{-1}, \qquad (2.40)$$

$$k'' = -\frac{\lambda^3}{2\pi c^2 d^2}\left[1 - \left(\frac{\lambda}{d} - \sin \gamma\right)^2\right]^{-1}. \qquad (2.41)$$

This expression demonstrates the possibility of selecting a set of parameters in such a way as to design a pair of gratings producing a positive or a negative group velocity dispersion. Therefore optical devices can be built that compensate a positive group velocity dispersion suffered by optical pulses travelling through a transparent material. Optical compressors have been a key to the development of various fields in which short optical pulses have been used as a primary tool. Figure 2.12 shows a typical arrangement for a reflective pulse compressor.

2.3 Nonlinear Optics

2.3.1 Second-Order Susceptibility

The second-order susceptibility $\chi^{(2)}$ governs several nonlinear optical effects, some of which will be considered in this section. Second-harmonic generation

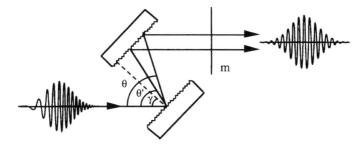

Fig. 2.12. Grating optical compressor. Two gratings are set in a subtractive diffraction geometry. Red components of a light pulse have a longer optical path than blue ones. The various components of a positively dispersed pulse can therefore be reset in phase

will be explored in some detail, while other effects will only be mentioned because of their importance in the physics of pulsed lasers.

2.3.1.1 Second-Harmonic Generation. Second-harmonic generation was the first nonlinear optical effect to be observed after lasers became available.

In this process, two photons having the same angular frequency ω, when propagating through a suitable material, can mix together, giving rise to a single photon with twice the original frequency. In the framework of classical physics, light propagation is described in terms of coherent emission by harmonic electronic dipoles, which have been set to oscillate by the light itself. This picture can be extended to the nonlinear regime by assuming that at high excitation intensities (meaning large oscillation amplitude of the electronic oscillators), the oscillations become strongly anharmonic. Under this assumption, the emitting dipoles can radiate energy at frequencies which are integer multiples of the original frequency.

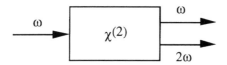

Fig. 2.13. When a noncentrosymmetric crystal is illuminated with photons with angular frequency ω some pairs of these photons disappear, replaced by single photons with angular frequency 2ω. This effect is called second-harmonic generation

The polarization of a material can be written as

$$P_i(2\omega) = \sum_{j,k} \chi_{ijk}^{(2)} E_j(\omega) E_k(\omega). \qquad (2.42)$$

In a dipole approximation of the electron–photon interaction, the second-order susceptibility $\chi^{(2)}$ is a third-order tensor, with components corresponding to the various possible orientations of the crystal axis and of light polarization. In the specific case in which the structure of the medium has an inversion centre the tensor is identical to zero: no second harmonic can be generated (the central symmetry of a material is relaxed on its surface and mixing of photons is possible there with a weak yield). In the ideal case of a plane wave the intensity of the generated second harmonic is as follows:

$$I(2\omega) = \frac{2^7 \pi^3 \omega^2 \chi_{\mathrm{eff}}^2 l^2}{n^3 c^3} I^2(\omega) \left(\frac{\sin(\Delta k l/2)}{\Delta k l/2} \right)^2. \tag{2.43}$$

The dependence on both the length of the material l and the incident intensity $I(\omega)$ is quadratic; n is the index of refraction and c the velocity of light. The intensity of the doubled frequency is proportional to the square of an effective susceptibility $\chi_{\mathrm{eff}}^{(2)}$, which depends on the material and reflects the mean doubling properties of this material in a given working direction. For the doubling yield to be maximum, the dephasing quantity $\Delta k = k_2 - 2k_1$ must be zero. This dephasing term accounts for the propagation effects during the doubling process. At any one point along the propagation axis the electronic oscillators radiate at angular frequency ω and 2ω in the same direction k. After propagation over a distance l in the medium the fundamental frequency generates the second-harmonic frequency with a time delay $k_1 l$, while the previously generated second harmonic reaches the same point with a time delay $k_2 l$. If the distance l is such that the dephasing quantity $\Delta k l = (k_2 - 2k_1)l$ is zero, then the two contributions add constructively; otherwise they interfere destructively.

The coherence length of a doubling crystal is defined as the characteristic distance along which the second-harmonic waves remain in phase, and it can be calculated using (2.43). To achieve efficient frequency-doubling devices it is now clearly apparent that dephasing must be overcome. The most common way to do this is to use birefringent crystals for which the ordinary and extraordinary index surfaces for the fundamental and doubled frequencies cross one another (Fig. 2.14).

In the propagation directions defined by these intersections, the indices of refraction at the fundamental and doubled frequencies are equal and the phase mismatch quantity $\Delta k = 0$. When this condition is fulfilled, the sinc function in (2.43) is equal to unity and the coherence length is infinite. Calculation of the phase-matching angle θ_c between the propagation direction, taken orthogonal to the surface of the crystal, and the c axis of the crystal can be made using

$$\left(\frac{1}{n_{\mathrm{e}}^{2\omega}(\theta)} \right)^2 = \frac{\sin^2 \theta}{(n_{\mathrm{e}}^{2\omega})^2} + \frac{\cos^2 \theta}{(n_{\mathrm{o}}^{2\omega})^2} = \left(\frac{1}{n_{\mathrm{o}}^{\omega}(\theta)} \right)^2. \tag{2.44}$$

Various situations can be encountered depending on the birefringent crystal: in positive crystals, like quartz, $n_{\mathrm{e}} > n_{\mathrm{o}}$, while in negative crystals, like calcite, $n_{\mathrm{e}} < n_{\mathrm{o}}$. For type I phase-matching crystals the two mixing photons have the

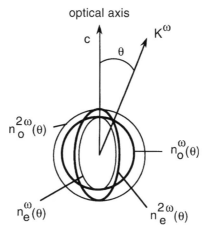

optical axis

Fig. 2.14. *Thin lines*: intersection of the plane containing the c axis of a bire-fringent crystal with its ordinary index surfaces. *Thick lines*: intersection with the extraordinary surfaces

same polarization direction, while in type II crystals they are orthogonally polarized.

Nowadays excellent doubling efficiencies are reached, close to 100 % in some cases.

Because of their wide spectral width, doubling of ultrashort light pulses needs special care. The doubling bandwidth of the crystal must accommodate the full spectral width of the pulse. For a negative uniaxial crystal, such as KDP, the wavelength bandwidth can be written as

$$d\lambda_1 = \frac{\pm 1.39\lambda_1}{2\pi l}\left(\frac{\partial n_o^\omega}{\partial \lambda_1} - \frac{1}{2}\frac{\partial n_e^{2\omega}}{\partial \lambda_2}\right), \qquad (2.45)$$

where λ_1 and λ_2 are the wavelengths of the light at the fundamental and doubled frequencies. Examination of 2.45 shows that to enlarge the doubling bandwidth of a crystal one has to decrease its thickness l, which of course is detrimental to the conversion yield, which is proportional to the square of that thickness (see (2.43)). The first ultashort pulses ever produced, 6 fs at 620 nm, have been characterized using a second-harmonic autocorrelation technique based on a 30 μm thick KDP crystal. Increasing the thickness of the crystal would have reduced its bandwidth, leading to an overestimation of the duration of the pulses.

Real life is always more complex: even if a perfect phase-matching is achieved, group velocity dispersion remains different for the fundamental and the doubled frequencies, inducing time broadening of the pulses.

2.3.1.2 Parametric Interactions. The second-order susceptibility is responsible for a large variety of other effects involving three photons and gener-

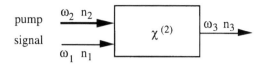

Fig. 2.15. Second-order susceptibility is responsible for various parametric effects, in which one photon is split into two other ones or two photons are mixed

ating new frequencies. These effects are called parametric effects, as a reference to forced oscillators and related techniques, which were previously developed in the field of radio-frequency electronics.

A classification of these parametric effects can be based on energy and momentum conservation for the three photons involved in the process. A nonlinear crystal with second-order susceptibility $\chi^{(2)}$ is illuminated with an intense pump pulse containing n_2 photons at angular frequency ω_2 and a weak signal pulse containing n_1 photons at angular frequency ω_1; n_3 photons are emitted with angular frequency ω_3. Such a classification is given in Table 2.2.

Photon number variation $\omega_1 > \omega_2$		$\omega_1 < \omega_2$	Photon number variation
$n_1 - 1$	Frequency difference	Frequency sum	$n_1 - 1$
$n_2 + 1$	$\omega_3 = \omega_1 - \omega_2$	$\omega_3 = \omega_2 + \omega_1$	$n_2 - 1$
$n_3 + 1$	$k_1 = k_2 + k_3$	$k_3 = k_1 + k_2$	$n_3 + 1$
$n_1 - 1$	Frequency sum	Parametric amplification	$n_1 + 1$
$n_2 - 1$	$\omega_3 = \omega_1 + \omega_2$	$\omega_3 = \omega_2 - \omega_1$	$n_2 - 1$
$n_3 + 1$	$k_3 = k_1 + k_2$	$k_2 = k_1 + k_3$	$n_3 + 1$

Table 2.2. Optical parametric interactions ordered according to energy and momentum conservation

$\omega_1 > \omega_2$. The pump frequency is smaller than the signal frequency. There are two cases to consider.

- *Case 1.* One photon with angular frequency ω_1 is annihilated (photon number becomes $n_1 - 1$) while two photons with angular frequencies ω_2 $(n_2 + 1)$ and ω_3 $(n_3 + 1)$ are created. Energy conservation implies the equality $\omega_3 = \omega_1 - \omega_2$. Every parametric interaction is a coherent effect which must respect a phase-matching condition between the various waves propagating in the crystal; this is implied by momentum conservation, $k_1 = k_2 + k_3$. In this case photons are emitted with a frequency which is equal to the frequency difference between the two incident photons. The signal (ω_1) intensity decreases (n_1 decreases), while the intensity of the pump (ω_2) and the frequency difference (ω_3) increase.

- *Case 2.* One photon with angular frequency ω_1 is destroyed ($n_1 - 1$) along with a pump photon (ω_2) ($n_2 - 1$), while one photon is created ($n_3 + 1$) with angular frequency ω_3. Energy conservation implies $\omega_3 = \omega_1 + \omega_2$ while momentum conservation implies $k_3 = k_1 + k_2$. This parametric effect is called "up-conversion". The pump and signal intensities decrease, favoring the resulting emission, contrary to Case 1, in which the pump intensity increases.

As in the case of second-harmonic generation, these processes can only take place in crystals having a large $\chi^{(2)}$, and where there are propagation directions allowing index matching and momentum conservation. When these prerequisites are fulfilled, waves produced in phase can add constructively, giving rise to significant light intensities. Selection of one effect or the other is realized by adjusting the angles of incidence of the incident light beams with respect to the crystal axes.

In a semiclassical model in which only the electronic system is quantized, these effects are described, in the transparency regime, via virtual electronic transitions. Therefore, as they are not related to any specific electronic resonance, tunability can be achieved by adjusting the incidence angles of the beam, on the crystal. However, the tunability bandwidth is limited by the phase-matching condition. On any timescale down to a few femtoseconds these effects can be considered as instantaneous.

$\omega_1 < \omega_2$. The pump frequency is greater than the signal frequency. Again, there are two cases.

- *Case 1* corresponds again to up-conversion with a reversal of the roles played by the pump and the signal, compared to the previous Case 2.
- *Case 2.* One photon with angular frequency ω_2 is destroyed ($n_2 - 1$) while two photons are created, one with the signal frequency $\omega_1(n_1 + 1)$ and one with the new frequency ω_3 ($n_3 + 1$), in such a way as to ensure $\omega_2 = \omega_1 + \omega_3$ and $k_2 = k_1 + k_3$. This is called "optical parametric amplification" (OPA) because the weak signal is amplified, in contrast to the strong pump intensity.

Other combinations could be added to the table which would not respect the realistic condition that a frequency must be positive. For example, one cannot expect a process to start with the destruction of a photon with angular frequency ω_3.

These effects are not totally independent. Let us consider, for example, the parametric frequency difference effect. For long enough propagation paths in matter, the implicit assumption that there is no depletion of the beams is not necessarily valid; the intensity of the signal can drop to 0 ($n_1 = 0$). The process then stops but the situation is such that up-conversion may start, one photon with angular frequency ω_3 mixing with one photon with angular frequency ω_2.

2.3.1.3 Spontaneous Parametric Fluorescence. Spontaneous parametric fluorescence is a special case of the parametric amplification process, in which the signal with angular frequency $\omega_1 < \omega_2$ is turned off. A $\chi^{(2)}$ nonlinear crystal spontaneously emits radiation with angular frequency ω_1 and ω_3 when irradiated with a strong pump beam made up of photons with angular frequency ω_2 such that $\omega_3 = \omega_2 - \omega_1$ and $k_1 + k_3 = k_2$. Although only a quantum-mechanical treatment can rigorously account for this effect, a rather good classical picture is to consider the effect as being started by the ambient noise, replacing the ω_1 signal photons. For a given orientation of the nonlinear crystal, numerous photons simultaneously satisfy the energy and momentum conservation conditions and therefore the spectral bandwidth of the parametric emission is large. This is being widely studied for the amplification of stretched ultrashort light pulses.

2.3.1.4 Applications of Parametric Effects. Second-harmonic generation is by far the most commonly used optical nonlinear effect in the field of lasers. Argon ion lasers are poor converters of electrical energy to light; a 10^{-3} conversion yield is a good figure. Solid-state lasers, on the contrary, are more efficient converters, for which a yield ten times larger is readily achievable. Nd:YAG, Nd:YLF, Nd:glass, etc. lasers emit light around $1.06\,\mu m$ and it is therefore of useful policy to double this frequency, since a 30 % doubling yield is easy to realize. This is clearly illustrated by the introduction onto the market of doubled frequency diode-pumped solid-state lasers, generating more than 5 W of CW green light, needing only a regular electrical socket as an energy source. Doubling is also commonly used for ultrashort pulse metrology.

Up-conversion, also used to increase the frequency of laser pulses, is useful as well for time resolution of photoluminescence. The spontaneous emission of a material can be mixed in a $\chi^{(2)}$ nonlinear crystal with a short reference, pulse allowing a spectro-temporal study of the emission. Detectors in the infrared spectrum are less sensitive than their equivalents in the visible range, so it can be of great interest to up-convert a weak signal to the visible range, using a strong pump. More applications are described in some detail in the following chapters.

OPA. Optical parametric amplification has already been described above and we have learned that the angular frequencies ω_1 and ω_3 are amplified via an intensity decrease of the pump. This amplification process should not be confused with the amplification by stimulated emission in a medium which has an inverted electronic population. Because of the instantaneous character of the process there is no time delay between excitation and emission.

The Manley–Rowe equation

$$-\Delta\left(\frac{I_2}{\omega_2}\right) = \Delta\left(\frac{I_1}{\omega_1}\right)\Delta\left(\frac{I_3}{\omega_3}\right) \tag{2.46}$$

relates the intensity variations of the various beams. Using a LiNbO$_3$ crystal having a susceptibility $\chi_{\text{eff}}^{(2)} = 0.5 \times 10^{-22}$ MKS, pumped in the vicinity of

$\lambda_2 = 1\,\mu$m with an intensity $I_2 = 10^6\,$W/cm^2, a gain of only $0.667\,$cm^{-1} is reached for the emitted wavelengths $\lambda_3 = 3\,\mu$m and $\lambda_1 = 0.5\,\mu$m. Parametric gains are always low and are of best use in laser cavities.

OPO. An optical parametric oscillator is a laser in which the optical gain is produced in a parametric crystal rather than by an electronic inverted population. The analysis of this application is beyond the scope of this short introduction and is usually described in quantum-electronics textbooks. We will merely point out that one can set up laser cavities oscillating at both amplified frequencies or only one, depending on needs.

These techniques are developing rapidly and are worth the work they imply, because of the nonresonant character of the parametric amplification process, which allows the tuning of the emitted light by adjustment of the phase-matching angles. In practice this means that tunability is achieved by orienting a nonlinear crystal with respect to the axis of an optical resonator.

The parametric techniques have already reached a point of development which makes them commercially available. Several companies are presently competing in niche markets.

2.3.2 Third-Order Susceptibility

Four photons are involved in the nonlinear optical effects due to the third-order susceptibility term of the expansion of the polarization of a material. In contrast with second-order effects, the various third-order effects do not vanish in centrosymmetric materials and they can be observed in liquids or amorphous materials, such as fused silica for instance. For all the possible effects, energy and momentum are conserved quantities and phase-matching conditions must be fulfilled.

One of the possible effects is the mixing of two photons of a laser beam in a $\chi^{(3)}$ nonlinear medium, generating two frequency-shifted photons such that $2\omega_L = \omega_1 + \omega_2$, with $\omega_1 > \omega_L$ and $\omega_2 < \omega_L$. When two laser beams are present a more general interaction occurs, such as, for instance, the mixing of three photons into one high-frequency photon. A single photon can also split into three lower-frequency photons. However, the efficiency of these effects is always poor. Degenerate and nondegenerate four-wave mixing have no popular application in the field of lasers; rather, they are used for the measurement of energy and phase relaxation dynamics in matter.

Stimulated fluorescence, amplification and Raman and Brillouin scattering, as well as two-photon absorption, belong to this class of $\chi^{(3)}$ effects. Obviously the number of observable effects is a rapidly growing function of the susceptibility order, and we will now focus only on the variations of the index of refraction induced by a strong pulse in a transparent isotropic medium.

2.3.2.1 Nonlinear Index of Refraction. When an intense electromagnetic wave passes into an isotropic medium its dielectric response is changed:

$$\varepsilon_t = \varepsilon + \varepsilon_2 \langle \boldsymbol{E} \cdot \boldsymbol{E} \rangle. \tag{2.47}$$

Where $\langle \boldsymbol{E} \cdot \boldsymbol{E} \rangle$ is the time average of the squared electric field. Therefore $\langle \boldsymbol{E} \cdot \boldsymbol{E} \rangle = \frac{1}{2} |\boldsymbol{E}|^2$. It can be demonstrated that such a dielectric function occurs when the polarization is written as

$$\boldsymbol{P} = \varepsilon_0 \chi^{(1)} + \varepsilon_0 \chi^{(3)} \langle \boldsymbol{E} \cdot \boldsymbol{E} \rangle \boldsymbol{E}; \tag{2.48}$$

therefore $\varepsilon = 1 + \chi^{(1)}$ and $\varepsilon_2 = 1 + \chi^{(3)}$. The $\chi^{(2)}$ term is zero because of the choice of a centrosymmetric medium.

As the index of refraction is given by the square root of the dielectric function, one easily finds that

$$n = \sqrt{\varepsilon_t} = \sqrt{\varepsilon + \varepsilon_2 \langle \boldsymbol{E} \cdot \boldsymbol{E} \rangle} \approx n_0 + n_2 \langle \boldsymbol{E} \cdot \boldsymbol{E} \rangle \tag{2.49}$$

and one eventually gets

$$n = n_0 + \tfrac{1}{2} n_2 I, \tag{2.50}$$

where $I = |\boldsymbol{E}|^2$ is the intensity of the light.

Numerous physical effects can yield a nonlinear polarization of the form $\varepsilon_0 \chi^{(3)} \langle \boldsymbol{E} \cdot \boldsymbol{E} \rangle \boldsymbol{E}$ and therefore an intensity-dependent index of refraction.

When an applied electric field is strong enough the electronic cloud of an atom or a molecule is strongly distorted. This happens when the mean electrostatic energy of the field $\frac{1}{2} \varepsilon_0 \varepsilon \langle \boldsymbol{E} \cdot \boldsymbol{E} \rangle EV$ (V being the interaction volume) becomes comparable to the energy of the electronic states (a few eV). Also the rotational Kerr effect can take place in liquids with anisotropic dissolved molecules whose polarisability is different along their large and small axes. A linearly polarized electric field favors the alignment of one of the axis of the molecules along its direction. This effect is generally stronger than the electric-field effect; it is also slower because it involves the full mass of the molecules instead of only the mass of the electrons.

From 2.50 we know that in the $\chi^{(3)}$ limit the change of index of refraction in a transparent medium is simply proportional to the intensity of the applied electric field. Closer examination also indicates that properties of the intensity distribution must be mapped onto the index change. Generally speaking, the light emitted by a pulsed laser is distributed in both space and time, $I = I(\boldsymbol{r}, t)$.

The analysis of the nonlinear propagation of a light pulse is a complex problem which needs numerical approaches; we will only consider situations which can be studied analytically.

2.3.2.2 Kerr Lens Effect. We will start with the spatial dependence of the light intensity and for the sake of simplicity we will study the propagation of a Gaussian laser beam in a $\chi^{(3)}$ material. This beam is described by a Gaussian function of its radius with shape parameter g. In this case the index distribution can be written as:

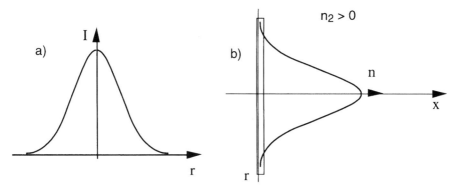

Fig. 2.16. (a) Intensity distribution of a Gaussian beam along one of its diameters. (b) Assuming the propagation of this intensity distribution along the x direction in a thin slab of $\chi^{(3)}$ material, the variation of the index of refraction follows the intensity distribution along the diameter. Depending on the sign of the nonlinear index of refraction, the index of refraction increases ($n_2 > 0$) or decreases ($n_2 < 0$) when going toward the center of the laser beam

$$n(r) = n_0 + \tfrac{1}{2}n_2 I(r), \quad \text{with } I(r) = e^{-gr^2}. \tag{2.51}$$

Figure 2.16a shows the intensity distribution along a diameter of a Gaussian beam. If this beam is propagating through a thin slab of $\chi^{(3)}$ transparent material (Fig. 2.16b), then the index change follows the intensity along that same diameter. For a positive n_2 the index of refraction is larger at the center of the slab than at the side. In the framework of geometrical optics, the quantity relevant to the propagation of a light ray is the optical path, i.e. the product of the index of refraction and the propagation distance e, $P(r) = n(r)e$. To visualize the Kerr lens effect we replace the constant thickness e by a variable one such that its product with a constant index of refraction leads to the same optical path:

$$P(r) = n(r)e = e(r)n_0. \tag{2.52}$$

Then

$$e(r) = \frac{en(r)}{n_0} \tag{2.53}$$

and one gets a Gaussian lens, which focuses the optical beam. During the propagation of a pulse through a thick material (Fig. 2.17) this process is enhanced along the path because focusing of the beam increases the focal power of the dynamical lens. This increase of the focusing stops when the diameter of the beam is small enough so that the linear diffraction is large enough to balance the Kerr effect. This effect, named self-focusing, is of prime importance in the understanding of self-mode-locking, or Kerr lens mode-locking (KLM) which occurs in titanium-doped sapphire lasers.

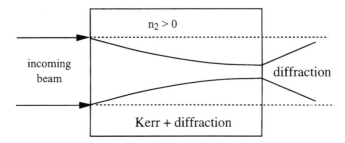

Fig. 2.17. Self-focusing of a laser beam crossing a medium with positive n_2

2.3.2.3 Self-Phase-Modulation. Similarly, the nonlinear index of a material depends on the time dependence of a light pulse intensity envelope, which can be expressed as

$$n = n_0 + \tfrac{1}{2}n_2 I(t), \qquad \text{with } I(t) = e^{-\Gamma t^2}. \tag{2.54}$$

What is the influence of this time-varying index on the frequency of the light? To simplify the analysis we only consider a plane wave propagating in a nonlinear medium:

$$E(t,x) = E_0\, e^{i(\omega_0 t - kx)}, \qquad k = \frac{\omega_0}{c} n(t). \tag{2.55}$$

The instantaneous frequency, being the time derivative of the phase, can be written as

$$w(t) = \frac{\partial}{\partial t}\Phi(t) = \omega_0 - \frac{\omega_0}{c}\frac{\partial n(t)}{\partial t}x \tag{2.56}$$

and the frequency variation as

$$\delta w(t) = w(t) - \omega_0 = -\frac{\omega_0 n_2}{2c}x\frac{\partial I(t)}{\partial t} \tag{2.57}$$

(Fig. 2.18).

As a very general consequence of the Fourier duality between time and frequency, any time a periodic amplitude or phase modulation is applied to a periodic signal new frequency components are created in its frequency spectrum. In the self-phase-modulation process, with n_2 positive, new low frequencies are created in the leading edge of the pulse envelope and new high frequencies are created in the trailing edge. These new frequencies are not synchronized, but are still created inside the original pulse envelope. Self-phase-modulation is not a dispersive effect in itself, but a pulse does not remain transform-limited when it crosses a transparent material. The transparent material in which the pulse propagates is dispersive, however, and therefore the frequencies are further chirped along the propagation.

Self-phase-modulation, because it opens the way to a spectral broadening of a light pulse, has been (and still is) the very basis of the optical compression

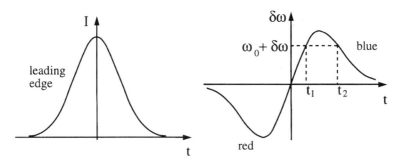

Fig. 2.18. *Left*, intensity dynamics of a Gaussian light pulse; the earlier times, i.e. the leading edge of the pulse, lie on the left side of the graph. *Right*, time variation of the central pulsation, which is proportional to the negative of the pulse envelope derivative when the nonlinear index of refraction is positive

technique used for producing ultrashort light pulses less than 10 fs in duration in the visible spectral range. When propagating light in a monomode silica fiber, only light with momentum along the fiber axis can be coupled; for that reason intense light pulses cannot undergo self-focusing, which would correspond to forbidden light momentum directions, and as a consequence self-phase-modulation is favored. Self-focusing down to the diffraction limit in sapphire leads to the generation of the large white light continua used as seeds in commercially available OPA.

Induced index variations can also be observed in degenerate or nondegenerate pump-and-probe experiments. A strong pump pulse can induce an index change in a medium, which can be experienced by a weak probe pulse. In the time domain the dynamical index change created by the pump shifts the central frequency of the probe. A record of the frequency shift versus time delay maps the derivative of the pump envelope. When n_2 is positive the probe frequency is red-shifted when it leads the pump pulse and blue-shifted when it trails the pump. This is the physical basis of the famous FROG technique used to simultaneously characterize phase and amplitude of short optical pulses (see Chap. 7). In the space domain the probe beam undergoes focusing, or defocusing as well as deflection, depending on the relative geometry of the pump and the probe beams.

Fourier transformation allows us to calculate the spectrum which results from the self-phase-modulation process. Figure 2.19 shows the spectrum of a self-phase-modulated pulse when the total nonlinearly induced phase shift is equal to 2π.

The spectral signature of self-phase-modulation is a "channeled" spectrum: as can be noticed from 2.57 and seen in Fig. 2.18, pairs of identical frequencies are created at two different time delays, t_1 and t_2, inside the pulse envelope. For values ω of the angular frequency such that the time delay $t_2 - t_1 = n\pi/\omega$, n an odd integer, there is a destructive interference between the two

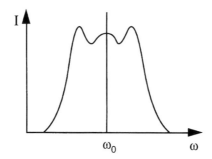

Fig. 2.19. Frequency spectrum of an originally Gaussian pulse which has suffered an induced phase shift equal to 2π

generated waves and a subsequent dip in the spectrum. At a time for which $t_2 - t_1 = n2\pi/\omega$ (n is an integer) the interference is constructive and the spectrum passes through a maximum. A simple examination of the spectrum therefore allows one to determine the maximum phase change and to estimate the absolute value of the nonlinear index of refraction.

Two other $\chi^{(3)}$ nonlinear effects are of great importance in the understanding of laser physics: gain saturation and absorption saturation. Gain saturation has already been examined in Chap. 1, so we will now only consider absorption saturation.

2.3.2.4 Saturable Absorbers. A saturable absorber is a material in which the transmittance increases with increasing light illumination. To describe the process we will consider simple qualitative arguments based on a two-level electronic model, for which saturable absorption is symmetrical to gain saturation. In this simple framework, a hyperbolic dependence of the absorption coefficient on the intensity is again a satisfactory approximation:

$$\alpha(I) = \alpha_0 \left(1 + \frac{I}{I_{\text{sat}}} \right)^{-1} , \qquad (2.58)$$

where α_0 is the low-intensity (linear) absorption coefficient and I_{sat} is the saturation intensity, a phenomenological parameter.

For increasing intensities the absorption coefficient decreases, which translates for the transmitted intensity as

$$T(I_{\text{i}}) = \frac{I_{\text{t}}}{I_{\text{i}}} = \exp \left[\frac{-\alpha_0 L}{1 + I_{\text{i}}/I_{\text{sat}}} \right] . \qquad (2.59)$$

Equation 2.59 is obtained by replacing the constant low-intensity absorption coefficient in the Beer–Lambert law $I_{\text{t}} = I_0 \exp(-\alpha_0 L)$ by expression 2.58. I_{i} is the intensity incident on the absorbing medium of length L, and I_{t} is the transmitted intensity. Figure 2.21 shows a numerical simulation of the transmission behavior as a function of the incoming light intensity.

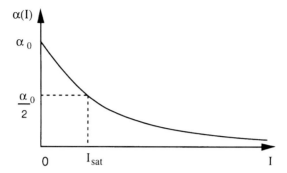

Fig. 2.20. Simple hyperbolic model of the absorption coefficient α of a two-level system versus incident intensity. When the intensity reaches the saturation value I_{sat} the low-intensity absorption α_0 is reduced by a factor of two

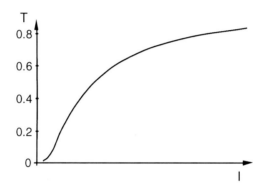

Fig. 2.21. Numerical simulation of transmission versus incoming intensity in a two-level electronic system

Saturable absorbers can be split into two classes depending on the speed of the change of transmission. Fast saturable absorbers have an instantaneous response under light illumination. Figure 2.22 shows a simulation of the effect of such a saturable absorber on a Gaussian pulse.

On the wings of the pulse, where the intensity is low, absorption takes place, while the summit of the pulse remains mostly unchanged. As a result of its propagation through the fast saturable absorber the pulse undergoes a temporal narrowing.

Saturation in a slow saturable absorber depends on the light intensity integrated over some characteristic time duration. As a result, the leading part of the pulse is strongly absorbed. Once saturation is reached, the saturation of the absorption decays with some other characteristic time and the trailing part of the pulse passes through the medium with no distortion. The result of the nonlinear absorption is asymmetric, as can be seen in Fig. 2.23.

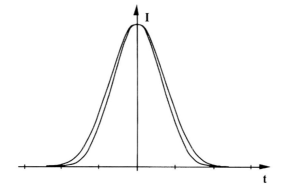

Fig. 2.22. Numerical simulation of the effect of a fast saturable absorber on a Gaussian light pulse. In order to make clear the narrowing of the pulsation of the pulse, the modified pulse has been normalized to the original one

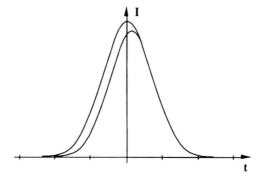

Fig. 2.23. Numerical simulation of the effect of a slow saturable absorber on a Gaussian light pulse

"Slow" and "fast" only have a meaning when comparing the characteristic time of the saturable absorber with the duration of the light pulses. For ultrashort light pulses, say less than a few picoseconds, all saturable absorbers are slow ones.

In a simple description, the photons from the leading part of a light pulse penetrating a saturable absorber disappear when promoting electrons from a ground to an excited state. As the pulse propagates further, more and more photons are absorbed, promoting more and more electrons, up to the point at which the ground-state and excited electronic populations are equalized (see Chap. 1) and no more absorption can take place. The dynamics of the saturation are under the control of the electronic relaxations.

2.4 Cascaded Nonlinearities

Cascading effects were discovered in 1986 when U. Osterberg and W. Margulis [2.1] observed the doubling of a Nd:YAG laser beam in a centrally symmetric optical fibre with a 10% yield. They had, however, been discovered in 1972 by E. Yablanovitch, C. Flytzanis, and N. Bloembergen [2.2]. The electric field E which appears in the development of the material polarization (expression (2.1)) is the total macroscopic field and, as the polarization modifies the field through Maxwell's equations, it is an unknown quantity. In the development of the optical nonlinearities we have considered so far, the effect of the medium polarization on the field in the medium was not considered: only the contribution $E^{(1)}$ to the field that would occur if the medium would have a linear response was taken into account. The total response of a material to an applied electric field is therefore a nonlinear function of this $E^{(1)}$ field, and it contains products of the nonlinear susceptibilities $\chi^{(i)}$. These terms are referred to as cascaded nonlinearities.

No cascading effect has to be considered including the linear susceptibility $\chi^{(1)}$ because this term depends only on the linear field. Therefore the lowest-order cascading term contains $\chi^{(2)} . \chi^{(2)}$ a fourth-order rank tensor which is therefore equivalent to a direct $\chi^{(3)}$ third-order optical nonlinearity.

To exemplify a cascading process we will examine the third-harmonic generation in a material with no central symmetry. For the sake of simplicity, scalar susceptibilities and no polarizations of the field will be considered. We start with the linear field $E^{(l)} = E_\omega = A_\omega e^{ik_\omega x}$ being a plane wave propagating in the x direction. This wave induces a material polarization at pulsation 2ω

$$P_{2\omega} = \frac{\varepsilon_0}{2} \chi^{(2)} E_\omega E_\omega, \tag{2.60}$$

as well as a polarization at pulsation 3ω

$$P_{3\omega} = \frac{\varepsilon_0}{4} \chi^{(3)} E_\omega E_\omega E_\omega. \tag{2.61}$$

The second-order polarization induces a second harmonic field

$$E_{2\omega} = A_{2\omega} e^{ik_{2\omega} x}, \tag{2.62}$$

which amplitude can be calculated in the nondepletive regime, when the decrease of the fundamental amplitude is negligible, from the light propagation equation

$$E_{2\omega} = \frac{\omega}{2n_{2\omega}c} \chi^{(2)} \frac{1}{\Delta k} \left(1 - e^{-i\Delta kx}\right) E_\omega E_\omega \tag{2.63}$$

where $n_{2\omega}$ is the index of refraction of light at the two-photon pulsation, and

$$\Delta k = 2k_\omega - k_{2\omega} = \frac{2\omega}{c}(n_\omega - n_{2\omega}) \tag{2.64}$$

is the phase mismatch between the fundamental and the doubled waves. The generated second harmonic wave mixes with the fundamental wave through up-conversion (sum-frequency generation)

$$P'_{3\omega} = \varepsilon_0 \chi^{(2)} E_{2\omega} E_\omega, \qquad (2.65)$$

which by using equation (2.63) leads to the following cascading contribution:

$$P'_{3\omega} = \varepsilon_0 \frac{\omega}{2n_{2\omega}c} \chi^{(2)} \chi^{(2)} \frac{1}{\Delta k} (1 - e^{-i\Delta kx}) E_\omega E_\omega E_\omega \qquad (2.66)$$

so that the real material polarization at the third harmonic pulsation contains a term depending on $\chi^{(2)}\chi^{(2)}$. In this simplified model the cascaded third-order polarization diverges when the phase-matching condition for the second-harmonic generation $\Delta k = 0$ is realised so that it becomes the main contribution to the third harmonic generation. The enlarged third-order susceptibility is written as:

$$\chi_t^{(3)} = \chi^{(3)} + \frac{\omega}{2n_{2\omega}c} \chi^{(2)} \chi^{(2)} \frac{1}{\Delta k} (1 - e^{-i\Delta kx}), \qquad (2.67)$$

which depends on the propagation length x. In the case when phase matching is realized for the pure third harmonic generation process (first contribution in Eq. (2.67)) the phase-matching condition for the second harmonic generation process is not fulfilled and the exponential term in (2.67) becomes negligible and the expression simplifies to

$$\chi_t^{(3)} = \chi^{(3)} + \frac{\omega}{2n_{2\omega}c\Delta k} \chi^{(2)} \chi^{(2)}, \qquad (2.68)$$

which is now independent of the length of propagation.

The simple appearance of the effect as described earlier is deceptive because cascading effects are vector dependent, they vary with the propagation vectors of the light beams as well as with their polarization. They cannot alway be distinguished from the direct-order nonlinear effects and their amplitude is generally of the same order of magnitude. Generally, an nth-order polarization has contributions arising from any set $\{n_i\}$ of multistep processes such that $\sum_i (n_i - 1) = n - 1$. General solutions for the optical cascading equations describing the $\chi^{(2)} : \chi^{(2)}$ process have been developed in reference [2.3].

The frequency doubling in a fibre mentioned in the introduction was tricky. It was shown that, because of surface symmetry breaking, a small amount of doubled frequency is created at the entrance of the fibre. The interference between the rectified field from the fundamental and doubled frequencies creates a static grating that in turn creates and organises macroscopic dipoles owed to the presence of doping atoms in the fibre (the original experiments were performed in germanium-doped fibres). This optical polling of the fibre breaks down the central symmetry and efficient frequency doubling can take place [2.4].

Considerable energy has been spent developing the cascaded Kerr with the perspective of developing optronic devices [2.5]. Non-phase-matched second harmonic is generated in a fibre that is down-converted to the fundamental frequency because of the non-phase-matched case. Then the phase shift between the fundamental light and the down-converted one translates as an intensity-dependent index change equivalent to a classical Kerr effect.

Cascading has been recognised as a way to efficiently generate fourth harmonic [2.6] of optical spatiotemporal solitons [2.7]; quantum optics should rapidly benefit from the recent advances of the field.

2.5 Problems

1. A light pulse has an intensity envelope shaped as a Gaussian function $I(t) = \exp(t/t_0)^2$. Calculate the full width at half maximum Δt (FWHM) for this pulse.

   The autocorrelation function of the pulse envelope can be measured and is expressed as $G(\tau) = \int_{-\infty}^{+\infty} I(t)I(t-\tau)\,dt$. Explicitly calculate the autocorrelation function of the pulse.

   What is the FWHM $\Delta t'$ of this pulse autocorrelation function? What is the value of the ratio $\Delta t'/\Delta t$? (Answer: $\sqrt{2}$)

2. Coherence length for frequency doubling in a quartz crystal.

   For quartz, the ordinary index n_o is isotropic. In a plane which contains the c axis and the propagation direction k, the ordinary index lies on a circle with radius n_o. In that same plane the extraordinary index lies on an ellipse described by $[1/n_e(\theta)]^2 = \cos^2\theta/n_o^2 + \sin^2\theta/n_e^2$, θ being the angle between c and k.

   Consider a fundamental wavelength in vacuum $\lambda_0^{(1)} = 620\,\mathrm{nm}$ and its doubled counterpart $\lambda_0^{(2)} = 310\,\mathrm{nm}$; the indices for quartz close to these wavelengths are given in the following table:

$\lambda\,(\mu m)$	n_o	n_e
0.62782	1.542819	1.551880
0.312279	1.57433	1.584485

   Show that under these circumstances there is no crossing between the index curves for the fundamental and doubled frequencies. Phase matching is not possible in quartz, at least for these wavelengths.

   From 2.43 it can be used that $I(2\omega) \propto \sin c^2\,\Delta kL/2$, L being the thickness of the crystal. This expression shows that the length $L_c = 2\pi/\Delta k$, called the coherence length, maximizes the intensity of the doubled frequency. Assuming type I propagation in a quartz crystal of a single beam with central wavelength 620 nm, the phase mismatch can be written as

   $$\Delta k = k_e^{2\omega} - 2k_o^{\omega} \quad \text{and} \quad k = 2\pi n(\lambda_0)/\lambda_0. \tag{2.69}$$

What is the value of the coherence length when the beam propagates along the c axes, $\theta = 0$? Using your favored mathematics software, plot the dependence of the coherence length on θ, under these conditions.

3. Show that the pulsation chirp law described by expression (2.16)

$$\omega(t) = \partial\phi/\partial t = \omega_0 + at$$

is a limit when the observation time t is smaller than the duration of the pulse $(\Gamma t^2 \ll 1)$. Deduce the value of the parameter a. (Answer: $a = \omega_0 n_2 x \Gamma/c$).

Further Reading

S.A. Akhmanov, V.A. Vysloukh, A.S. Chirkin: *Optics of Femtosecond Laser Pulses* (American Institute of Physics, New York 1992)

R.R. Alfano (ed.): *The Supercontinuum Laser Source* (Springer, Berlin, Heidelberg 1989)

M. Born, E. Wolf: *Principles of Optics* (Pergamon, Oxford 1993)

H. Haken: *Light*, Vol. 1, "Waves, photons, atoms" (North-Holland, Amsterdam 1986)

J.D. Jackson: *Classical Electrodynamics*, 2nd edn. (Wiley, New York 1974)

L. Landau, E. Lifschitz: *Field Theory* (Mir, Moscow 1966)

S.L. Shapiro (ed.): *Ultrashort Light Pulses*, Topics in Applied Physics, Vol. 18 (Springer, Berlin, Heidelberg 1977)

A.E. Siegman: *Lasers* (University Science Books, Mill Valley, CA 1986)

A. Yariv: *Quantum Electronics*, 3rd edn. (Wiley, New York 1989)

References

[2.1]   U. Osterberg, W. Margulis: *Optics Lett.* **11**, 516 (1986)

[2.2]   E. Yablanovitch, C. Flyzanis, N. Bloembergen: *Phys. Lett.* **29**, 865 (1972)

[2.3]   S. Lafortune, P. Winternitz, C.R. Menuyk: *Phys. Rev. E.*, **58**, 2518 (1998)

[2.4]   B.P. Antonyuk: *V.B. Antonyuk Physics-Uspekhi*, **44**, 53 (2001)

[2.5]   R. Schiek: *J. Opt. Soc. Am.* **10**, 1848 (1993)

[2.6]   A.A. Sukhorukov, T.J. Alexander, Y.S. Kivshar, S.M. Saltiel: *Phys. Lett. A*, **281**, 34 (2001)

[2.7]   M. de Sterke, S.M. Saltiel, Y.S. Kivshar: *Opt. Lett.*, **26**, 539 (2001)

[2.8]   X. Liu, L.J. Qian, F.W. Wise: *Phys. Rev. Lett.*, **82**, 4631 (1999)

3

Methods for the Generation of Ultrashort Laser Pulses: Mode-Locking

A. Ducasse, C. Rullière and B. Couillaud

With 26 Figures

3.1 Introduction

After the considerations developed in the preceding chapter, it seems contradictory, a priori, to generate ultrashort pulses with a laser source, because of the frequency selection imposed by the laser cavity. Indeed, the Fourier transform of an extremely short light pulse is spectrally very broad. Yet, a laser cavity will allow oscillation in only a few very narrow frequency domains around the discrete resonance frequencies $\nu_q = qc/2L$ (where q is an integer, c the speed of light and L the optical length of the laser cavity). Therefore a laser cannot deliver ultrashort pulses while functioning in its usual regime, in which the cavity plays the part of a frequency selector. However, it has been shown in Chap. 1 that when a laser operates in its most usual regime, it oscillates simultaneously over all the resonance frequencies of the cavity for which the unsaturated gain is greater than the cavity losses. These frequencies make up the set of longitudinal modes of the laser. While operating in the multimode regime, the output intensity of the laser is no longer necessarily constant with time. Its time distribution depends essentially on the phase relations existing between the different modes, as illustrated by the simulation in Fig. 3.1. Figure 3.1a shows the intensity of oscillation of a single mode, Fig. 3.1b that of the resultant intensity of two modes in phase, and Figs. 3.1c and d that of eight modes. In the case of Fig. 3.1c, where the phase differences between the modes were chosen randomly, the time distribution of the intensity shows a random distribution of maxima. In the case of Fig. 3.1d, the eight modes oscillate with the same initial phase, and the time distribution shows a periodic repetition of a wave packet resulting from the constructive interference of the eight modes.

This very simple simulation, which can be done with a home computer, points out the importance of phase relations to the time distribution of laser intensity. This role can be very simply understood with Fresnel visualization, as shown in Fig. 3.2.

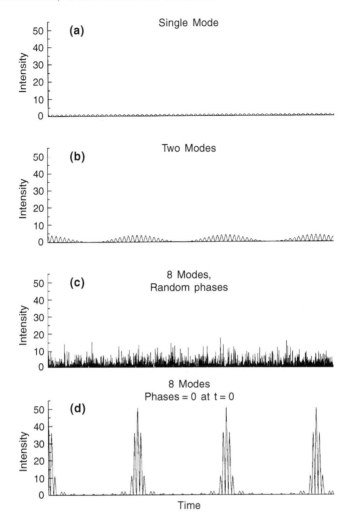

Fig. 3.1. Illustration of the influence of the phase relation between the modes on the resultant intensity of oscillation. (**a**) One mode, (**b**) two modes in phase, (**c**) eight modes with random phases, (**d**) eight modes with the same phase

Let us assume n modes, with sinusoidal oscillation at angular frequencies ω_i, with identical phases at time $t = 0$ and with equal amplitude E ($E_i = E \sin \omega_i t$). Further, let us assume that $\omega_{i+k} - \omega_i = k\Delta\omega$, k being an integer and $\Delta\omega$ a fixed spectral interval. At $t = 0$, the resultant amplitude $E_T = nE$, since all the components are aligned along the x axis of the diagram (Fig. 3.2a). At some later time Δt, the representative vector has rotated through an angle equal to $\omega_i \Delta t$ and the angle difference θ between two adjacent modes will be (Fig. 3.2b):

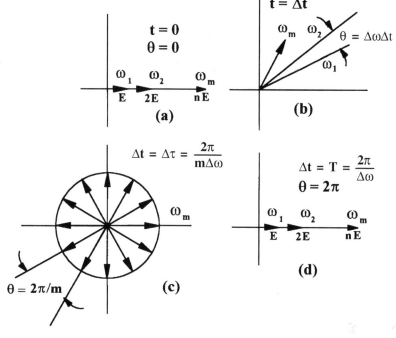

Fig. 3.2. Fresnel representation of m modes. The sum of the vectors represents the amplitude of the field inside a laser cavity at different times, when all the modes are supposed to be in phase at time $t = 0$ (see text)

$$\theta = \Delta\omega \, \Delta t. \qquad (3.1)$$

As shown in Fig. 3.2c, when $\theta = 2\pi/m$ the resultant amplitude E_T goes to zero. This occurs at time $\Delta t = \Delta\tau$, where

$$\Delta\tau = \frac{2\pi}{m \, \Delta\omega}. \qquad (3.2)$$

The larger the number of modes m, the shorter is the time $\Delta\tau$ for the amplitude to go from its maximum to zero.

Each time $\theta = 2k\pi$, E_T will again reach a maximum ($E_T = mE$), and this will occur at time $T = 2k\pi/\Delta\omega$ (Fig. 3.1d). To summarize, for a large number of modes m and a spectral interval $\Delta\omega$, the resultant amplitude will periodically reach its maximum, with period T, and will go to zero very fast at time $kT + \Delta\tau$. This very simple representation illustrates the role of the phase in constructive and destructive interference between the different modes. Note that if $\Delta\omega = 2\pi c/2L$ then $T = 2L/c$.

We see that the laser output will consist of a periodic sequence of pulses instead of just a single pulse. The width of each pulse will be inversely pro-

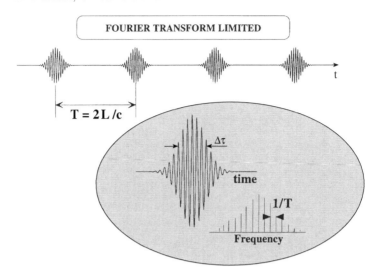

Fig. 3.3. Illustration of Fourier-transform-limited pulses

portional to the number of modes contributing to the oscillation. The value of the period T is given by $T = 2L/c$.

Figure 3.3 shows a calculated simulation of this kind of behavior. Pulses which are obtained by assuming that the initial phases are rigorously equal are said to be Fourier-transform-limited and the laser is said to be "mode-locked".

The aim of this chapter is, first of all, to examine the mode-locked regime of the laser in detail, as compared to its usual free multimode regime. We shall pay special attention to the various methods actually used to lock the mode phases relative to one another. In the rest of the chapter, we shall implicitly assume that all longitudinal oscillation modes are transverse fundamental TEM_{00} modes. In this book, we shall not go into the phase-locking of other transverse modes, which may also give rise to interesting laser behavior [3.1].

3.2 Principle of the Mode-Locked Operating Regime

The longitudinal modes which are able to self-oscillate in the free multimode regime of the laser are those for which the unsaturated gain is greater than the cavity losses, as shown in Fig. 3.4. The number of these modes, N, can vary from just a few (in He–Ne lasers for example) to some 10^4 (in dye lasers and in Ti:sapphire lasers for example). Under these conditions, the instantaneous intensity shows fluctuations, with lines whose mean width $\Delta\tau \approx 1/\Delta\nu_L$ is of the order of the inverse of the width of the gain curve [3.2]. The larger the number of modes concerned, the smaller the value of $\Delta\tau$. If now we assume the modes to have constant phase differences, the laser output will consist of a periodic succession of single pulses, each lasting $\Delta\tau$, the repetition rate

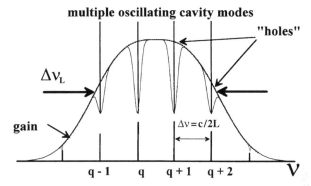

Fig. 3.4. Allowed oscillating modes in a cavity of optical length L, with net gain $G > 0$

being $T = 2L/c$. Under these conditions, the laser is said to operate in the mode-locked regime.

The fundamental difference in behavior between these two cases is due to the beat terms of modes with different frequencies. These terms influence the fluctuations only slightly in the random-phase case, but they completely determine the time distribution in the fixed-phase case. Let us try to give a physical meaning to this remark. First, let us consider the case of just two locked modes [3.3]. The two modes are assumed to be linearly polarized in the same direction so that a scalar description of their fields can be used. Let e_1 and e_2 stand for these fields. We then have

$$e_1 = E_1 \cos[\omega_1 t + \varphi_1(t)],$$
$$e_2 = E_2 \cos[\omega_2 t + \varphi_2(t)]. \tag{3.3}$$

The operating regime of the laser – whether with two free modes or with two locked modes – will depend on whether or not a time relation exists between the two initial phase factors $\varphi_1(t)$ and $\varphi_2(t)$.

No matter what type of detector is placed at the output of the laser, it will only react to the intensity I of the field, that is, to the square of the total electric field amplitude:

$$I = (e_1 + e_2)^2 = E_1^2 \cos^2[\omega_1 t + \varphi_1(t)] + E_2^2 \cos^2[\omega_2 t + \varphi_2(t)]$$
$$+ 2E_1 E_2 \cos[\omega_1 t + \varphi_1(t)] \cos[\omega_2 t + \varphi_2(t)]. \tag{3.4}$$

However, the response time τ_D of all existing actual detectors is much greater than the period of optical oscillation ($\sim 10^{-14}$ s). Therefore, the detector will respond with a signal proportional to the mean value $\langle I \rangle$ of the intensity I over a time τ_D, and this response will represent the experimental laser intensity. We have

$$\langle I \rangle = \frac{E_1^2}{2} + \frac{E_2^2}{2} + \frac{E_1 E_2}{\tau} \int_0^{\tau_D} \cos\left[(\omega_1 - \omega_2)t + \varphi_1(t) - \varphi_2(t)\right] dt. \tag{3.5}$$

In (3.5), the first two terms correspond to the intensity of each of the modes taken separately. The third term expresses the beating between the two modes. If the phase factors $\varphi_1(t)$ and $\varphi_2(t)$ have random and independent fluctuations with time and if, moreover, the characteristic time of these fluctuations τ_f is much shorter than τ_D, the effect of the beating term on the detector averages out to zero; this is true even if the detector is very fast and should, in theory, be able to record the beating of the two modes at $\omega_1 - \omega_2$, that is, when we have $2\pi/\omega_0 \ll \tau_f \ll \tau_D < 2\pi/(\omega_1 - \omega_2)$ with ω_0 the central frequency corresponding to the gain maximum. When this condition is true, the laser operates in a multimode (two-mode in this case) regime and the detector records an intensity which is constant in time.

Let us now consider the case in which the two modes still fluctuate independently, but with a characteristic time slower than that of the intermode beats. We thus have the situation where $2\pi/\omega_0 \ll \tau_D < 2\pi/(\omega_1 - \omega_2) < \tau_f$. On top of the invariant component, this fast detector will now record a beating of period $2\pi/(\omega_1 - \omega_2)$, and the phase of this beating varies at random with a characteristic time τ_f. Generalizing this to the case of multimode operation with N modes will lead, for large N, to the recording of extremely strong intensity fluctuations of the kind shown in Fig. 3.1c. Two characteristic times describe the fluctuations: a slow period of $2\pi/(\omega_1 - \omega_2)$ corresponding to the intermode beating, and a time $\Delta\tau \sim 2\pi/N(\omega_1 - \omega_2)$ which is due to the phase fluctuations of the large number N of modes, as shown in Figs. 3.2 and 3.4.

Next, let us consider the case where the phases $\varphi_1(t)$ and $\varphi_2(t)$ are totally correlated so that $\varphi_1(t) - \varphi_2(t) = 0$. Then (3.5) shows us that, for all τ_f characterizing the phase fluctuations, the recorded intensity will show a sinusoidal modulation with time, of period $2\pi/(\omega_1 - \omega_2)$ (Fig. 3.1b). In fact, saying that $\varphi_1(t) - \varphi_2(t) = 0$ is saying that the phases of the modes corresponding to ω_1 and to ω_2 are strictly constant with respect to each other. So we have rediscovered a very general result about the sinusoidal beating arising from two interfering pure sine waves recorded by a quadratic detector. If $\varphi_1(t) - \varphi_2(t)$ fluctuates, this fluctuation of the instantaneous phase difference will partially or totally destroy the beating.

If, instead of two modes, N modes interfere, with N a large number, the two-wave beating term has to be replaced by a multiple-wave beating term. The sine-shaped beating of the two-mode case then becomes an intensity variation with very intense periodic maxima, at least in the case where the N phases are constant with respect to each other, i.e. when $\varphi_i(t) - \varphi_j(t) = 0$, $\forall i, j$. Note that this progression from a sinusoidal curve to a much narrower curve as one goes from two-wave beating to multiple-wave beating illustrates a very general principle in physics. In optics, for instance, the response curve of a Michelson interferometer – which is a two-wave interferometer – is a sine function, while a Fabry–Pérot interferometer, which is a multiple-wave interferometer, has a response curve, called the Airy curve, with a very narrow structure.

To illustrate the influence of multiple beating on the time distribution of the intensity, let us consider the case of a laser with an infinite number of phase-locked modes: $\varphi_i(t) = \varphi_j(t)$, $\forall i, j$. It can be shown that the exact fluctuations of the phases are not important, provided they fluctuate in the same way, and therefore we may write $\varphi_i(t) = 0$, $\forall i, j$. In complex notation, the total electric field $E(t)$ can then be written as:

$$E(t) = \sum_{-\infty}^{+\infty} E_n \exp i\omega_n t, \tag{3.6}$$

where the frequencies ω_n of the different modes are distributed around a central frequency ω_0:

$$\omega_n = \omega_0 + n\,\Delta\omega. \tag{3.7}$$

$\Delta\omega$ is the frequency interval between two adjacent modes ($\Delta\omega = 2\pi c/2L$). In order to simplify the calculations, we choose a Gaussian distribution for the amplitudes E_n centered around ω_0:

$$E_n = E_0 \exp\left[-\left(\frac{2n\,\Delta\omega}{\Delta\omega_0} \right)^2 \ln 2 \right]. \tag{3.8}$$

E_0 is the amplitude of the central mode at ω_0 and $\Delta\omega_0$ the total width at half maximum of the total field distribution. $E(t)$ can also be written as

$$E(t) = \exp i\omega_0 t \sum_{-\infty}^{+\infty} E_n \exp(in\,\Delta\omega t). \tag{3.9}$$

Therefore, $E(t)$ can be considered as a carrier wave with the optical frequency ω_0 which is modulated by a function $k(t)$ varying slowly with time; $k(t)$ is defined by

$$k(t) = \sum_{-\infty}^{+\infty} E_n \exp(in\,\Delta\omega t). \tag{3.10}$$

Writing $k(t)$ like this as a Fourier series shows that $k(t)$ is a real-valued periodic function with period $T = 2\pi/\Delta\omega = 2L/c$. Hence, the laser intensity $\langle I \rangle$ is also periodic:

$$\langle I \rangle = E(t)E^*(t) = k^2(t). \tag{3.11}$$

In the case of a Gaussian amplitude distribution, the shape of the laser intensity curve is easy to calculate. Remembering that the Fourier transform of a Gaussian is a Gaussian, and approximating the Fourier series by an integral over n, we find that this integral has the following value:

$$k'^2(t) \propto \exp\left[-\left(\frac{2t}{\tau_{\mathrm{p}}} \right)^2 \ln 2 \right] \tag{3.12}$$

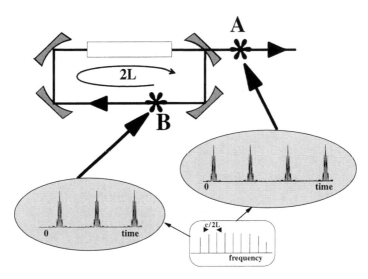

Fig. 3.5. Illustration of a pulse building up in a ring laser cavity

where τ_{p}, the total pulse duration at half maximum, is given by

$$\tau_{\mathrm{p}} = \frac{2\sqrt{2}}{\pi\,\Delta\omega_0}\ln 2. \tag{3.13}$$

However, upon approximating the Fourier series by an integral, the periodicity of the function $k^2(t)$ was lost for mathematical reasons. Actually, the detector will record a periodic series of pulses, each with a Gaussian shape described by $k'^2(t)$, with a period of $T = 2L/c$.

Though a Gaussian amplitude distribution of the modes was chosen for the sake of simplifying the mathematics, the actual shape of the amplitude distribution only influences the shape of the pulse. Any other amplitude distribution of modes with the same frequency spacing and with an equivalent spectral width $\Delta\omega_0$ would lead, in a similar way, to a periodic series of pulses of non-Gaussian shape, but having an approximate pulse duration of $\tau_{\mathrm{p}} \propto 1/\Delta\omega_0$ and a period of $T = 2L/c$. In order to understand the cause of the periodicity of the laser pulses, we can analyze the space–time distribution of the stationary electromagnetic wave which exists inside the laser cavity when it operates with phase-locked modes. It is possible to do a similar calculation for the beating of N axial modes, but taking into account the specific space–time distribution of each mode inside the cavity. This calculation shows that if constructive beating occurs at a specific instant and at a specific point B of the cavity, then at this same instant, the modes beat destructively at all other points of the cavity (Figs. 3.5 and 3.6). This is due to the presence of a large number of modes whose wavelengths λ_n all differ slightly from each other. If all the waves have their maximum amplitude at point B, that implies that

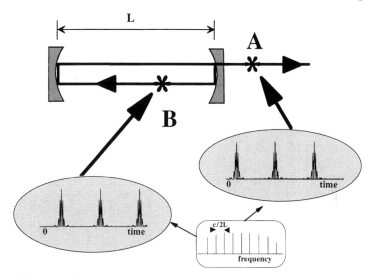

Fig. 3.6. Illustration of pulse building up in a linear laser cavity

their beating is completely destructive everywhere inside the cavity, except at a short distance from point B.

The greater the number of modes N, the shorter this distance is. Therefore, at each instant t, the electromagnetic wave inside the cavity is concentrated in a small spatial domain of length $l_p = c\tau_p$. A picosecond pulse ($\tau_p = 10^{-12}\,\mathrm{s}$), for instance, is represented at each instant t by a small wave packet of length $l_p = 0.3\,\mathrm{mm}$. Interference calculations also show that this wave packet travels back and forth inside the cavity (Figs. 3.5 and 3.6). Every time the packet is reflected off the output mirror, part of its energy is transmitted into the output beam of the laser. Thus it is easy to understand why this beam consists of a periodic series of pulses whose period $T = 2L/c$ is determined by the transit time of the wave packet through the cavity.

The pulse duration τ_p is always inversely proportional to the spectral width $\Delta\omega_0$ of the amplitude distribution of the modes (and therefore to the number N of phase-locked modes). It is clear that the laser whose amplifying medium presents the broadest emission band will emit the shortest pulses. An ionized-argon laser has a bandwidth of $\Delta\omega_0 \sim 0.7 \times 10^{-2}\,\mathrm{nm}$; for a ruby laser $\Delta\omega_0 \sim 0.2\,\mathrm{nm}$, for a $\mathrm{Nd}^{3+}/\mathrm{YAG}$ laser $\Delta\omega_0 \sim 10\,\mathrm{nm}$, for a dye laser $\Delta\omega_0 \sim 100\,\mathrm{nm}$ and for a Ti:sapphire laser $\Delta\omega_0 \sim 400\,\mathrm{nm}$. These data show that the smallest pulse duration one can hope to obtain directly at the output of a laser with-phase locked modes varies from 150 ps for an ionized-argon laser to 3 fs ($3 \times 10^{-15}\,\mathrm{s}$) for a Ti:sapphire laser. The proportionality factor $2\sqrt{2}\log 2/\pi$ which relates τ_p to $1/\Delta\omega_0$ in (3.13) depends on the specific choice of a Gaussian distribution for the amplitudes of the modes. For another distribution, the proportionality factor would be different, but it would still be of the order of unity.

A pulse is said to be Fourier-transform-limited if τ_p and $\Delta\omega_0$ are related to each other by an equation of type (3.13) with a proportionality constant of the order of unity. To verify experimentally whether or not a pulse is Fourier-transform-limited, τ_p can be measured with the help of a fast photodiode, a streak camera or an autocorrelator. The width of the spectral distribution can be measured by means of a spectrograph or with a Fabry–Pérot interferometer. In both cases, since the detector is quadratic, the measurement will give the spectral distribution of the intensity; let us call the spectral width thus measured $\Delta\omega_0'$. The exact relation between $\Delta\omega_0'$ and the spectral width of the field $\Delta\omega_0$ depends on the shape of the distribution. However, it can be shown that, for a Gaussian distribution, $\Delta\omega_0' = \Delta\omega_0/\sqrt{2}$ and that this ratio remains of the same order of magnitude for all distributions encountered in practice. If there is a large discrepency between the measured value of $\Delta\omega_0$ and the value obtained by calculation using (3.13) along with the measured value of τ_p, this proves that the mode-locking is not optimal. For instance, the laser output can consist of a series of recurrent pulses while the central frequency ω_0 of the carrier wave wanders within a band of width $\Delta\omega_0$. For a precise analysis of the laser output signal corresponding to this more complicated mode-locking regime, more complex testing is needed. If the phase-locking of the modes is perfect, knowing the spectral distribution is equivalent to knowing the modulating function $k(t)$ since one can go from one to the other by a simple mathematical transformation. Therefore, in the phase-locked regime, the situation can be described in an equivalent way either in the frequency domain or in the time domain without loss of generality. We shall use this property in the following paragraphs, in which we describe how the conditions for phase-locked modes, i.e. $\varphi_1(t) = \varphi_2(t)$, may be obtained in practice.

3.3 General Considerations Concerning Mode-Locking

When the laser operates in the free multimode regime, there is usually a competition going on among the different modes to be amplified by stimulated emission of the same atoms, molecules or ions. This competition causes big fluctuations in the relative phases and amplitudes of the modes, which explains the big fluctuations of the instantaneous intensity observed in this type of laser (Fig. 3.1c). The purpose of locking the modes is to organize the competition between modes in such a way that the relative phases stay constant or, alternatively (but this is equivalent through Fourier transformation), in such a way that the output intensity of the laser consists of a periodic series of pulses resulting from the shuttling back and forth of a wave packet within the laser cavity. The first idea which comes to mind is best explained in the time domain. Since in a free-mode regime, spatial concentration occurs within the cavity at several points, it should be possible to concentrate the laser energy even more by inserting a nonlinear medium – or any other system promoting stronger intensities – inside the cavity. As the wave travels back and forth in the cavity, maxima that are initially slightly stronger will grow much stronger at the expense of lesser maxima because of the competition within the amplifying medium. If the conditions are well chosen, the situation may arise where all the energy of the cavity is concentrated in one single pulse. This is the mode-locked situation: selecting a single intensity maximum in the time domain is equivalent to establishing a phase relation between the longitudinal modes in the frequency domain. A second fundamental idea is best formulated in the frequency domain. If something is inserted inside the cavity which modulates the modes at a frequency close to the intermode frequency interval $c/2L$, competition for maximum gain inside the amplifying medium will result in a coupling between each mode and the sidebands created by the modulation of its neighboring modes. The phases of the modes could then lock onto each other. One must bear in mind that selecting an intensity maximum inside the cavity by way of a nonlinear medium is equivalent to modulating the wave packet at frequency $c/2L$ (the frequency corresponding to the back-and-forth movement of the packet in the cavity). The first method indeed induces a self-modulation of the modes at frequency $c/2L$ and thus establishes a phase relation between these modes.

These two ideas form the starting point of the two main mode-locking methods developed so far:

- *passive mode-locking* resulting from the insertion of a saturable absorbing medium into the cavity in order to select one single pulse;
- *active mode-locking* resulting from an external modulation at frequency Ω either of the cavity losses (by inserting an acousto-optical crystal inside the cavity, for instance) or of the gain of the amplifying medium (for example by pumping this medium with another mode-locked laser).

Lately, however, the advent of the Ti:sapphire laser, in which mode-locking is especially simple to obtain, has made an old method popular again. This method was already well known, but it has now found new and fruitful applications. We are speaking about a self-locking mechanism of the modes by which the nonlinear properties of the amplifying medium naturally enhance the intensity maxima arising within the cavity.

Let us now look at the details of these various methods, starting with the one which is easiest to understand and easiest to put into practice: the active method.

3.4 The Active Mode-Locking Method

Let us insert an element inside the cavity which modulates its losses (Fig. 3.7). This element will induce a modulation of the amplitude of each longitudinal mode.

If we assume the modulation to be sinusoidal, of angular frequency Ω, and to have a modulation depth α, the time dependency of mode n of frequency

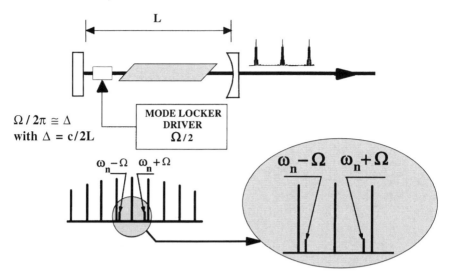

Fig. 3.7. Illustration of an actively mode-locked laser cavity

ω_n can be written as

$$e_n(t) = E_n \cos(\omega_n t + \varphi_n)[1 - \alpha(1 - \cos(\Omega t + \varphi))]. \qquad (3.14)$$

This expression can also be written as follows, showing that in the frequency domain two sidebands show up at either side of mode $e_n(t)$:

$$e_n(t) = E_n(1 - \alpha) \cos(\omega_n t + \varphi_n) + E_n \frac{\alpha}{2} \cos[(\omega_n - \Omega)t + \varphi_n - \varphi]$$
$$+E_n \frac{\alpha}{2} \cos[(\omega_n + \Omega)t + \varphi_n - \varphi]. \qquad (3.15)$$

Now if the modulation frequency $\Omega/2\pi$ is close to the intermode frequency separation $c/2L$, the two sidebands will be very close to the two neighboring modes $n + 1$ and $n - 1$, as shown in Fig. 3.7 (a modulation frequency of the order of $mc/2L$, where m is an integer, can also be used to couple mode n to modes $n + m$ and $n - m$ [3.4]). The sidebands and the longitudinal modes themselves therefore compete inside the amplifying medium for maximum gain. The situation in which the medium is used most efficiently is that in which the longitudinal modes lock their phases onto the sidebands, causing a global phase-locking over the whole spectral distribution. This global phase-locking, in turn, is the condition under which the competition will give rise to the concentration of all the electromagnetic energy of the cavity into a single pulse traveling back and forth inside it.

This technique goes by the name of mode-locking by amplitude modulation (AM) [3.5, 3.6] and it is used to lock the modes of ionic lasers and Nd^{3+}:YAG lasers, for example. Usually, an acousto-optical modulator is placed inside

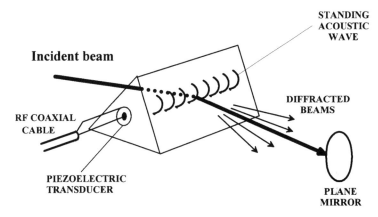

Fig. 3.8. Acousto-optical modulator generally used for active mode-locking of lasers

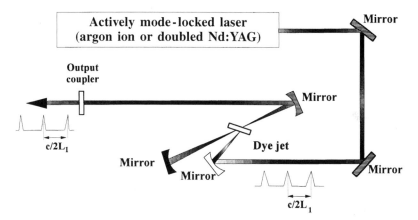

Fig. 3.9. Illustration of the synchronous pumping method for mode-locking of a laser

the cavity to modulate the cavity losses. For example, in Fig. 3.8 an acousto-optical prism is shown; it can be used in an ionic laser, where it serves to lock the modes and at the same time to select the desired mean wavelength. Similar mode-locking may also be achieved by frequency modulation (FM) [3.7–9]. Whether using an AM or an FM method, an external radio-frequency power source, whose angular frequency Ω should be as well defined as possible, must be imposed upon the system. These AM and FM methods make up one kind of active mode-locking method. Of course, one could also modulate the gain of the amplifying medium at a frequency $\Omega/2\pi$. We shall now go into this second type of active method, most often referred to as the synchronous pumping method.

Usually, synchronous pumping is obtained by pumping the amplifying medium of the laser with the output beam of another mode-locked laser.

Fine tuning of the length L of the cavity makes it possible to adapt the intermode frequency separation $c/2L$ to the frequency of recurrence $\Omega/2\pi$ of the pump laser pulses. Figure 3.9 shows the general layout of a synchronously pumped dye laser. The pump is usually either a mode-locked argon laser or a frequency-doubled mode-locked YAG:Nd^{3+} laser. The mode-locked regime of the pump laser is most often obtained by modulating its cavity losses with an acousto-optical modulator set at frequency $c/2L_1$ (L_1 is the length of the pump cavity). The dye laser will go into the mode-locked regime if the interval between its modes, $c/2L_2$, is adjusted to the frequency $c/2L_1$ of the gain modulation. The purpose of this example to illustrate synchronous pumping is twofold. First, it shows that it is possible to combine two kinds of active pumping – modulation of the cavity losses and modulation of the gain – within the same laser system. But second, and more importantly, it shows how one can use a mode-locked laser as a source of a continuous train of 150 ps pulses to pump a second laser and thus obtain a series of subpicosecond pulses. Indeed, the mode-locking of the pump laser leads to pulse durations τ_p which are limited by the bandwidth of the gain (about 10^{-2} nm for an argon laser). However, the bandwidth of the dye laser is much broader (typically 100 nm). Therefore the output of the second synchronously pumped laser will consist of a continuous train of picosecond or subpicosecond pulses with tunable wavelengths, provided one stays within the gain band.

Let us look at the behavior of a mode-locked dye laser when its cavity length L_2 varies around the value of L_1. First, let us consider the case of a stationary regime: a narrow pulse travels back and forth inside the cavity. The oscillation condition implies that the pulse is crossing the center of the amplifying medium when the gain induced by the pump pulse is greater than the cavity losses. Figure 3.10 shows how the gain changes as the pump pulse and the dye laser pulse travel through the amplifying medium, i.e. through the dye.

The horizontal axis of the figure represents time in arbitrary units and the gain is measured at a given position inside the amplifying medium. At $t = 0$, the pump pulse arrives at this position, where it excites the dye molecules. The gain grows proportionally to the integral (i.e. the total accumulated sum) of the pulse intensity until the arrival of the dye laser pulse in the medium. At that instant, stimulated emission due to the dye laser photons will diminish the number of excited molecules in the dye, thereby decreasing the population inversion of the dye. This decrease shows up as a narrow dip (hole) in the gain curve during the time the dye laser pulse passes through the medium. When the gain becomes smaller than the intracavity losses, the long pulse duration of the pump makes it possible for the gain to restore itself after the passage of the dye laser pulse.

What influence would a change of the cavity length of $\Delta L = L_2 - L_1$, say, have on this behavior? One must not forget that the transit time of a pulse in the cavity is determined by the repetition frequency of the pump pulses. Therefore, it is very critical that the cavity lengths are well adapted. However,

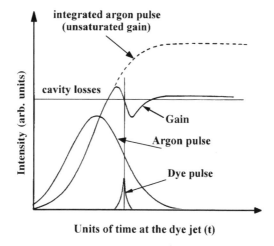

Fig. 3.10. Illustration of the generation of a pulse in a dye laser pumped by a mode-locked pumping laser. The pulse is created by competition between population of the emitting level by the pump and depopulation by stimulated emission (see text)

a distortion of the dye pulse inside the amplifying medium turns out to be able to compensate small transit-time changes automatically, so that changes of length ΔL of the order of $100\,\mu$m can occur without hampering the stability of the system. The causes of pulse distortion can be understood by looking at the gain saturation induced by the dye pulse inside the amplifying medium. Before the pulse arrives, the gain is unsaturated, so the pulse front feels a larger gain than the tail of the pulse. This differential amplification of the wavefront results in a pulse which seems to be ahead of where it should have been, if it had not been distorted. The transit time of the pulse in the amplifying medium is shortened. The shortening of the transit time is directly proportional to the difference between the saturated and the unsaturated values of the gain, and moreover, it always occurs, no matter what the cavity length corresponding or not to a stable regime is. This explains, among other things, why the perfect mode-locked regime occurs for very small positive values of $\Delta L = +\varepsilon$ [3.10].

Let us now shorten the cavity length of the dye laser L_2 with respect to its optimal length. The dye pulse which travels in the cavity arrives in the amplifying medium slightly earlier than it would have done in the optimum condition. The wavefront thus feels a slightly weaker unsaturated gain than it would otherwise have done. The pulse is somewhat less distorted, so that its transit time through the medium is longer than it would have been. This produces an effective lengthening of the total transit time. A new steady state arises in which the shortening of the transit time due to the shorter length of the cavity is compensated by the longer time needed to go through the amplifying medium. A similar reasoning can be applied to a slightly lengthened cavity. In this case, the dye pulse is late because of the longer cavity, but

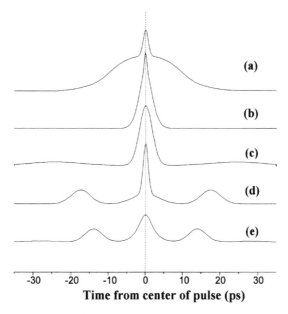

Fig. 3.11. Autocorrelation curve shape as a function of the mismatch between the cavity optical lengths of the pump laser and of the pumped laser (see text)

it feels a stronger unsaturated gain than it would have done in the optimum condition, so its transit time through the amplifying medium is proportionally shortened.

Thus, small changes in the length of the cavity in stable operating conditions translate into changes in the shape of the dye pulse. More generally, most of the behavior of the pulse as a function of cavity length can be obtained qualitatively from the model of gain variation shown in Fig. 3.10. Figure 3.11 shows the autocorrelation traces of the pulses for different values of the dye cavity length L_2. Figure 3.11b shows the trace of the pulse associated with an optimal cavity length. Figure 3.11a shows what happens in a longer cavity. Lengthening the cavity delays the arrival of the pulse in the amplifying medium, so that on the gain curve it shows up further from the point where the gain curve cuts the loss curve.

The maximum intensity of the pulse is stronger because of the larger unsaturated gain. Simultaneously, the pulse broadens in order to adapt its transit time to the pumping period. When the change in length is sufficiently large, greater than $50\,\mu$m, the intensity maximum decreases after having reached a maximum, because the unsaturated gain decreases owing to spontaneous emission. Finally, lengthening the cavity makes total mode-locking impossible since the characteristic form of the autocorrelation trace of the pulse, reminiscent of a Prussian helmet, shows that the pulses are no longer Fourier-transform-limited.

On the other hand, if one shortens the cavity length with respect to the optimal length (Fig. 3.11c, d), the pulse shows up closer to the point where the gain curve cuts the loss curve. The maximum pulse intensity decreases since the unsaturated gain is smaller, and the pulse duration decreases slightly in order to stay synchronized with the pump pulse. However, if the gain decrease induced by the dye pulse occurs early enough, the amplifying medium, which still feels the pump pulse, may retrieve enough gain for it to overcome the cavity losses. The reconstitution of the oscillation condition allows a second pulse to form in the cavity, and this second pulse is time-delayed by several tens of picoseconds with respect to the main pulse. This satellite pulse is the cause of the two secondary maxima in the autocorrelation trace of Fig. 3.11c–e. If the cavity length is shortened even more, the delay between the two pulses can decrease so much that one may observe a third and even a fourth pulse inside the cavity. If the pulse arrives just before the point where the gain curve crosses the loss curve, it is damped. Still, the laser effect continues in the form of a very broad pulse whose parameters are determined by the pump pulse and by the saturation characteristics of the amplifying medium. Only very slight mode-locking occurs under these conditions, and no short pulse is formed.

All this shows that the pulse structure depends critically on the adjustment of the two cavity lengths. For instance, a Rhodamine 6G laser whose dispersive element is a three-slide birefringent filter and which is synchronously pumped by a mode-locked Ar^+ laser displays the whole range of pulse shapes shown in Fig. 3.11 for a total length variation of only $130\,\mu m$. It is not always easy to obtain, or even to identify, the proper mode-locked regime of the synchronous pump source. However, a qualitative approach is possible by producing a pulse with an autocorrelation trace similar to that of Fig. 3.11b, and then reducing the pulse width experimentally until a second pulse arises. For a dye laser with a three-slide birefringent filter and with an output of 6 ps pulses, the tolerance in the change in cavity length is about $10\,\mu m$. If only one birefringent slide is used to obtain subpicosecond pulses, the tolerance drops to about $1\,\mu m$.

In fact, the second pulse must be suppressed in order to obtain subpicosecond pulses with a synchronously pulsed laser. Two methods have been suggested to achieve this. As was said earlier, the second pulse arises from the retrieval of the gain after the passage of the first pulse. There is a critical value of the intracavity power density such that if the power density is greater than this value, a second pulse is generated provided the cavity lengths are properly adapted. If, however, the power density is lower than this critical value, and if the system is designed to operate near enough to the functioning power threshold, the gain curve cannot increase above the loss curve even after the passage of the dye pulse [3.11]. One way to avoid the formation of a second pulse is therefore to widen the beam waist in the amplifying medium so that the power density remains below the critical density. A second method involves the simultaneous use of a passive mode-locking method and is called the hybrid mode-locking method. It will be described in the next section. The

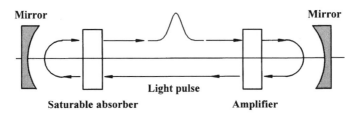

Fig. 3.12. Round-trip pulse in a laser cavity including saturable absorber and amplifying medium

extreme sensitivity of active mode-locking methods to the relative fluctuations of the imposed frequency Ω and the intermode frequency interval $c/2L$ makes it very difficult to establish subpicosecond regimes with just a single active locking procedure. These regimes are easier to obtain with hybrid methods or with passive methods in which no exterior frequency Ω intervenes.

3.5 Passive and Hybrid Mode-Locking Methods

The dynamics of the gain saturation in the amplifying medium are responsible for distorting the pulse shape, as was discussed in the previous section. Now we want to show that if an absorbing medium with a saturable absorption coefficient (Fig. 3.12) is placed inside the cavity, the association between this saturable absorber and the saturable amplifying medium leads to a natural mode-locking of the laser, without any need for external monitoring. The most commonly used saturable absorbers usually consist of liquid dye solutions. However, especially since the advent of Ti:sapphire lasers, an active search is going on to find solid media which could serve as saturable absorbers, especially among the semiconductors.

This mode-locking process is simpler to explain in the time domain [3.12]. Let us look at the transmission characteristics of a saturable absorber. At low incident intensity, the transmission T stays practically constant, with a value of T_0 which is almost independent of the incident intensity (Fig. 3.13). But if the incident intensity increases, the population of the upper level involved in the absorption process increases, as does the stimulated emission from this level. At the same time, the population of the lower level decreases. The combination of these two effects results in a nonlinear behavior of the transmission coefficient, as shown in Fig. 3.13. The absorber is characterized by its saturation intensity I_s^{abs}. This parameter is defined as the intensity at which the population difference that exists between the two levels at low intensity is reduced by a factor of two. This definition thus implies that the absorption coefficient α must be proportional to the population difference.

Moreover, as was said earlier, the amplifying medium also possesses saturation properties. At low intensity, the gain G has a constant value G_0, which

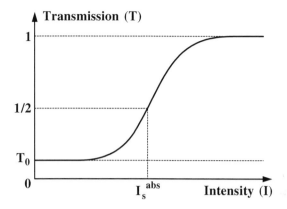

Fig. 3.13. Transmission through a saturable absorber as a function of the input intensity

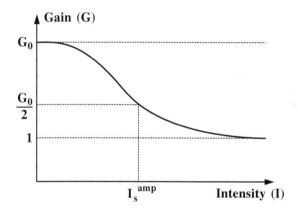

Fig. 3.14. Gain through an amplifying medium as a function of the input intensity

is rather large (Fig. 3.14). It is said to be unsaturated. At higher input intensities, there is a lessening of the population inversion between the two levels involved in the process, and the gain G decreases. As can be seen in Fig. 3.14, we can define I_s^{amp}, the saturation intensity for G, in an equivalent way to the saturation intensity I_s^{abs} for the absorption.

Using these parameters, we can describe how a pulse arises in a passively mode-locked laser comprising the elements shown in Fig. 3.12. At the instant $t = 0$ the pump beam is applied to the amplifying medium. The power inside the cavity is initially zero and the unsaturated gain G_0 is greater than the sum of all the cavity losses. An oscillation arises inside the cavity with a characteristic electromagnetic field showing very strong fluctuations at low power (Fig. 3.1c). As the power increases, the strongest intensity maxima start to saturate the absorbing medium. These stronger maxima thus suffer smaller

Saturable absorber

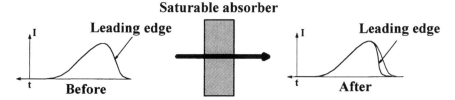

Fig. 3.15. Illustration of pulse shape modification after crossing a saturable absorber

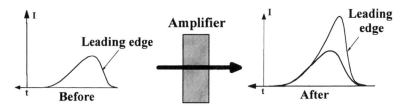

Fig. 3.16. Illustration of pulse shape modification after crossing an amplifying medium

losses than the lesser intensity maxima of the fluctuating field. Therefore the strongest maximum will eliminate the others in the competition for gain which occurs inside the amplifying medium, since it will grow faster and will stifle all competing processes. If the conditions are favorable, it will end up by being the only intensity maximum in the cavity and will contain all the energy of the wave.

Let us now follow the pulse along its round trip through the cavity. Let us start at the point where the pulse which will eventually take over is already formed, but has not reached its final shape nor its final duration. As the pulse travels through the saturable absorber, the pulse front is strongly absorbed (Fig. 3.15), but if the maximum of the pulse saturates the absorber medium and if – as is the case in a dye laser or in a Ti:sapphire laser – the relaxation time of the medium is longer than the pulse duration, the tail of the pulse will benefit from the induced transparency of the medium and travel through it without being attenuated. When the pulse reaches the amplifying medium, the pulse front will come upon the unsaturated gain G_0 and will be strongly amplified while the tail of the pulse will feel a much weaker gain, which has just been saturated by the front of the pulse, and thus it will be much less amplified (Fig. 3.16). It is clear that after many back-and-forth trips, the resulting pulse will have narrowed and will have a very strong maximum, since the center of the initially broad pulse is not affected by the absorber but is amplified by the amplifying medium. This process is illustrated in Fig. 3.17. The fact that the saturation dynamics of the absorber are more rapid than those of the amplifying medium explains why only the center of the pulse is amplified, the wings, on the contrary, being attenuated.

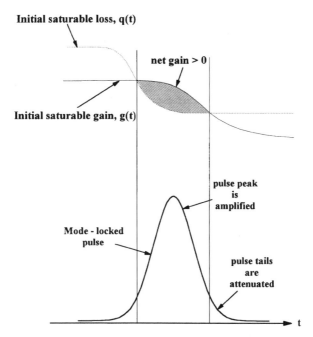

Fig. 3.17. Illustration of the pulse-shortening process by simultaneous action of a saturable absorber and amplifying medium, due to saturation effects of the absorption and of the gain

The pulse reaches its final shape when it becomes self-consistent in the cavity, that is, when the system reaches a steady state. For the pulse to be self-consistent, it must keep the same shape after a round trip through the cavity. However, the previous paragraph suggests that the pulse should grow ever narrower upon traveling constantly back and forth. Of course, if applied with sufficient care, the laws of physics will show that there is a limit to this process. As said earlier, the pulse duration under perfect mode-locked conditions is inversely proportional to the spectral width of the amplitude distribution. Therefore, each element of the cavity which tends to limit the width of the oscillation band will tend to lengthen the pulse duration. The element which is determining from this point of view may be an external optical element such as a prism, a grating or a Lyot filter. The amplifying medium itself may also be the determining element.

This broadening effect is best explained in the frequency domain [3.13]. Just before entering the filtering element, the pulse has a spectral distribution with certain amplitudes. The transmission filter, which is frequency-dependent, changes the distribution by decreasing the amplitudes of the wing frequencies while leaving the central frequencies unchanged (Fig. 3.18). The narrowing of the spectrum explains the lengthening of the pulse duration as it passes through the filtering element. Similarly, each dispersive element

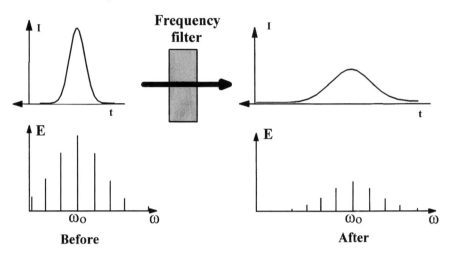

Fig. 3.18. Influence of a spectral filter on the spectral distribution and duration of a pulse

of the cavity will also affect the pulse duration by time-delaying the various frequency components of the spectral distribution by different amounts. The pulse reaches steady state when the narrowing effect due to the saturation properties of the absorber and of the amplifying medium is exactly compensated by the broadening effects of the various elements in the cavity, of which the main two have been described above. In fact, the actual study of the propagation of the pulse through the various elements of the cavity is extremely complex, especially if one remembers that amplifying media and absorbers are nonlinear media. To tackle the problem in a reasonably realistic way, one should take into account, among other things, the variation of the refractive index with intensity inside these media. This variation results in a self-modulation of the phase of the pulse. This point has been covered in Chap. 2.

Figure 3.19 shows an example of a passively mode-locked laser. It is a dye laser whose amplifying medium consists of Rhodamine 6G pumped by a continuously emitting argon ion laser. The saturable absorber is a DODCI (3,3′-diethyl oxadicarbo cyanine iodide) jet. In order to reach the saturation threshold of the absorber, the laser beam is strongly focused inside the absorber medium by means of spherical mirrors of short focal length. The cavity shape is calculated so that the astigmatism introduced by the mirrors, which are used off axis, is compensated by the jets placed at the Brewster angle. One should remember, as said before, that by passing through the absorbing medium and the amplifying medium every $2L/c$ seconds, the pulse creates a modulation of each of its longitudinal modes at frequency $c/2L$. Lateral bands therefore arise at the intermode frequency distance. These in turn induce mode-locking of the phases, and the frequency spectrum tends to broaden

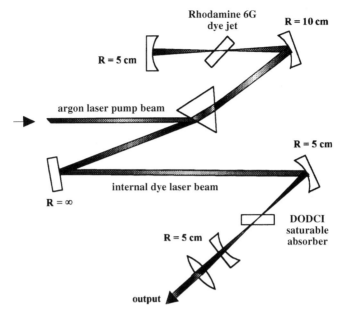

Fig. 3.19. Typical laser cavity of a passively mode-locked laser using a saturable absorber. This example is for a dye laser pumped by a CW argon laser

while the pulse narrows. Here we have presented the frequency description of the passive mode-locking process.

The problems which arise in the passive mode-locking method are, first of all, the fact that there are not many compatible pairs of saturable absorbing and amplifying media with the correct properties, and, second, the fact that the pulses obtained by this method are not very powerful and that their wavelength is only slightly tunable. A hybrid method was devised to try to overcome the drawbacks of both the passive and the active mode-locking methods. It consists of inserting a saturable absorbing medium inside the synchronously pumped cavity [3.14]. This hybrid locking method allows a wider choice of wavelengths and powers than a simple passive mode-locking method would. On the other hand, when compared with the simple active mode-locking method, hybrid locking makes it easier to obtain subpicosecond pulses and it avoids the formation of secondary pulses.

Figure 3.20 shows an example of a mode-locked laser using the hybrid method. The system is similar to the one shown in Fig. 3.19 to illustrate the passive method, but now the pumping is done with an Ar^+ laser whose modes are locked using the active method. In this case, the cavity length must be adjusted to the cavity length of the Ar^+ laser, as explained in Sect. 3.4. One can see that the pumping inside the amplifying medium is not colinear with the laser wave of the cavity. This kind of pumping, which avoids the inconvenience of having a prism inside the cavity, is also frequently used in the passive

Pump Beam

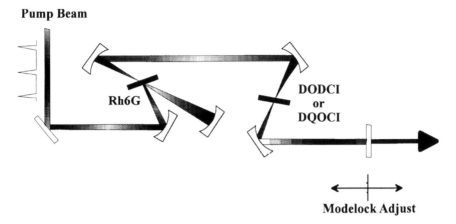

Modelock Adjust

Fig. 3.20. Typical cavity design of a hybrid mode-locked laser

method instead of the pumping method shown in Fig. 3.19. Some examples of pairs of amplifying dyes and saturable absorbers, covering a wide spectral range between 575 nm and 860 nm, are shown in Table 3.1. In Figs. 3.19 and 3.20, the dispersive element (a Lyot filter for example) which determines the mean wavelength of the pulse has not been shown. The cavity also usually contains a device which compensates for the dispersive effects of the group velocity of the pulse, which limit the pulse duration, as explained earlier. All these elements (series of prisms for instance) will be discussed in detail in the following chapters. Figure 3.21 shows how a series of prisms is used in a hybrid pumping setup. In Fig. 3.22 one can see that such a system, compensating the dispersion of the group velocity, is needed in order to obtain subpicosecond pulses with a hybrid pumping method.

Table 3.1. Some examples of dye-amplifier–dye-absorber pairs for generation of short pulses in different spectral ranges

Gain dye	Saturable absorber	Wavelength in nm
Rh6G	DODCI, DDI	575–620
Kiton Red	DQOCI	600–655
DCM	DODCI, DTDCI	620–660
Pyridine 1	DTDCI, DDI	670–740
LD 700	DTDCI, DDI, IR 140	700–800
Pyridine 2	IR 140, HITC	690–770
Styryl 9M	DDI, IR 140	780–860

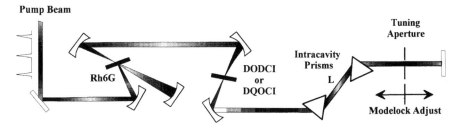

Fig. 3.21. Cavity design of hybrid mode-locked laser with inserted prisms for group velocity dispersion (GVD) compensation

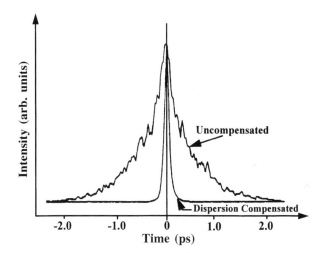

Fig. 3.22. Illustration of the effect of GVD compensation on the pulse duration in a hybrid mode-locked laser

3.6 Self-Locking of the Modes

We have seen that the nonlinear properties of the amplifying medium are always very important for the locking process, whether the locking method is active or passive. In some types of lasers, these properties are so fundamental that the modes may lock, partially or totally, without any need for an external modulation (active locking) or for a saturable absorbing medium (passive locking). This situation is called self-locking of the modes. For such a situation to arise, the amplifying medium must induce a narrowing of the pulse at each of its round trips through the cavity. The dynamics which were described for the passive locking method show that saturation of the gain is not sufficient. There needs to be an associated effect which favors strong intensity maxima at the expense of weak ones. This effect is provided by the saturable absorber in the passive mode-locking method. We shall now describe a specific self-locking situation in which the amplifying medium decreases the losses of the stronger intensity peaks of the cavity by modifying the transverse structure of the laser wave selectively with respect to intensity. This situation exists in the Ti:sapphire laser, explaining why the interest in self-locking of modes suddenly flared up recently.

Historically speaking, the self-locking of modes was first observed accidentally in a laser whose amplifying medium consisted of a Ti:sapphire crystal pumped by an Ar^+ laser, which, in principle, operated in a continuous regime, in a cavity without a saturable absorber. Scottish scientists [3.15] then noticed that the laser went into a pulsed regime when they jerked the table on which the laser was mounted. The pulsed regime consisted of very short pulses, and once initiated, maintained itself. It was first called "magic mode-locking", but it stayed magic only for a few months, an explanation having been found meanwhile. In fact, it was a case of self-locking by a Kerr lens effect [Kerr lens mode-locking (KLM)], a self-locking process which clearly reveals that the following conditions are needed for this kind of behavior to appear:

- The pulsed regime must somehow be favored over the continuous regime.
- The overall system must possess the property of shortening the pulses.
- Some mechanism must initiate the self-locking process.

Figure 3.23 shows the classical configuration of a self-mode-locked Ti:sapphire laser. The $Ti:Al_2O_3$ crystal is pumped by the output of a continuous Ar^+ laser through M_2, a dichroic mirror which is transparent at $0.5\,\mu m$ and which reflects at the emission wavelengths of Ti:sapphire around $0.8\,\mu m$. The birefringent filter (B.R.F.) determines the central wavelength of the oscillation. The two prisms P_1 and P_2 compensate for the dispersion of the group velocity inside the cavity, as was explained above. The mode-locking process arises exclusively from the amplifying medium and an associated pinhole with adjustable diameter. This process results from the Kerr effect and can be summarized in the following way.

The fact that the amplifying medium is nonlinear implies that its refractive index is a function of the intensity (Kerr effect): $n = n_0 + n_2 I$. The Gaussian wave therefore does not feel a homogeneous refractive index as it passes through the medium. If n_2, the nonlinear coefficient of the refractive index, is positive, the refraction is stronger on the axis of the beam than away from it. So the amplifying medium behaves like a converging lens and focuses the beam just like a lens (Kerr lens). We are speaking of the phenomenon of self-focusing, which has been known for a long time in nonlinear optics and is explained in Chap. 2. However, self-focusing is proportionally more important for strong intensities. This means that the strong intensity maxima of the laser cavity will be much more strongly focused than the weaker ones, for which focusing will be negligible (Fig. 3.24).

These strong intensity maxima, whose transverse structures have now been reduced in size, are usually less subject to losses in the cavity than the weaker intensities, which occupy a large volume, so they are enhanced. Clearly, the intensity-differentiated self-focusing associated with the natural cavity losses

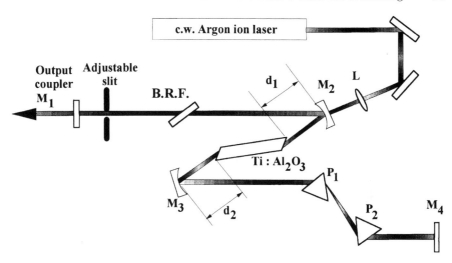

Fig. 3.23. Typical cavity design of a self-mode-locked Ti:sapphire laser using the KLM (Kerr lens mode-locking) process (see text)

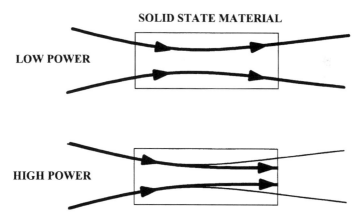

Fig. 3.24. Illustration of the self-focusing effect by the optical Kerr effect on the beam waist of a laser beam at high and low intensity

plays a part similar to that of the saturable absorber in the passive mode-locking method, and, indeed, self-mode-locking of the modes arises. A slit can be placed inside the cavity to help the self-locking process since it increases the difference between the losses undergone by the weak intensities and those undergone by the intensity maxima. The exact position, diameter, and shape of the slit must be calculated so that it lets through most of the power of the stronger intensity maxima while it stops the greater part of the weaker intensities. Figure 3.25 illustrates the passage from a continuous regime (with open slit) to a mode-locked regime. The experimental data of Fig. 3.25 clearly show the broadening of the spectrum and its transformation into a spectrum

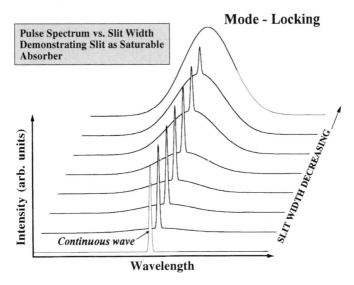

Fig. 3.25. Spectral distribution of laser output for different widths of the slit inserted in the cavity, which controls the mode-locking by the optical Kerr effect

corresponding to a correctly mode-locked regime for a slit width of 0.47 mm. It should be noted that the amplifying medium must be relatively thick to observe a strong lens effect. This explains why the lens effect, which is so important in the Ti:sapphire laser, is negligible in dye lasers.

The laser illustrated in Fig. 3.23 has the drawback that it does not go into the pulsed regime spontaneously. As long as no important intensity fluctuations arise to create sufficiently strong Kerr lens effects, the continuous regime prevails over the pulsed regime, as shown in Fig. 3.25. To start the pulsed process, one can insert a rapidly rotating optical slide so as to create a changing optical path, or one can simply give a quick jolt to one of the mirrors of the cavity to create an intensity pulse which triggers off the process.

We see therefore that in a Kerr-lens mode-locked (KLM) laser, the discrimination between a continuous regime and a pulsed regime is due to the self-focusing in the amplifying medium, and this discrimination becomes effective after a rapid transient change in the optical length of the cavity. We still have to explain how the pulses shorten and stabilize their duration by following their journey back and forth through the cavity in the proper self-locking conditions defined above. In fact, on top of its influence on the spatial evolution of the wavefront (self-focusing), the change of refractive index as a function of intensity in the amplifying medium also has very important consequences for the time structure of the laser wave. Indeed, the intensity $I(t)$ being a rapidly varying function of time, the change of refractive index as a function of the intensity, following the equation $n = n_0 + n_2 I(t)$, implies a rapid change of the phase of the wave as a function of time. This

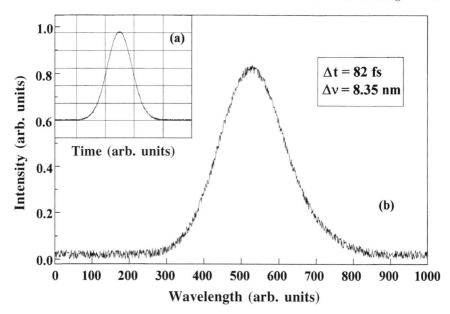

Fig. 3.26. (a) Autocorrelation trace (b) and spectrum of the pulse a few hundreds of femtoseconds wide obtained with the laser shown in Fig. 3.23 after starting the mode-locking process

self-modulation of the phase broadens the spectrum of the wave and therefore shortens its duration. As was the case for the passive mode-locking method, an equilibrium is reached between this process of pulse shortening and the dispersion of the group velocity, which tends to lengthen the pulse. The compensation between the self-modulation of the phase and the dispersion of the group velocity gives rise to a pulse which travels back and forth in the cavity while keeping its shape. This pulse is called a quasi-soliton. Here again, it is essential to be able to keep control of the group velocity inside the cavity, which is the reason for the two prisms in the system shown in Fig. 3.23.

This kind of mode-locking leads to excellent results for the pulse duration as well as for its stability. In Fig. 3.26 we show the autocorrelation trace and the spectrum of the pulse a few tens of femtoseconds wide obtained with the laser shown in Fig. 3.23. The ease with which these systems are made to work, their great reliability and the great stability from pulse to pulse place them among today's most popular mode-locked systems. At present, they are chosen routinely each time a good-quality femtosecond oscillator is required.

Further improvements of this mode-locking can produce even shorter pulses in the sub-10-fs range as a routine [3.17]. These improvements are obtained using cavity chirped mirrors and/or SESAMs (semiconductor saturable absorber mirror).

a) Chirped Mirrors: The main limitation of the pulse duration in a Kerr lens mode-locking cavity is due to the difficulty of obtaining exact compensation of the GVD and SPM effects using pair prisms. Indeed, higher-order dispersion introduced by the prisms is the main limitation factor. Chirped mirrors make it possible to overcome this difficulty [3.18–3.20].

Such mirrors consist of stacks of different transparent layers with high and low refraction indexes. Tuning the thickness of layers to a corresponding $\lambda_B/4$ phase delay makes constructive interferences of the different waves reflected at λ_B (Bragg wavelength).

If the thickness of the layers is made different along the mirror structure, it will be possible to reflect different waves at different λ_B wavelengths to accommodate the bandwidth of the laser cavity. Then the different waves at λ_B will have different optical pathways inside the mirror before it is reflected. These different waves will be more or less delayed after reflection on this mirror, depending on the thickness of the different layers. In such a way, a chirp is created by this type of mirror which can be tailored, adjusting the thickness layers, to compensate exactly the highter-order dispersion effects and obtain shorter pulses near the theoretical limit (3 fs) of Ti/Sa material.

b) SESAM mirrors: As the pulses become shorter, self-starting of mode-locking by KLM is more difficult. One interesting solution is to use a saturable absorber designed for such lasers. A SESAM makes it possible to help the self-starting of Ti/Sa lasers when an ultimate pulse duration below 10 fs is required [3.21].

Different designs of SESAMs have been employed. Four parameters have to be taken into acount for proper design of SESAMs: the modulation depth introduced by the saturable absorber, the spectral bandwidth to be accommodated, the saturation energy, and the recovery time constant. For this last parameter, the time response should be biexpoential with a fast answer for ultrashort pulse generation and a slow answer for making it possible for the mode-locking to sef-start [3.21].

Semiconductors as absorbers are well suited because they absorb in a large spectral bandwidth compatible with Ti/Sa emission range and because they have fast and slow responses correlated with the thermalized process inside the band (fast response) and carrier trapping and recombination (slow response).

SESAMs use different designs to account for the different parameters. Sandwiches of saturable absorber between a dielectric reflector and a semiconductor Bragg mirror are, for example, a simple solution [3.22–3.23]. To account for large bandwidth supporting very short pulses, the Bragg mirror can also be replaced by a silver mirror [3.22, 3.24–3.26].

A panel of solutions is described in [3.21], and cited references make possible to tailor the pulse duration in the range of 50 fs–sub-10 fs with self-starting mode-locking and stable laser running.

References

[3.1]   D.H. Auston: IEEE J. Quant. Electr. **QE 4**, 420 (1968)
[3.2]   D.J. Bradley, G.H.C. New: Proc. IEEE **62**, 313 (1974)
[3.3]   B. Couillaud, A. Ducasse: Rev. Phys. Appl. **14**, 331 (1979)
[3.4]   M.F. Becker, D.J. Kuizenka, A.E. Siegman: IEEE J. Quant. Electr. **QE 8**, 687 (1972)
[3.5]   L.E. Hargrove, R.L. Fork, M.A. Pollack: Appl. Phys. Lett. **5**, 4 (1964)
[3.6]   O.P. McDuff, S.E. Harris: IEEE J. Quant. Electr. **QE 3**, 101 (1967)
[3.7]   S.E. Harris, B.J. McMurtry: Appl. Phys. Lett. **7**, 265 (1965)
[3.8]   S.E. Harris, O.P. McDuff: IEEE J. Quant. Electr. **QE 1**, 243 (1965)
[3.9]   D.J. Kuizenga, A.E. Siegman: IEEE J. Quant. Electr. **QE 6**, 709 (1970)
[3.10] A. Scavennec: Opt. Commun. **17**, 14 (1976)
[3.11] J. Kluge, D. Wiechert, D. von der Linde: Opt. Commun. **45**, 278 (1083)
[3.12] U. Keller, N.H. Knose, H. Roskos: Opt. Lett. **15**, 1377 (1990)
[3.13] H.A. Haus: IEEE J. Quant. Electr. **QE 11**, 736 (1975)
[3.14] J.P. Ryan, L.S. Goldberg, D.J. Bradley: Opt. Commun. **27**, 127 (1978)
[3.15] U. Keller, W.H. Knose, G.W. 'tHooft, H. Roskos, T.R. Woodward, J.E. Cunningham, D.L. Sivco, A.Y. Cho: Adv. Sol. State Lasers **10**, 115 (1991)
[3.16] G.R. Jacobovitz-Veleska, U. Keller, M.T. Asom: Proceedings of Conference on Lasers and Electro-Optics (CLEO), 1992, p. 188
[3.17] G. Steinmayer, L. Gallmann, F. Helbing, U. Keller: C.R. Acad. Sci. Paris t.2, Serie IV, 1389 (2001)
[3.18] R. Szipöcs, K. Ferencz, C. Spielmann, F. Krausz: Opt. Lett. **19**, 201 (1994)
[3.19] A Stingl, C. Spielmann, F. Krausz, R. Sipöcs: Opt. Lett. **19**, 204 (1994)
[3.20] M. Zalevani-Rossi, G. Cerullo, S. De Silvestri, L. Gallmann, N. Matuschek, G. Steinmeyer, U. Keller, G. Angelow, V. Scheuer, T. Tschudi: Opt. Lett. **26**, 1155 (2001)
[3.21] I.D. Jung, F.X. Kärtner, N. Matuschek, D.H. Sutter, F. Gorier-Genoud, Z. Shi, V. Scheuer, M. Tilsch, T. Tschudi, U. Keller: App. Phys. **B 65**, 137 (1997)
[3.22] U. Keller, K.J. Weingarten, F.X. Kärtner, D Kopf, B Braun, I.D. Jung, R. Fluck, C. Hönninger, N. Matuschek, J. Aus der Au: IEEE J. Selected Topics in Quantum Electronics **2**, Sept. (1996)
[3.23] U. Keller, D.A.B. Miller, G.D. Boyd, T.H. Chu, J.F. Ferguson, M.T. Asom: Opt. Lett. **17**, 505 (1992)
[3.24] R. Fluck, I.D. Jung, G. Zhang, F.X. Kärtner, U. Keller: Opt. Lett. **21**, 743 (1996)
[3.25] C. Hönninger, G. Zhang, U. Keller, A. Giesen: Opt. Lett. **20**, 2402 (1995)
[3.26] I.D. Jung, L.R. Brovelli, M. Kamp, U.Keller, M. Moser: Opt. Lett. **20**, 1559 (1995)

4

Further Methods for the Generation of Ultrashort Optical Pulses

C. Hirlimann

With 23 Figures

4.1 Introduction

Up to the beginning of the sixties, the shortest measurable time duration was of the order of one nanosecond (10^{-9} s). Short pulses were produced through the generation of short electrical discharges. After the laser was invented in 1960, the situation quite rapidly changed. In 1965, the picosecond (10^{-12} s) regime was reached by placing a saturable absorber inside a laser cavity. Twenty years of continuous progress led to the production of light pulses of less than 10 femtoseconds. In the race towards ever shorter pulses, recent developments in the generation of tabletop X-ray lasers have opened the way to dynamical studies in the attosecond (10^{-18} s) regime [4.1–2]. In the meantime, progress was made on the tunability of the pulsed-laser sources. Today's tunability extends from the near ultraviolet to the near infrared [4.2–6].

Periodic trains of ultrashort light pulses, with duration less than 10^{-12} s, can be generated using mode-locking techniques in laser cavities: active, passive or self-mode-locking. A large variety of lasers have been designed but only a few of them have been made commercially available. In this chapter we will discuss some of the lasers which have been of key importance in the field of short pulse generation. Although mode-locking is the most popular way to have a laser generate light pulses, we will also describe some other, more exotic techniques, based on the specific dynamical behavior of amplified spontaneous emission (ASE). A short section will also be devoted to the development of ultra-intense lasers, which has led to the arrival of table-top X-ray lasers.

4.1.1 Time–Frequency Fourier Relationship

Following L. Euler, who gave in 1748 the cosine representation of an analytical function, Joseph Fourier introduced in 1807 the more general trigonometric analysis, named after him, which was to become one of the most powerful mathematical tools used in physics. Fourier analysis can be applied to periodic pulse trains. If one defines $\Delta\nu$ as the full spectral width at half maximum and Δt as the duration of a light pulse at half maximum, then these quantities obey in the following inequality (see Chaps. 2 and 3):

$$\Delta\nu\,\Delta t \geq K, \tag{4.1}$$

where K is some constant number depending on the profile of the light pulse [4.7]. The Fourier relationship (4.1) has important consequences in the field of the generation and transfer of ultrashort light pulses.

To generate a short duration one has to be able to manipulate wide spectra. For a Gaussian pulse shape $K = 0.441$, and the spectral width of a one picosecond pulse (10^{-12} s) is at least $\Delta\nu = 4.41 \times 10^{11}$ Hz, while it is already 4.41×10^{13} Hz for a 10 fs pulse. Such large spectra need special care because they suffer group velocity dispersion when propagating through transparent materials. As an example, a 100 fs transform-limited pulse, crossing a 50 cm long cell filled with ethyl alcohol [4.8], broadens to 400 fs, a situation easily found early in dye amplifiers and now in modern large scale glass amplifiers. The instantaneous frequency of the pulse is then not constant any more; the pulse is chirped along the propagation direction.

Another dispersion effect takes place when light is reflected from a multilayered dielectric mirror like the ones used in laser cavities [4.9]. Reflection of light from such mirrors is the result of a constructive interference effect between rays which have been partially reflected at the interfaces of $\lambda/4$ dielectric layers. The frequency dispersion introduced by such mirrors is turned to advantage in Gires–Tournois interferometric optical compressors, but has proved to be particularly treacherous in broadband multistack multilayer dielectric mirrors [4.10]. The group velocity dispersion which takes place in these mirrors has also been turned to advantage in the so-called chirped mirrors [4.11–12].

Optical compressors are used to correct positively-group-velocity-chirped pulses by introducing negative group velocity dispersion. In a grating optical compressor, for instance, the various frequency components of the pulse are spatially separated and the device can be set up in such a way that their optical path is shortest when their frequency is highest.

The wide spectral width of short optical pulses also implies that each and every optical element used in an experimental setup must have a large enough bandwidth to be able to accommodate all the frequency components of the pulses. Ultrashort light pulses are fragile entities: from (4.1) it can be deduced that any spectral narrowing induces a broadening in the time domain. As a consequence these pulses can only be generated by laser oscillators having a wide gain bandwidth; a 100 fs pulse needs a gain spectrum 10 nm wide in the mid-visible spectral range. Outside the oscillator, ultrashort light pulses are better transferred from one point to another using aluminum- or silver-protected mirrors in order to avoid spurious phase distortions.

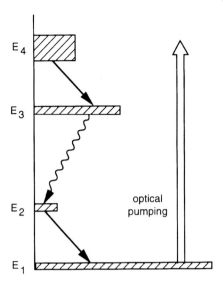

Fig. 4.1. Schematic diagram showing the orbital electronic states involved in a gas laser. Direct or collision excitation moves electrons from a ground state E_1 to an excited state E_4. Nonradiative relaxation of electrons from E_4 to E_3 and from E_2 to E_1 is faster than the spontaneous radiative relaxation of electrons between states E_3 and E_2. An inversion of the electronic population may grow between states E_3 and E_2

To provide the structure for this chapter, the arbitrary choice has been made to classify the laser oscillators according to the density of their gain medium, from gases to solids.

4.2 Gas Lasers

In gas lasers inverted populations are created between electronic levels in either gases, ion plasmas or metal vapors. The following discussion will be restricted to the visible and near-infrared regions of the electromagnetic spectrum, so that molecules as gain media will not be discussed.

Figure 4.1 shows a very general sketch of the electronic transitions involved in a gas laser. Most often, energy is transferred to the gas through an electrical discharge inside a sealed tube. Excitation may be direct, as in ionized argon or ionized krypton lasers, or indirect as in He–Ne lasers, where the energy is transferred through collisions from electrically excited helium to inert neon. From a highly excited state the electrons relax nonradiatively down to an intermediate metastable state, creating an inverted population.

Atomic-like states have energetically narrow bandwidths, but in a gas the atoms or ions move around with a distribution of speeds which depends on the working condition of the laser. For each atom velocity, the electronic state

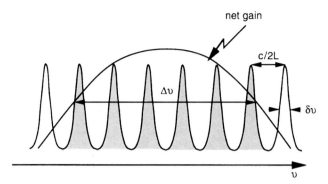

Fig. 4.2. The axial modes of a laser cavity are equally spaced with intermode frequency spacing $c/2L$. The effective gain curve (gain minus losses) is Doppler-broadened up to Δv and in this sketch only six modes may participate in the laser action [4.13]. The spectral width Δv of the axial modes is related to the finesse of the Fabry–Pérot cavity

is shifted in energy according to the Doppler effect; therefore the emission radiation of a gas is inhomogeneously broadened to a width $\Delta \nu$ (Fig. 4.2). Figure 4.2 shows the Dirac's comb of the Doppler-broadened axial modes of a laser cavity, which are equally spaced in frequency with a frequency separation equal to $c/2L$ (see Chaps. 1 and 3), c being the velocity of light and L the optical length of the cavity. Only the modes for which the gain is larger than the losses are involved in the coherent laser emission. For a typical gas laser the Doppler broadening is of the order of 1500 MHz, while the frequency spacing, for a 1 m cavity length, is of the order of 150 MHz, so that there are only a few tens of modes really involved in the laser process. With this kind of spectral bandwidth the shortest pulse achievable with gas lasers is of the order of a few hundreds of picoseconds; these lasers are not well suited for generating ultrashort pulses.

4.2.1 Mode-Locking

Active mode-locking (e.g. [4.14]) of ionized-argon [4.15] or krypton [4.16] lasers is widely used, as commercial lasers are available. Mode-locking is achieved by placing in the laser cavity an acousto-optical device which periodically modulates the losses of the laser at frequency $c/2L$ (see Chap. 3). In this way pulses lasting for about 150 ps can be generated, with a repetition rate of the order of 80 MHz and an average power of several watts or more.

4.2.2 Pulse Compression

Gas lasers, even if they are not good short-pulse generators, have been widely used as primary light sources in optical setups which allow the generation of

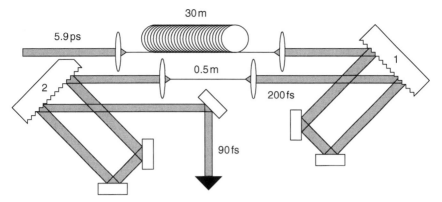

Fig. 4.3. Compound optical compressor. Two optical compressors made of a length of single-mode optical fiber followed by a dual-pass-grating time compressor are used in series. As the self-phase-modulation process broadens the spectrum of the pulses in a manner proportional to the intensity derivative of the pulses, the fiber needs to be longer in the first stage, where the peak intensity is smaller, than in the second stage, where it is larger

really ultrashort pulses. One basic idea is to use nonlinear optical effects to enlarge the spectral content of the pulses and to synchronize all the frequency components through an optical compressor.

The first demonstration of pulse shortening was achieved using a mode-locked He–Ne laser generating 500 ps pulses [4.17]. The spectrum of the pulses was broadened by up to 2.8 MHz by externally applying a strong acousto-optical modulation. The modulation creates new frequencies in the same way as in mode-locking. The frequency-broadened pulses were then time-compressed, using a specially designed Gires–Tournois interferometer [4.18], down to 250 ps.

Self-phase-modulation (see also Chap. 2) in single-mode optical fibers was developed at the end of the 1970s as a powerful tool for generating short optical pulses. Figure 4.3 shows the optical path of a dual-stage pulse compressor [4.19]. A mode-locked argon ion laser generates light pulses which are coupled into a 30 m long single-mode optical fiber. After propagation along the fiber, the spectral bandwidth is enlarged by a factor of 30 because of self-phase-modulation. The pulses are then chirp-corrected in a dual-pass grating compressor and coupled to a second but shorter optical fiber. After final time compression in another compressor the pulse shortening is as large as 65 times.

This technique is totally general and can be used with any kind of picosecond or subpicosecond laser source (see also Sect. 4.3.3). Its use is of particularly great importance when the original pulses are generated by a tunable picosecond laser, because tunability is maintained in the process.

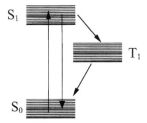

Fig. 4.4. In a dye molecule the electronic states are split into numerous substates corresponding to deformations of the electron clouds of the molecule induced by vibration and rotation of various parts of the molecule relative to each other. S_0 and S_1 are singlet states, T_1 is the lowest triplet state

4.3 Dye Lasers

Liquids are of higher density than gases, so they give rise to better yields when used as gain media, even though the real active media are organic molecules dissolved in polar solvents such as water, alcohols and glycols. These molecules absorb light in a specific wavelength range and emit light in some other, red-shifted range (Stokes shift).

Figure 4.4 shows a simplified electronic structure for a dye molecule (see e.g. [4.20]). This structure is much more complex than the electronic structure of gases because of interactions of the electron clouds with various rotations and vibrations inside the molecule, as well as interactions of the molecule with the solvent. These movements distort the electronic structure, each state being split into a great many rotation–vibration substates. The speed of the molecules in their solvent is much smaller than the speed of atoms in a gas, and therefore Doppler broadening may be totally neglected. Contrary to gases again, the energy positions of states and substates in a molecule are strongly dependent on the molecular environment. Electronic states are inhomogeneously broadened because of the interactions of the electron clouds with the neighbouring solvent molecules. For the purpose of understanding dye lasers, it is legitimate to assume that the electronic structure in these media is made of quasi-continuous energy bands. Between these bands one finds a quasi-continuous distribution of four-level electronic systems from which laser emission may take place. The spectral width of these bands is of the order of a few tens of terahertz (a few tens of nanometers for the wavelength), which opens direct access to the generation of ultrashort light pulses.

4.3.1 Synchronously Pumped Dye Lasers

The technique of synchronously pumping a dye laser (see Chap. 3 and [4.21]) is commonly used to generate short light pulses in the visible and in the near infrared. The primary source of energy for optically pumping the dye in these devices is an actively mode-locked laser generating pulses lasting for about

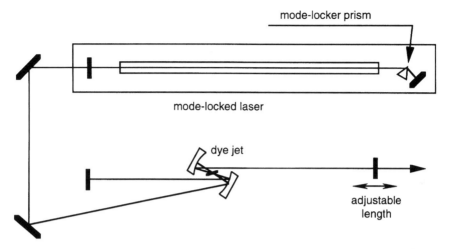

Fig. 4.5. Dye laser cavity, optically pumped by a mode-locked laser. The lengths of the laser cavities are made equal in order for the build-up of the pulse in the cavity to be synchronous with the pumping of the population inversion in the gain medium

100 ps. This primary oscillator may be either a gas laser or a solid-state laser (Fig. 4.5).

The mode-locked laser source excites dye molecules dissolved in a solvent; the liquid gain medium is either contained in a cell or dynamically shaped as a jet [4.22]. The dye laser is set up with the same optical length as the source laser to ensure the equality of the times of flight in both cavities. That way, the light pulse traveling in the dye cavity encounters the gain medium at the exact time it is optically pumped. Because of gain saturation (Fig. 4.6) and because the spectral bandwidth of a dye is wider than the bandwidth of the mode-locked laser, much shorter pulses may be generated.

Gain saturation has been explained in Chaps. 1 and 2; this nonlinear effect is such that for large incoming intensities the population inversion in the gain medium is depleted and stimulated emission stops. During the amplification of short pulses, the pumping rate of the inverted population is generally at least one order of magnitude slower than the duration of the pulse to be amplified. Therefore the incoming intensity rapidly grows, followed quickly by the stopping of the stimulated emission. By this mechanism the leading front of the pulse is preferentially amplified, compared with the trailing edge; this leads to a duration narrowing of the pulse from its trailing part.

These dual-cavity lasers generate light pulses with a duration of the order of a few picoseconds, a repetition rate of the order of 80 MHz and an average power ranging from 10 mW to a few hundreds of milliwatts. Various designs have been proposed, using a large number of dye molecules, allowing the generation of pulses with wavelengths ranging from 500 nm to the infrared.

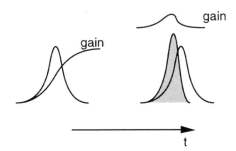

Fig. 4.6. Pulse shortening through gain saturation. *Left*, the time envelope of the pump pulses and the corresponding gain increase in the dye (in arbitrary normalized units). *Right*, result of the synchronized propagation of the dye oscillator pulse through the gain medium. The leading part of the pulse is strongly amplified up to the point when the gain saturates. The trailing part is not amplified, which results in a narrowing of the pulse [4.23–24]

4.3.2 Passive Mode-Locking

For general information, see for example [4.25–26]. The most famous member of this family is the so-called colliding-pulse mode-locked (CPM) dye laser, of which one variation is shown in Fig. 4.7 [4.27–28].

We will first consider the operation of a CPM laser without any compensation element. In a ring laser cavity two light pulses can counter-propagate, clockwise and counterclockwise. The saturable absorber is chosen to have a concentration such that it only saturates when the two pulses are simultaneously present. As a consequence, the two pulses automatically synchronize their flight in the cavity so as to cross in the saturable absorber. The superposition of the two pulses results in a transient stationary wave, creating a transient phase and amplitude modulation. Part of the light of each of the pulses is diffracted back into the other pulse by these modulations, improving the locking of the modes [4.29]. For the amplification of the two pulses to be symmetrical the gain medium and the saturable absorber are placed in the cavity according to the geometry shown in Fig. 4.8.

When the gain medium is placed at a quarter of the total cavity length L from the saturable absorber, the total pumping duration $L/2c$ is the same for both pulses. This geometrical rule is not very stringent as long as the gain recovers its full value between consecutive saturation processes [4.30–31].

Figure 4.9 illustrates the three mechanisms responsible for the final duration of the pulses. Gain saturation has already been discussed as a time-shortening mechanism in which the trailing edge of a pulse is clipped off. Symmetrically, in the saturable absorption process, the leading part of the pulse is absorbed (see Chaps. 2 and 3), which time-shortens them from the front edge. Time-shortening by gain and absorption saturation is balanced by group velocity dispersion. Along their propagation the pulses encounter the two jets and the coatings of the cavity mirrors, which introduce time-

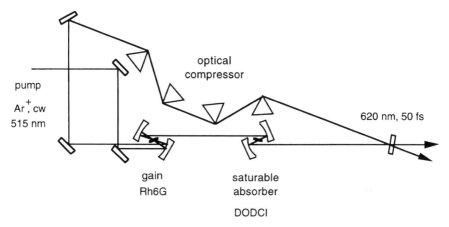

optical
compressor

pump

Ar⁺, cw

515 nm

620 nm, 50 fs

gain
Rh6G

saturable
absorber

DODCI

Fig. 4.7. Sketch of a balanced colliding-pulse mode-locked dye laser. The geometry of the cavity is that of a ring laser with Rhodamine 6G as a gain medium. These molecules are optically pumped by a CW argon ion laser delivering 5 W in its 515 nm green line. To ensure mode-locking, a second jet, dyed with DODCI (3,3′-diethyloxadicarbocyanine iodide), is placed at the focal point of a short-focal-length telescope. The emission maximum of Rhodamine 6G is close to 590 nm. This wavelength is strongly absorbed by the DODCI molecules, for which the absorption maximum is close to 580 nm in the ground state. Absorption of photons induces a conformation change of the DODCI molecules to a photoisomeric state, for which the absorption is red-shifted down to 620 nm. The resulting photoisomer is the actual saturable absorber acting as a mode-locker in the cavity. This photoisomerization explains why, without a dispersive element in the cavity, CPM lasers emit light at 620 nm rather that at 590 nm

broadening by group velocity dispersion. The strong focusing of the light in the saturable-absorber jet induces self-phase-modulation, which introduces phase distortion. When the laser runs in a steady regime the balance of these various effects leads to the generation of 60 fs pulses. The repetition rate of such a laser is of the order of 80 MHz, the central wavelength is 620 nm and the average power is of the order of 20 mW.

Note that the generation of ultrashort light pulses is not based on the use of ultrafast phenomena. Both the gain medium and the saturable absorber recover from saturation on timescales which are of the order of one nanosecond!

Up to this point, a temporal picture has been used to describe the pulse-shortening effects acting in a passively mode-locked laser. The Fourier duality between time and frequency implies that there must be an equivalent frequency description of the working condition of the laser. Let us consider now such a frequency description. It has been shown in Chap. 3 that the application of an amplitude modulation to one of the axial modes of a laser cavity through, for instance, an acousto-optic modulator induces sideband frequencies. The frequency shift of these sidebands is equal to the modulation frequency, and mode-locking occurs when the modulation frequency is properly

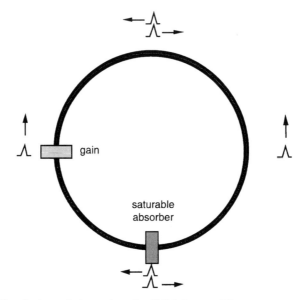

Fig. 4.8. Circulation of the pulses in CPM lasers. The two counterpropagating pulses automatically cross in the saturable absorber because the value of its optical density is chosen so that twice the intensity of one pulse is necessary to reach saturation. With the saturable absorber as origin, the gain medium is placed at a distance which is a quarter of the total length of the cavity. This ensures symmetrical amplification of the two pulses

adjusted to the axial-mode frequency spacing. In a passively mode-locked dye laser, the amplitude modulation responsible for the locking of the modes is produced by the light pulses themselves. When the light pulses make their way inside the gain and saturable-absorber media they induce a time-periodic amplitude modulation through gain and absorption saturation. The period of the modulation is exactly coincident with the time of flight of the pulses around the cavity, L/c (here L is the perimeter of the ring cavity). These two modulations create, for each light component, sidebands with a frequency shift $\pm c/L$, which is exactly coincident with the intermode frequency of the cavity c/L, leading to efficient mode-locking. This frequency description teaches us that there is no basic difference from active mode-locking.

One of the major advances in the technique of short-pulse generation has been to introduce a prism optical compressor inside the cavity of a CPM laser in order to compensate for group velocity dispersion [4.32].

Even when properly blazed, gratings are lossy optical devices because of their numerous diffraction orders. As at least two gratings are necessary for phase corrections, a Treacy compressor [4.33] (grating compressor) cannot be used in a laser cavity because it would introduce too much loss. Prism optical compressors were originally designed as intracavity devices. This is

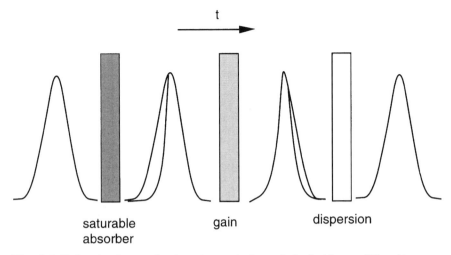

Fig. 4.9. Pulse-shaping mechanisms in passively mode-locked lasers. When it crosses the saturable absorber, the leading edge of a pulse disappears, while the trailing edge is clipped off in the gain medium. The final duration of the pulses is the result of the balance of these effects with the time-broadening induced by group velocity dispersion

made possible by the fact that the apex angle of a prism with a given index of refraction can be chosen so that a ray which passes through the prism at the minimum-deviation angle also falls onto the faces at the Brewster angle. At the minimum-deviation angle there is no geometrical distortion of a beam refracted through a prism. Not being at the minimum-deviation angle would distort the beam and this would be deleterious to the cavity quality factor. When in-plane linearly polarized light is used, the reflection from the faces at the Brewster angle is almost zero, therefore limiting additional losses in a cavity. A pair of prisms set in a subtractive refraction geometry with a long enough separation distance induces a negative group velocity dispersion in a light pulse. At the same time the various frequency components of a short pulse are spread across the diameter of the refracted beam, but this is easily corrected for by using a second pair of symmetrical prisms, which recollimates the rays. A four-prism device does not introduce much loss, so that it can be used harmlessly in a laser cavity (Fig. 4.7). The distance between the prisms must be chosen to be sufficiently long to induce net negative group velocity dispersion when the laser beam passes through the apices of the prisms. If one of the prisms can be translated in a direction perpendicular to the beam then the sign and the amplitude of the group velocity dispersion may be adjusted: introducing more glass into the beam path, i.e. introducing positive group velocity dispersion, balances or overcomes the negative dispersion of the device.

The dispersion parameter of the prism compressor is related to the second derivative of the optical path with respect to the wavelength:

$$D = \frac{\lambda_0}{cL} \frac{\mathrm{d}^2 P}{\mathrm{d}\lambda^2}, \tag{4.2}$$

L being the geometrical path of the light and P its optical path. It can be shown that $\mathrm{d}^2 P/\mathrm{d}\lambda^2 = 1.0354 - l(7.481 \times 10^{-3})$ for silica glass, with l being the distance between the two prism apices (measured in cm) and λ the wavelength (measured in μm). When $l = 13.8$ cm, the four prisms in the sequence shown in Fig. 4.7 are dispersionless, while for $l = 25$ cm the negative group velocity dispersion is enough to compensate for the addition of a 6.5 mm thick slab of silica inside the laser cavity [4.34].

Prisms have been introduced into CPM lasers with the primary purpose of balancing the linear dispersion effect and therefore reaching shorter duration. The shortest measured duration obtained in this way has been 27 fs [4.35]. Now what is the mechanism which prevents the collapse of the pulse to zero duration? Bandwidth limitation is the answer. The gain of Rhodamine 6G is 62 nm wide; from the time–frequency relationship one can deduce that pulses cannot be shorter than 30 fs. This limitation is dramatically illustrated by the experimental fact that the shortest duration may only be achieved with fresh dyes. In one hour of running, Rhodamine 6G undergoes a bandwidth narrowing which shifts the duration of the pulses toward the more familiar 50 fs figure.

Beyond reducing the duration of the pulses, the addition of a prism compressor greatly changed the operation of CPM lasers [4.36–37], introducing a quasi-soliton way of working. In a simplified description, solitons are functions one obtains as solutions to nonlinear propagation equations. They are found in a large variety of physical systems which have lost at least one space dimension; a "tsunami" (the solitary wave caused by an underwater earthquake in the Sea of Japan) is a good example of a solitary wave in a two-dimensional space. In their simplest expression, they can be seen as a unique intensity envelope which propagates without any deformation. Optical solitons are generated by propagating optical pulses in silica optical fibers [4.38]. To do so, the central wavelength of the pulse must be chosen to be on the high-energy side of the OH absorption band of silica, around 1.5 μm. In this configuration the linear group velocity dispersion is negative. For a high enough intensity, during the propagation of the pulse a self-phase-modulation is induced in which frequencies are positively chirped. With the right amount of self-phase-modulation, the linear and nonlinear effects completely compensate each other and the pulse propagates with no temporal distortion. The total propagation length is only limited by the residual absorption in the fiber and by light scattering. A light pulse traveling in a balanced CPM laser alternately encounters the saturable absorbing medium and the optical compressor. Self-phase-modulation, with positive chirp, occurs in the former, while negative group velocity dispersion is induced in the later. When the laser is in good working order, these two effects exactly compensate each other. Because of the periodic nature of the group velocity dispersion and self-phase-modulation, this propagation mode is not a solution of a continuous nonlinear propagation equation, and is called a quasi-soliton mode. This kind of propagation is very stable and the addition of prisms in the laser cavities has mainly improved the intensity stability of passively mode-locked lasers.

One of the drawbacks of the CPM laser is the weakness of the generated pulses. A few tens of milliwatts is a typical value of the average power for a repetition rate of the order of 100 MHz. This corresponds to only a few tens of picojoules for the energy per pulse, and a peak power not larger than a few hundred watts.

Amplification by stimulated emission has been widely used for the purpose of generating far more intense light pulses. An alternative way is to introduce a pulse-picker inside the cavity, to extract pulses at a repetition rate ranging from a few kilohertz to a few megahertz, with an energy of one nanojoule [4.39].

4.3.3 Really Short Pulses

The ultrashort light pulses described in Sect. 4.1 were generated using the optical fiber compression technique; we will describe here the first generation of 6 fs pulses [4.40]. A CPM laser is used as the primary source, generating 50 fs pulses. Some of these pulses are amplified in a multipass dye amplifier pumped by a copper-vapor laser, with a repetition rate of 8 kHz.

The pulse duration of copper-vapor lasers or Q-switched Nd:YAG lasers is of the order of 10 ns, which makes them last five orders of magnitude longer than the pulses to be amplified. This quasi-steady-state situation suggests the use of a thin gain medium to minimize group velocity dispersion and making the pulse cross the gain medium repeatedly for as long as there is gain available. During each pass of the amplified pulse the gain saturates, and then recovers during the flight of the pulse inside the device. Pulses of 9–10 microjoules can be generated in this way.

After being amplified, the pulses are first corrected for residual chirp down to 40–50 fs and then coupled into an optical fiber. The fiber is a single-mode fiber, in order to favor self-phase-modulation, and only a few millimeters long, to avoid Brillouin and Raman stimulated emission. The spectrum of the exiting pulses spans almost the full visible spectrum.

The real difficulty in generating ultrashort pulses is to properly compensate the phase distortion in the broadened spectrum. It has long been recognized that compensation up to the cubic phase distortion is a key ingredient for taking full advantage of the full spectral width. The solution to this problem came from the discovery that cubic phase distortion generated by a cubic prism compressor can be made negative, while the equivalent distortion in a grating compressor remains positive. This change in sign indicates that the use of both type of optical compressor allows one to compensate for phase

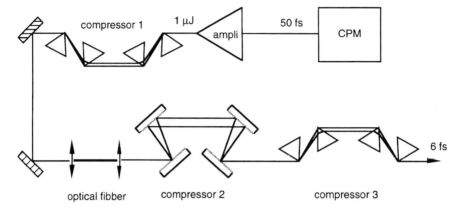

Fig. 4.10. Sketch of the setup used to generate 6 fs light pulses

distortion up to the third order. In this way, in 1987, 6 fs pulses were generated at 620 nm, which corresponds to only three oscillations of the electric field at half intensity maximum. 10 fs pulses were also produced between 690 and 750 nm [4.41]. More recently, a similar compression technique based on a hollow fiber filled with noble gases and a Ti:sapphire laser generating 20 fs pulses has been used to reduce the old record slightly to 4.5 fs [4.42].

4.3.4 Hybrid Mode-Locking

Various laser cavities have been designed in an attempt to take advantage of both the good duration shortening achieved with passively mode-locked dye lasers and the rather high intensities reached with synchronously pumped dye lasers.

Figure 4.11 shows one of these variations. The primary source is a Nd:YAG solid-state laser. The population inversion between Nd^{3+} states is produced with CW flashlights and the axial modes are actively mode-locked (CW mode-locked YAG laser). This laser emits 100 ps pulses at 532 nm when its 1.064 μm fundamental wavelength is frequency-doubled; it optically pumps a synchronous dye laser. The dye laser subcavity also contains a saturable-absorber jet and a prism compressor. When only two prisms are used for the compressor, they must be placed in front of the high reflector of the cavity so as to ensure that spatial frequency spreading is cured when recrossing the prisms. In front of the output coupling mirror, four prisms should be used to avoid any spurious spatial dispersion in the transmitted beam.

When pulse duration of 0.5–1 ps are short enough for an envisioned application, a bandpass filter can be added to the cavity to provide tunability of the central wavelength of the pulses. When pumped with 70 ps pulses at 532 nm, such lasers generate subpicosecond pulses in the range 560–974 nm with an average power of 650 mW [4.43]. The covering of such a large tunability range

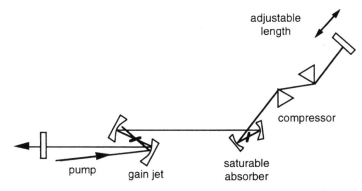

Fig. 4.11. Hybrid synchronously pumped dye laser. A saturable-absorber jet and a prism compressor are added to the cavity shown in Fig. 4.6 or 4.7

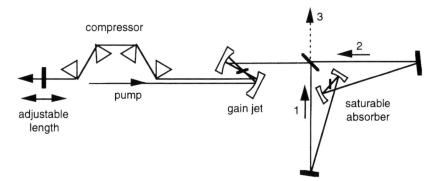

Fig. 4.12. Antiresonant cavity in a hybrid synchronously pumped dye laser. The saturable-absorber jet is placed in the middle of the Sagnac interferometer which closes the laser cavity. Fine geometrical tuning is necessary to achieve collision of pulses in the saturable absorber

requires the use of several pairs of gain and saturable-absorber dyes. On the other hand, generation of ultrashort pulses of less than 200 fs needs to be done with a large gain bandwidth, and tunability must be sacrificed. The shortest published duration with these hybrid mode-locked dye lasers is of the order of 50 fs [4.44]. This family of dye lasers is probably the most versatile in the field of short and ultrashort pulse generation.

At the price of added complexity, pulse collision can also be added to hybrid lasers [4.45–47]. For this purpose a Sagnac interferometer is placed at one end of the laser cavity. It consists of a ring subcavity coupled by a 50 % beam-splitter to the main cavity (Fig. 4.12). The saturable-absorber jet is placed exactly midway in the subcavity, so that the two counterpropagating pulses can cross there, as they do in CPM lasers. Dephasing at the beam-splitter is such that the lossy reflection (beam 3 in Fig. 4.12) is interferentially destroyed [4.48].

A good spatial superposition of the beams at the beam-splitter is mandatory for minimization of losses. Such a configuration generates 50 fs pulses at 620 nm, with an average power larger than 100 mW and a repetition rate of 100 MHz.

As already mentioned, synchronously pumped cavities must have exactly the same length, as the shortening of pulses is based on the time coincidence between gain and amplification. The spectral and temporal shapes of the pulses are very sensitive to minute length differences between the two cavities. As an example, in the picosecond regime a 1 μm difference is enough to disturb the working condition of the setup; the shorter the pulse the greater the necessary accuracy [4.49]. Many methods have been tested to create error signals allowing a feedback control of the cavity length. A simple one takes advantage of the fact that a 1 μm difference between the cavity lengths shifts the central wavelength of the emitted light by 10 nm. Placing a grating and a pair of photodiodes in one of the mirrors' leakage is enough to generate an error signal, which activates a PZT translator to move the end mirror of the cavity. The pump laser must be carefully stabilized because its phase jitter has also been recognized as a limiting factor in generating ultrashort light pulses.

4.3.5 Wavelength Tuning

Most of the lasers we have considered up to here emit light at a fixed central wavelength, at least in the femtosecond regime. By using another pair of dyes, pulses have been generated at 800 nm [4.50]; doubling of the pulses' frequency gives access to near-UV wavelengths around 300 nm [4.51].

The setup shown in Fig. 4.13 has been used to generate blue–green 10 fs pulses with the help of continuum generation [4.52]. In this technique the primary pulse generator is a CPM laser delivering 50 fs pulses at 620 nm. These pulses have a low energy content, only a few tens of picojoules, so that the first part of the device is devoted to the amplification of the pulses up to a few microjoules. The gain medium is a 2 mm thick cell filled with sulforhodamine-640 dye dissolved in water. A multipass amplifier is set up, with plane mirrors and lenses in order to adjust the beam diameter for intensity control on each pass of the pulses through the gain medium. The population inversion is realized by optical pumping of the dye with the 351 nm UV light of a XeF excimer laser. The duration of the pumping pulses is 10 ns at a repetition rate of 400 Hz. Multipass amplifiers [4.53] with longitudinal pumping, by taking advantage of long-lasting gains, give access to large amplification yields, and also they do not produce too much phase distortion. After amplification, phase distortions are corrected by a folded dual-pass prism compressor.

The red pulses amplified in the first amplification stage are used to generate a white-light continuum. When pulses shorter than 100 fs are focused on a transparent material, a strong spectral broadening starts because of self-phase-modulation. The newly created frequencies act as seeds for a large variety of

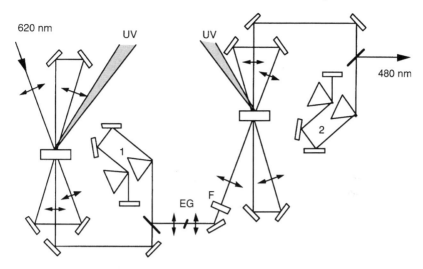

Fig. 4.13. External wavelength tunability using dual-stage multipass amplifier pumped in the UV by an excimer laser. The dyed solvent circulates in 2 mm thick cells, crossed three times by the pulses to be amplified. The primary pulse generator is a CPM laser. Each amplifier is followed by a folded dual-pass prism compressor (labeled 1 and 2). A white-light continuum is generated by focusing the amplified pulses on an ethylene glycol jet (EG). Part of the continuum is selected using an interference filter (F) and then amplified

high-order nonlinear optical processes, such as wave mixing and parametric effects, resulting in a wide spectrum which spans from the near infrared to the near ultraviolet [4.54–55], as will be shown in Chap. 6. In the setup described on Fig. 4.13 the continuum is generated by tightly focusing the pulses on a pure ethylene glycol jet. Blue pulses, 450 nm, are selected by an interference filter having a 17 nm wide spectral width. These blue pulses are then amplified in a second multipass amplifier in which the gain medium is made of coumarin dyes pumped again by the light of the excimer laser. When compressed, the blue pulses last for only 20 fs with an energy of a few microjoules.

The technique we have just described is quite general; by changing the interference filter, as well as the dye acting as a gain medium in the second-stage amplifier, one can generate ultrashort light pulses at any convenient central wavelength in the visible spectrum [4.53]. The white-light continuum generated in ethylene glycol has the specific temporal behavior shown in Fig. 4.14. In the blue and red parts of the continuum spectrum the frequency chirp is close to linear and can easily be compressed down to 10 fs. This technique can therefore be recognized as another time-shortening technique.

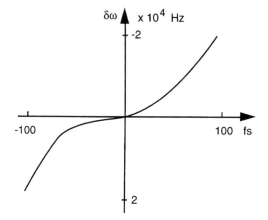

Fig. 4.14. Dynamics of angular frequency in a white-light continuum generated by tightly focusing 620 nm, 100 fs pulses on a jet of ethylene glycol. The overall time-spreading of the resulting pulses is not dramatically large. In the blue and red parts of the spectrum the frequency chirp is close to linear. The pulsation variation $\delta\omega$ is measured by sum-frequency mixing selected parts of the continuum with a reference pulse [4.54–55]

4.4 Solid-State Lasers

The first laser ever built was a ruby laser and it has always been a strong trend in laser physics to try to substitute solids for gases and dyes as gain media. This has been done for the purpose of increasing the generated power density and therefore the optical yield of lasers, and to minimize the amount of effort necessary to operate them. In the following, we will discuss some of the applications of solid-state media relevant to the generation of ultrashort pulses.

4.4.1 The Neodymium Ion

Neodymium, in its three-times ionized state Nd^{3+}, has been widely studied and used as a gain medium. This ion can be dissolved, with a concentration by weight of the order of 1%, in a large variety of solid-state matrices: phosphate glasses, yttrium–aluminum garnets (YAG), yttrium orthoalumi-nate perovskites $YAlO_3$ (YAP) and lithium–yttrium fluorides $LiYF_4$ (YLF). In all cases, as far as the 1.06 μm emission is concerned, the electronic structure of the Nd^{3+} ion can be schematized as shown in Fig. 4.1. Because of interactions of the ion with its surrounding solid matrix, the absorption and emission bands are wider for Nd^{3+} than for gases, despite the fact that these bands are only homogeneously broadened. Shorter pulse durations can there-fore be expected from Nd^{3+}-doped matrices than from gases. For short pulse durations these gain media are most often continuously optically pumped and used in cavities which are actively mode-locked (CW mode-locked lasers). The typical duration of the generated pulses is of the order of or less than 100 ps in neodymium-doped glasses [4.56], Nd:YLF [4.57] and Nd:YAG [4.58]. The repetition rate of these lasers is of the order of 80 MHz due to their length ($f = c/2L$), and they emit, close to $1.06\,\mu m$, pulse trains for which the average power can be as high as several watts. The most popular versions of these lasers are based on Nd:YAG and Nd:YLF; they deliver pulses with durations of the order of 100 ps when nonlinearly frequency-doubled to 532 nm. As already said, they are commonly used as primary light sources to excite synchronously pumped dye lasers running in the subpicosecond regime.

The larger the pump intensity in a gain medium, the larger the number of modes for which the gain is greater than the losses. When one increases the gain in a laser cavity the number of modes involved in a mode-locking process becomes larger, thus allowing shorter pulses to be generated. Gain increase can be achieved by Q-switching a mode-locked cavity, leaving more time for the gain to grow. Typical Q-switch frequencies range from a few hertz to a few kilohertz. Combining active and passive mode-locking with Q-switching leads to the generation of 1 ps pulses with Nd:YAG, Nd:YLF and Nd:YAP as gain media [4.59]. Since the local disorder of a glass matrix is greater than the disorder in crystal matrices, linewidths are inhomogeneously broadened in Nd-doped glasses and subpicosecond pulses may be generated [4.60].

4.4.2 The Titanium Ion

In the past few years, the most spectacular advances in laser physics and, particulary, in the field of ultrashort light pulse generation [4.61–62] have been based on the development of titanium-doped aluminum oxide (Ti:Al_2O_3, Ti:sapphire) [4.63–64] as a gain medium.

Titanium is only weakly soluble in aluminium oxide, but progress in crystal growth now allows doping up to a few percent. The Ti^{3+} ion substitutes for the Al^{3+} ion in the sapphire structure. The absorption band in the blue–green part of the visible spectrum is unusually wide. This is due to the fact that the ionic radius of the titanium ion is 26 % larger than that of the aluminum ion, which induces a strong distortion of the local environment of the titanium ion. This distortion, in turn, creates a strong local electric field. The absorption of visible wavelengths promotes electrons from a ^2T_g ground state to a ^2E_g excited state, which splits into two sublevels 50 nm apart under the influence of the local electric field. The ground and excited states are strongly coupled to vibrational modes of the sapphire matrix, inducing strong homogeneous broadening (Fig. 4.15).

The emission band of Ti:sapphire is shifted towards low energies ($\sim$ 750 nm) and its width is as large as 200 nm [4.65]. Titanium-doped sapphire, used as a gain medium, beats dye gain media as far as CW tuning and short-pulse generation are concerned. It is also of prime importance to recognize that synthetic sapphire has a thermal conductivity as high as metals [4.66]

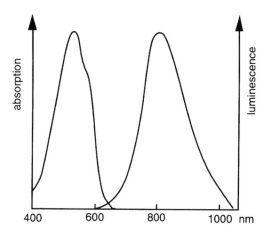

Fig. 4.15. Normalized absorption and emission spectra of Ti^{3+} ions embedded as impurities in a sapphire matrix $(Ti:Al_2O_3)$

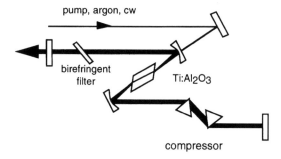

Fig. 4.16. Simplified sketch of a Ti:sapphire laser

at low temperature, a simple fact that explains why very high CW optical pumping powers (≈ 20 W) are achievable in this material.

A large number of mode-locking techniques have been developed to generate short pulses with Ti:sapphire as a gain medium: active mode-locking [4.67] and passive mode-locking with a dye jet [4.68] or with a coloured glass [4.69]. Self-mode-locking (or Kerr lens mode-locking) has proved to be the best way to achieve this. Kerr lens mode-locking (see Chap. 3) is based on the spatial phase-modulation effect, which favors coherent short-pulse emission. This effect starts at an intensity fluctuation in the laser cavity and various ways have been explored to generate the starting fluctuation in a reproducible way, e.g. synchronous pumping [4.70] and a moving mirror [4.71]. These techniques do not keep the laser in a steady working condition, but the addition of an external cavity containing a nonlinear optical element does [4.72–73].

Figure 4.16 shows a simplified sketch of a possible design for a Ti:sapphire laser cavity. The primary light source is an all-line blue-green argon-ion laser generating about 10 W of CW emission or a diode pumped green emitting laser. The length of the gain crystal being of the order of 1 cm, the intracavity laser beam and the pumping beams must be kept collinear. For this purpose, pumping is generally done through one of the gain collimating mirrors. If this colinearity is not respected, then the spatial phase modulation is inefficient. In Fig. 4.16, the device which keeps the laser running steadily is not shown, because there is no unique solution.

A pulse duration of 50 fs from a Ti:sapphire laser was rapidly published [4.74], and nowadays durations are close to 10 fs. This is achieved directly from the laser cavity with no external treatment of the pulses, because of the extra-large gain bandwidth of Ti:sapphire crystals. Commercially available lasers routinely generate pulses of less than 100 fs at a repetition rate of 80 MHz, with an average power of the order of one watt. Peak powers close to 100 kW are easily and directly reached. A great many nonlinear optical effects therefore come into the hands of more experimentalists, since no amplification is necessary, and this dramatically simplifies the experimental setups.

4.4.3 *F*-Centers

In alkali halides, when an electron fills a halide vacancy it creates an F-center defect. The trapped electron is responsible for wide-spectral-width absorption and emission bands in the near infrared. F-center defects in alkali halide crystals have been used as gain media in synchronously pumped lasers. Gain levels are low in these materials and high pumping intensities are required, which require one to cool the crystal down to liquid-nitrogen temperature. The first laser in this family was made using the LiF:F^{2+} color center pumped in the red by an actively mode-locked krypton-ion laser [4.75]. Pulses lasting for a few picoseconds were generated, which were tunable in the wavelength range 0.82–1.07 μm. Passive mode-locking in a balanced CPM-like cavity using the dye IR 140 as a saturable absorber has allowed the generation of 180 fs pulses [4.76].

4.4.4 Soliton Laser

The development of optical communications has lead to the design of lasers specifically devoted to the study of optical fibers. We will now briefly discuss the soliton laser [4.77]. In this experimental setup, a color-center laser is coupled to an external cavity containing a single-mode optical fiber. The losses in the chosen fiber are as low as 0.2 dB/km for wavelengths larger than 1.3 μm, a region where group velocity dispersion is negative. As we have already discussed in Sect. 4.2.2, solitons can propagate in such a fiber, and self-phase-modulation broadens the spectrum of the pulses, so soliton propagation is a pulse-shortening technique.

Figure 4.17 shows a sketch of a soliton laser. The first part is a color-center $(Tl^0(1):KCl)$ laser synchronously pumped by a mode-locked krypton-ion laser. A de Lyot filter inside the KCl laser tunes the emitted wavelength to 1.55 μm,

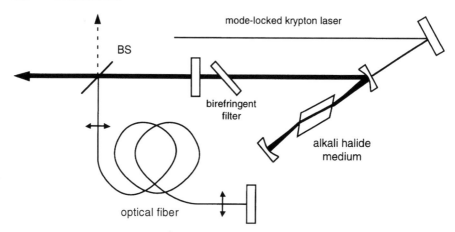

Fig. 4.17. Sketch of a soliton laser. The main oscillator is an alkali halide cavity, synchronously pumped by an actively mode-locked krypton-ion laser. This cavity is coupled to an optical-fiber cavity, which accommodates the propagation of a second-order soliton

in the negative-group-velocity-dispersion spectral region of the fiber. The main cavity is coupled through a 50 % beam-splitter to an external cavity which contains the optical fiber. Again, as in the antiresonant cavity (Fig. 4.12), the interference conditions are such that no light is lost at the beam-splitter when the superposition of the beams is properly adjusted.

The length of the optical fiber in the secondary cavity is chosen to be half of the characteristic propagation length of a second-order soliton. A second-order soliton splits into two pulses that recombine as a single pulse after some specific propagation length. In this way, the soliton pulse-shortening process takes place along one complete back-and-forth propagation cycle inside the secondary cavity.

Unlike to what happens in the Sagnac interferometer, the destructive interference condition must be actively monitored. In this way, 60 fs pulses at 1.5 μm have been externally compressed down to 20 fs, which corresponds to only 4 cycles of the electric field of the light. This kind of laser, being designed around specific optical properties of silica fibers, is not tunable, and it is in no way easy to set up or to service. Three coupled cavities must be actively kept to the same length, an interference condition must be continuously monitored, the coupling of the pulses into the fiber is quite delicate and coupling losses must be kept low. Using a better gain medium, $Tl^0(1)$:NaCl, 200 fs pulses have been generated with an average power of 300 mW at 1.56 μm [4.78].

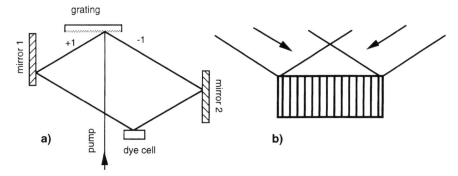

Fig. 4.18. Distributed feedback dye laser (DFDL). (**a**) The pump beam is split into two beams by means of reflections in the two first-order diffraction beams of a grating. The two beams interfere in a dye cell. (**b**) The interference pattern creates a gain modulation which behaves like a giant, amplifying, multilayered mirror

4.5 Pulse Generation Without Mode-Locking

From what has been previously said several times, one should not associate light-pulse generation only with mode-locking. Mode-locking remains the most widely used technique for the generation of periodic series of pulses, but is unable to generate single or low-repetition-rate pulses.

In what follows we will briefly consider other methods which have been used to generate short and ultrashort light pulses.

4.5.1 Distributed Feedback Dye Laser (DFDL)

Distributed feedback dye lasers are a quite simple and inexpensive way to enter the world of ultrashort pulse generation [4.79–80].

DFDLs have no mirror cavity, the feedback necessary for a laser resonance to take place being produced by a periodic spatial modulation of the gain. The two halves of a pumping beam are superposed in a dye cell, where they create an interference pattern. Bright interference fringes create periodic gain regions in the amplifying medium. One of the best ways to split the pump beam is to use symmetrical diffraction orders of a grating, as shown in Fig. 4.18a. This way the gain is structured as a single, thick, amplifying, multilayered mirror. As in a Lippmann optical filter, the gain modulation partly reflects waves for which half the wavelength is equal to the optical thickness of the modulation period. Amplified spontaneous emission is partially reflected back and forth in the distributed structure and further amplified by stimulated emission.

Coherently poor lasers, such as nitrogen or exciplex lasers (e.g. XeCl, XeF), can be used as the pump. The light emitted by a DFDL has complex dynamics. When the excitation intensity is slightly above the laser action threshold, a short, 100 ps pulse is generated. For 10 ns excitation pulses the compression

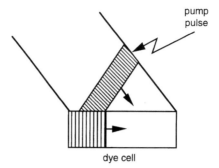

Fig. 4.19. Traveling-wave excitation. The excitation pulse is linearly time-delayed across a diameter of the light beam. This pulse creates a traveling gain region which amplifies traveling ASE (amplified spontaneous emission)

ratio is a factor of 100. The pulse narrowing is based on the fast decline of the distributed quality factor induced by gain saturation. Gain saturation diminishes the gain modulation contrast, which in turn kills the laser action. When the excitation intensity is increased, the gain can recover between two saturation processes: DFDLs emit trains of 10 ps pulses from which a single pulse can be extracted. Subpicosecond emission has been obtained by combining several DFDLs [4.81–83].

4.5.2 Traveling-Wave Excitation

In this technique the spatial phase of an excitation pulse is prepared in such a way that the pulse group delay is linearly distributed across the diameter of the linearly focused beam (Fig. 4.19). Transverse excitation of a gain cell with such a tailored pulse creates a moving gain front that can be made to propagate in the cell with the same group velocity as the spontaneous emission. When it first hits the cell, the excitation pulse generates a burst of spontaneous emission which travels along the cell, accompanying the gain packet and being continuously amplified. By this technique, poor-quality pulses are generated but short-lived, low-quantum-yield dyes can be used. Picosecond pulses in the wavelength range 1.2 to 1.8 μm have been generated in this way [4.84–86]. Even though it is not used any more at present, this simple technique could prove to be useful for generating far-UV or near-X-ray short pulses.

4.5.3 Space–Time Selection

Space–time selection allows one to directly generate picosecond pulses using a nanosecond laser. A dye cell is transversely pumped on one face and the two faces orthogonal to the excited volume constitute a low-quality optical laser cavity (Fig. 4.20). Such a low-Q-factor cavity, when excited by a long-lasting pulse, emits a strongly chirped broadband burst of light. This frequency chirp

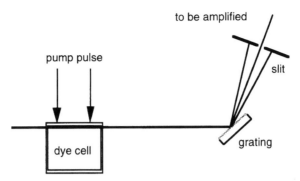

Fig. 4.20. Space–time selection of part of the emission from a low-Q-factor cavity generates 100 ps pulses

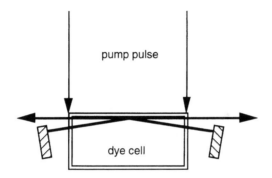

Fig. 4.21. Quenched-cavity laser. The natural reflections in the dye cell act as a low-quality-factor cavity. An external, folded, high-quality-factor cavity competes for the same gain with a short delay due to its longer length

is changed into a space sweep by dispersing the emission with a grating. A simple slit selects a 100 ps pulse of light when placed in the dispersed frequencies. This source is easily tunable by translating the slit or rotating the grating. As in most of these unconventional pulse generators, the emitted light pulses need to be amplified and multipass amplifiers are well suited for this purpose [4.87–89].

4.5.4 Quenched Cavity

Figure 4.21 shows another way to generate short optical pulses without mode-locking. In this technique the gain in a cavity is dynamically killed (quenched) after the rise of a laser emission in order to tailor explicitly a pulse shape. For this purpose two optical cavities are setup to compete one with the other. As in the previous setup, the natural reflections in a transversely pumped dye cell are used as a low-quality-factor cavity (weak feedback). A second external cavity is built, making a small angle with the same gain volume and having a longer length. This cavity has a high quality factor and high-reflectivity multilayered dielectric mirrors. The laser effect first starts in the short, low-Q-factor cavity, and then because of its longer length the laser effect starts with some time delay in the high-Q-factor cavity. This later effect is more efficient than the former and rapidly saturates the gain, ending the laser emission [4.90–92]. To some extent, this quenched-cavity technique allows some control of the pulse duration by simply adjusting the length of the high-Q-factor cavity.

4.6 New Developments

4.6.1 Diode Pumped Lasers

Over the past decade the most important advance in the field of short optical pulse generation has been the replacement of ion gas lasers as a primary source of energy by solid-state diode pumped lasers. This has been made possible by the continuous improvement of the technology of III–V semiconductor heterostructures and multiquantum wells based on Ga, As, Sb, In, As, and P. The driving force behind this move is the need for new technologies in the fields of optical communications and computer processors and memories. Nowadays, very sophisticated material structures allow the mass production of diode laser arrays, with CW emitting up to 40 Watts in the far-red or near infrared. In a very popular design, laser diodes are used to optically pump ND^{3+}:YVO_4 crystals, in a Q-switched cavity running at a repetition rate of a few kilohertz, at a wavelength close to 1.04 μm. Like in YAG lasers, the light is frequency doubled and used as a light source for short-pulse laser cavities. Despite a two-step generation process the overall optical yield is better than in ion gas lasers. Following this line, the need for electrical power and water cooling has dramatically decreased, saving maintenance. This new concept was introduced by Wilson Sibbett and his collaborators from St Andrews University in 1994. In these all solid-state laser sources, the diode pumping is indirect and the need was pointed out for a direct diode pumping oscillator [4.93].

Figure 4.22 shows a schematic of such a laser in which the gain medium is directly pumped through the subcavity mirrors by a couple of laser diodes. The difficulty in this scheme is related to the optical adaptation of the foci volume of the laser cavity and the pumping diodes.

4.6.2 Femtosecond Fibber Lasers

As has been seen, fibbers lasers were designed in the course of optical telecommunication technology development. They now are able to produce light pulses in the femtosecond regime [4.94], and their development is such that they are commercially available in niche markets. Because of their robustness and compactness, these lasers might have a bright commercial future for various industrial applications, ranging from mechanics to medicine.

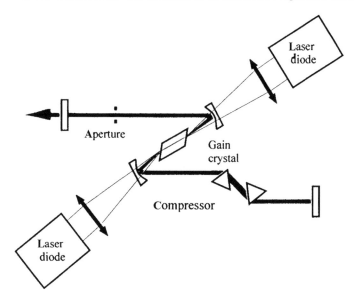

Fig. 4.22. Schematics of a laser diode pumped solid-state laser

Femtosecond fibber lasers are based on fibbers doped with rare earths, mainly erbium and ytterbium, used for optical amplification and regeneration processes in telecommunication repeaters. The gain bandwidth of these fibbers is large enough (≈ 6 THz) to ensure via the Fourier relationship a minimum duration of the order of 100 fs [4.95] and small enough for a large family of applications. Because of their simplicity, semiconductor saturable absorbers are mainly used to ensure mode-locking in these lasers. The main advantage of this technological method is the concomitant development of adapted diodes for the optical pumping of the ion-doped fibbers.

4.6.3 Femtosecond Diode Lasers

Studies of diode lasers are still very active to further decrease the duration of the pulses that can be directly generated. The aim of these studies is to use femtosecond light sources in compact opto-electronic devices on the one hand, and to dispose of the tiniest possible laser oscillators one could use as a seed for large-scale amplifying chains generating intense pulses on the other hand. Two ways are actively explored for the generation of short light pulses from laser diodes: Q-switching and mode-locking. In both techniques, picosecond pulses are first directly produced by a diode and then recompressed in the femtosecond regime.

In the Q-switched regime, optical pulses as short as 20 fs have been demonstrated [4.97] at a central wavelength of 1541 nm, with a repetition rate of 2 GHz and an energy per pulse of 0.5 nJ. The starting point of this generator

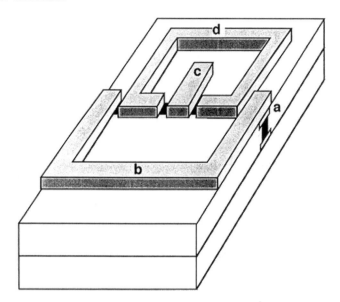

Fig. 4.23. Schematics of a femtosecond CPM laser diode. A multiquantum well acts as a gain medium. Electrons are confined and light is guided in the structure (a). Three electrodes are used: (b) gain pumping, (c) saturable absorber command, (d) active light guiding [4.96]. Light pulses are generated on both ends of the gain strip and collide inside the saturable absorption region

is a high-frequency gain-modulated diode, CW pumped by a tunable source. The first relaxation oscillation in the diode cavity on each modulation cycle is then launched in a series of optical fibbers, then propagating as high-order solitons. This experiment clearly demonstrated the potential of diode lasers for generating ultrashort light pulses.

Passive mode-locking is also applied to laser diodes by placing a saturable absorber inside the cavity of the laser. The saturable absorber can be either a special multiquantum well structure or a perturbed layer of material. Broadband gain diodes can emit highly chirped light pulses that can be nonlinearly recompressed down to 150 fs [4.98]. Nonlinear pulse compression can be realized, for example, by using a multipass Treacy compressor and deformable mirrors or a phase modulator. The ultimate improvement in passive mode-locking of diode lasers would be to integrate the compression function inside the laser cavity and to have it run in the quasisoliton mode. Along this line, a principal demonstration has been made [4.99] in which it was shown that during its propagation through a semiconductor a short optical pulse could undergo self-narrowing. This happens under the specific condition that the central wavelength of the pulse lies in the upper energies of the band gap. In this situation the self-phase modulation process does create new bluer frequencies on the leading edge of the propagating pulse while new redder frequencies are crated on the trailing edge. As the group velocity dispersion of semiconductors in their gap is positive (blue travels faster than red) the newly created frequencies recompress along the propagation in a manner that mimics a soliton propagation process. Therefore one could imagine that part of a semiconductor laser is made of the right material or concentration to ensure self-compression.

4.6.4 New Gain Materials

One of the most active fields of the past few years has been the search for new gain materials. Despite the fact that Ti:sapphire lasers have become a research and industrial standard, the need for new gain materials is still high for specific applications such as direct generation of specific wavelength ranges in the infrared of the near UV, more efficient use of available devices (diodes, fibbers ...), better yield, etc.

4.6.4.1 Chromium Doping

Forsterite. Large emission band crystals that had been intensively scrutinized in the 1970s for their potential as rivals to tunable dye lasers have been re-visited during the last decade for their potential as short pulse generators. One of the best examples is given by forsterites. Forsterities are a crystalline phase of magnesium and/or iron silicates (Mg, Fe) SiO_4 that can be doped with chromium. A low repetition rate operated Cr^{4+}:Mg_2SiO_4 cavity produced pulses as short as 14 fs [4.100] at 1.3 μm. As forsterites have a low gain value, low repetition rate cavities have been developed to allow more time for the gain to grow between two pulses. This is achieved by folding the laser cavity between mirrors. That way, high energy (17 nJ) pulses have been generated at a repetition rate of 26.5 MHz. In this specific case the cavity end mirror was a SESAME that insured the locking of the cavity modes [4.101].

YAG. Suffering spectral and thermal conductivity limitations, Cr^{4+}:YAG cannot be used for the generation of ultrashort pulses. But because of the excellent quality of the substate material, original applications such as high repetition rate (2.6 GHz) pulse generation at 1.5 μm, have been developed [4.102].

Colquirites. Chromium doped LiSAF, LiSGaF, and LiCaF do belong to that family [4.103–104] of gain media. They can be laser diode pumped at 670 nm.

4.6.4.2 Rare Earth Doping.
The aspect regarding the use of telecommunication devices has been only hinted at in the fibber laser part; it will now be developed more.

The Ytterbium ion has a simple electronic structure so that there is no absorption starting from excited states, no up-conversion, and no decline of the fluorescence time for high concentrations. This minimizes the losses at pumping time. The fluorescence gain width is of the order 20 nm, allowing for the generation of 60-fs duration pulses and even less in the vicinity of 1040 nm. A possible drawback of this ion is its quasi-three-level structure that induces reabsorption of the emitted light.

One of the most important points, in terms of experimental setups, is the commercial availability of powerful pumping laser diodes at 980 nm, designed for the pumping of ytterbium doped fibbers used for amplification and re-shaping of signals in all optical long-distance telecommunication cables. A developing market is favorable to both continued production and a decrease of prices.

Glasses. Glasses have been worked out to accommodate ytterbium doping and have allowed for the generation of femtosecond pulses [4.105]. Tunable diode pumped cavities produced 60 fs pulses around 1.06 μm [4.106]. Poor thermal conductivity certainly slows down the development of ultrafast lasers.

Oxoborates. Calcium and gadolinium oxoborate (GdCOB) doped with yt-terbium is a promising new gain material: doping as high as 27% is possi-ble and the thermal conductivity is of the order of 3 times larger than in glasses. This new material is a good compromise between the large fluores-cence band of glasses and the good thermal conductivity of crystals. Light pulses of 90 fs at 1045 nm have been produced with a 15% doped Gd-COB ($Yb^{3+}:Ca_4GdO(Bo_3)_3$) crystal directly pumped by a 976 nm laser diode [4.107]. In the same family the newly developed Yd:BOYS ($Yb^{3+}:Sr_3Y(Bo_3)_3$) sounds very promising. Because two different sites are available for the Yb ion in the crystal structure, disorder is induced in the crystals and large absorption and emission bands are available. Pulses of 70 fs have already been generated in one or two laser diode pumped cavities around 1062 nm [4.108].

YAG. Because of emission bandwidth limitations Yb:YAG does not allow for the generation of ultrashort light pulses. Despite this limitation, it can be of interest to use such a gain medium when quite long-lasting ($\approx$ 100 fs) pulses are enough, because the setup could be based on well-mastered technologies [4.109].

Tungstates. Because of their large emission cross section, tungstates are also explored as possible gain candidates for ultrashort light pulse generators. Ref-erence materials are: $Yb:KY(WO_4)_2$ and $Yb:KGd(WO_4)_2$ [4.110–111].

4.7 Trends

Evolution toward the generation of ever-shorter duration has been driven by the desire to gain knowledge of the fastest events that occur in nature. The study of fast events has long been dominated by the science of electricity and then by electronics. For forty years now, optics, through the science of lasers, has been leading the quest. Lasers generating short pulses have become commercially available instruments, which have proved to be valuable tools for many scientific fields including physics, chemistry, and biology.

No physical argument can be given to predict that diving towards shorter duration should stop. The shorter a pulse, the wider its spectrum; when the duration of a pulse comes close to the period of the central wavelength the spectral width is close to the value of that central frequency. In the visible it is therefore difficult to generate pulses shorter than a few femtoseconds. This is not a limitation of principle, but rather a technical one. To go to even shorter pulses it is evidently necessary to increase the central frequency of the pulse to keep the spectral bandwidth smaller than this central frequency. Two ways are being actively explored. In the first one, X ray bursts, 300 fs in duration, are generated by shining relativistic electrons with amplified 50 fs visible pulses [4.112], while other groups try to phase-lock very high-order harmonics generated by short optical pulses in rare gases. Recent progress have been made along these directions, and even an experiment has been published in the attosecond regime [4.113].

As short optical pulses in the visible have probably reached their shortest practical duration, large efforts are devoted to the generation of very intense pulses. For a given reasonable amount of energy, the narrower the time envelope of this energy amount the higher the peak power. One hundred mJ delivered in 100 fs correspond to a 10-terawatt peak power. What could these powerful pulses be useful for? When focused down, the intensity can reach a value as high a 10^{20} W/cm^2 and the electric field 2×10^{14} V/cm. In such electric fields the velocity of excited electrons comes close to the velocity of light and relativistic effects are to be expected. The effect of the magnetic field of an electromagnetic wave cannot be neglected anymore. New tabletop techniques for accelerating electrons are actively explored. Ultrashort powerful lasers in the visible allow the fast development of tabletop X ray laser physics [4.114], nuclear physics [4.115–116], and further openings in the field of nuclear fusion [4.117].

References

[4.1]   N.A. Papadogiannis, B. Witzel, C. Kalpouzos, D. Caralambidis: Phys. Rev. Lett. **83**, 4289–92 (1999)
[4.2]   M. Hentschel, R. Kienberger, Ch. Spielmann, G.A. Reider, N. Milosevic, T. Brabec, P. Corkum, U. Heinzmann, M. Drescher, F. Krausz: Nature **414**, 509–13 (2001)
[4.3]   A. Brun, P. Georges, G. Le Saux, G. Roger, F. Salin: Pour La Science **135**, 46 (1989)
[4.4]   J.D. Simon: Rev. Sci Instrum., **60**, 3597–3624 (1989)
[4.5]   W.H. Knox: IEEE J. Quantum Electron. **QE 24**, 388–97 (1988)
[4.6]   A.E. Siegman: *Lasers*, University Science Books, Mill Valley, CA (1986)
[4.7]   W.H. Lowdermilk: Technology of Bandwidth-Limited Ultrashort Pulses Generation, In *Laser Handbook*, 3, M.L. Stick, ed. North-Holland, Amsterdam (1979)

[4.8]   O. Seddiki: Thesis, University Pierre et Marie Curie, Paris (1986)

[4.9]   E. Spiller: Appl. Opt. **10**, 557–66 (1971)

[4.10]   W.H. Knox, N.M. Pearson, K.D. Li, C. Hirlimann: Opt. Lett. **13**, 574–76 (1988)

[4.11]   R. Szipöcs, C. Spielmann, F. Krausz: Opt. Lett. **19**, 201 (1994)

[4.12]   F.X. Kärtner, N. Matuschek, T. Schibili, U. Keller, H.A. Haus, C. Heine, M. Morf, V. Scheuer, M. Tilsch, T. Tschudi: Opt. Lett. **22**, 831–3 (1997)

[4.13]   cf. [4.6] p. 42

[4.14]   See, for instance, A. Yariv. *Quantum Electronics*, 2nd ed. (John Wiley & Sons, New York, 1975) chap. 11

[4.15]   P.G. May, W. Sibbett, K. Smith, J.R. Taylor, J.P. Wilson: Opt. Commun. **42**, 285 (1982)

[4.16]   L.L. Steinmetz, J.H. Richardson, B.W. Wallin: Appl. Phys. Lett. **33**, 163 (1978)

[4.17]   M.A. Duguay, J.W. Hansen: Appl. Phys. Lett. **14**, 14 (1969)

[4.18]   F. Gires, P. Tournois: C. R. Acad. Sci. (Paris) **258**, 6112 (1964)

[4.19]   B. Nikolaus, D. Grischkowsky: Appl. Phys. Lett. **43**, 228–30 (1983)

[4.20]   cf. [4.14], p. 238

[4.21]   A. Dienes, E.P. Ippen, C.V. Shank: Appl. Phys. Lett. **19**, 258 (1971)

[4.22]   C.K. Chan, S.O. Sari: Appl. Phys. Lett. **25**, 403 (1974)

[4.23]   D.J. Bradley: *Ultrashort Light Pulses*, S.L. Shapiro, Ed., Topics in Applied Physics, vol. 18 (Springer-Verlag, Berlin (1977), p. 60

[4.24]   cf. [4.4], p. 3602

[4.25]   cf. [4.14], p. 277

[4.26]   cf. [4.6], p. 1057, chap. 28, p. 1104

[4.27]   R.L. Fork, C.V. Shank, R.T. Yen, C. Hirlimann: IEEE J. Quantum Electron. **QE-19**, 500 (1983)

[4.28]   C.V. Shank, C. Hirlimann: Helv. Phys. Acta **56**, 373–81 (1983)

[4.29]   M.S. Stix, E.P. Ippen: IEEE J. Quantum Electron. **QE-19**, 520 (1983)

[4.30]   C.V. Shank, E.P. Ippen: Appl. Phys. Lett. **24**, 373 (1974)

[4.31]   cf. [4.27]

[4.32]   J.A. Valdmanis, R.L. Fork, J.P. Gordon: Opt. Lett. **10**, 131–3 (1985)

[4.33]   E.B. Treacy: IEEE J. Quantum Electron **QE-5**, 456 (1969)

[4.34]   R.L. Fork, O.E. Martinez, J.P. Gordon: Opt. Lett. **9**, 150 (1984)

[4.35]   [4.32]

[4.36]   F. Salin, P. Grangier, G. Roger, A. Brun: Phys. Rev. Lett. **56**, 1132 (1986)

[4.37]   F. Salin, P. Grangier, G. Roger, A. Brun: Phys. Rev. Lett. **60**, 569 (1988)

[4.38]   L.F. Mollenauer, R.H. Stollen, J.P. Gordon: Phys. Rev. Lett. **45**, 1095 (1980)

[4.39]   R.L. Fork, B.I. Green, C.V. Shank: Appl. Phys. Lett. **38**, 671 (1981)

[4.40]   R.L. Fork, C.H. Brito Cruz, P.C. Becker, C.V. Shank: Opt. Lett. **12**, 483 (1987)

[4.41]   C.H. Brito Cruz, A.G. Prosser, P.C. Becker: Opt. Commun. **86**, 65 (1991)

[4.42]   Ch. Spielmann, S. Sartania, F. Kransz, K. Ferencz, M. Nisoli, S. de Silvestri, O. Svelto: Laser Focus World **33**, 127 (1997)
         Ch. Spielmann, S. Sartania, F. Kransz, R. Szipöcs, K. Ferencz, M. Nisoli, S. de Silvestri, O. Svelto: QELS'97, OSA Technical Digest Series *12*, JTuA3 (1997)

[4.43]   D. Dawson, T.F. Boggess, D.W. Gawey: Opt. Commun. **60**, 79 (1986)

[4.44] H. Kubota, K. Kurokawa, M. Nakazawa: Opt. Lett. **13**, 749 (1988)
[4.45] A.E. Siegman: Opt. Lett. **6**, 334 (1981)
[4.46] T. Norris, T. Sizer II, G. Mourou: J. Opt. Soc. Am. **B 2**, 613 (1985)
[4.47] J. Chesnoy, L. Fini: Opt. Lett. **11**, 635 (1986)
[4.48] cf. [4.6]
[4.49] M.C. Nuss, R. Leonhardt, W. Zinth: Opt. Lett. **10**, 16 (1985)
[4.50] A. Migus, A. Antonetti, J. Etchepare, D. Hulin, A. Orszag: J. Opt. Soc. Am. **B2**, 584 (1985)
[4.51] D.C. Edelstein, E.S. Wachman, L.K. Cheng, W.R. Bosenberg, C.L. Tang: Appl. Phys. Lett. **52**, 221 (1988)
[4.52] R.W. Schoenlein, J.-Y. Bigot, M.T. Portella, C.V. Shank: Appl. Phys. Lett. **58**, 801 (1991)
[4.53] S. Petit, O. Crégut, C. Hirlimann: Opt. Commun. **124**, 49 (1996)
[4.54] R.L. Fork, C.V. Shank, R.T. Yen, C. Hirlimann, W.J. Tomlinson: in *Picosecond Phenomena III*, Chemical Physics series 23, ed.: K.B. Eisenthal, R.M. Hochstrasser, W. Kaiser, A. Laubereau Springer-Verlag, Berlin, (1982)
[4.55] R.L. Fork, C.V. Shank, R.T. Yen, C. Hirlimann, W.J. Tomlinson: Opt. Lett. **8**, (1983)
[4.56] P. Heinz, M. Fickenscher, A. Laubereau: Opt. Commun. **62**, 343 (1987)
[4.57] J. Weston, P.H. Chin, R. Aubert: Opt. Commun. **61**, 208 (1987)
[4.58] H.P. Kortz: IEEE J. Quantum Electron **QE-19**, 578 (1983)
[4.59] P. Heinz, A. Laubereau: J. Opt. Soc. Am. **B7**, 182–6 (1990)
[4.60] P. Heinz, A. Laubereau: J. Opt. Soc. Am. **B6**, 1574 (1989)
[4.61] P. Lacovara, L. Estorovitz, M. Kokta: IEEE J. Quantum Electron. **QE-21**, 1614 (1985)
[4.62] C.E. Byvik and A.M. Buoncristiani: IEEE J. Quantum Electron. **QE 21**, 1619 (1985)
[4.63] D.E. Spence, P.N. Kean, W. Sibbett: Opt. Lett. **16**, 42 (1991)
[4.64] D.E. Spence, J.M. Evans, W.E. Sleat, W. Sibbett: Opt. Lett. **22**, 1762 (1991)
[4.65] Lasers and Optronics, December 1989, p. 49
[4.66] C. Kittel: *Introduction to Solid State Physics*, 3rd ed. John Wiley, New York (1967), p. 194
[4.67] P.F. Curley, A.I. Ferguson: Opt. Lett. **16**, 1016 (1991)
[4.68] N. Sarukura, Y. Ishida, H. Nakano: Opt. Lett. **16**, 153 (1991)
[4.69] N. Sarukura, Y. Ishida, T. Yanagawa, H. Nakano: Appl. Phys. Lett. **57**, 229 (1990)
[4.70] C. Speilmann, F. Krausz, T. Brabec, E. Wintner, A.J. Schmidt: Opt. Lett. **16**, 1180 (1991)
[4.71] P.M.W. French, D.K. Naske, N.H. Rizvi, J.A.R. Williams, J.R. Taylor: Opt. Commun. **83**, 185 (1991)
[4.72] U. Keller, G.W. t'Hooft, W.H. Knox, J.E. Cunningham: Opt. Lett. **16**, 1022 (1991)
[4.73] G. Gabetta, D. Huang, J. Jacobson, M. Ramaswamy, E.P. Ippen, J.G. Fujimoto: Opt. Lett. **16**, 1756 (1991)
[4.74] N.H. Rizvi, P.M.W. French, J.R. Taylor: Opt. Lett. **17**, 279 (1992)
[4.75] L.F. Mollenauer, D.M. Bloom, A.M. Del Gaudio: Opt. Lett. **3**, 45 (1978)
[4.76] N. Langford, R.S. Grant, C.I. Fohnston, K. Smith, W. Sibbett: Opt. Lett. **14**, 45 (1989)

[4.77]   cf. [4.38]

[4.78]   C.R. Pollock, J.F. Pinto, E. Georgiou: J. Appl. Phys. B **48**, 287 (1989)

[4.79]   C.V. Shank, J.E. Bjorkholm, H. Kogelnik: Appl. Phys. Lett. **18**, 396 (1971)

[4.80]   Z. Bor, B. Rácz, G. Szabó: Helv. Phys. Acta **56**, 383 (1983)

[4.81]   S. Szatmári, B. Rácz: Appl. Phys. B **43**, 93 (1987)

[4.82]   G. Szabó, Z. Bor: Appl. Phys. B **47**, 299 (1988)

[4.83]   F. Raksi, W. Hener, H. Zacharias: Appl. Phys. B **53**, 97–100 (1991)

[4.84]   H.J. Polland, T. Elsaesser, A. Seilmeier, W. Kaiser: J. Appl. Phys. B **32**, 53 (1983)

[4.85]   P.O. Scherer, A. Seilmeier, W. Kaiser: J. Chem. Phys. **83**, 3948 (1985)

[4.86]   Z. Bor, S. Szatmári, A. Muller: J. Appl. Phys. B **32**, 101 (1983)

[4.87]   M.M. Martin, E. Breheret, Y.H. Meyer: Opt. Commun. **56**, 61 (1985)

[4.88]   F. Nisa, M.M. Martin, Y.H. Meyer: Opt. Commun. **75**, 294 (1990)

[4.89]   N.D. Hung, Y.H. Meyer: Appl. Phys. B **53**, 226 (1991)

[4.90]   S. Szatmári, and F.P. Schäfer: Opt. Commun. **48**, 279 (1983)

[4.91]   S. Szatmári, and F.P. Schäfer: Appl. Phys. B **33**, 219 (1984)

[4.92]   P. Simon, J. Klebniczki, G. Szabó: Opt. Commun. **56**, 359 (1986)

[4.93]   F. Falcoz, F. Balembois, P. Georges, A. Brun: Opt. Lett. **20** 1874 (1995)

[4.94]   M.E Fermann, A Galvanauskas, G. Sucha, D. Harter: Appl. Phys. B**65**, 259 (1997)

[4.95]   M.E. Fermann, A. Galvanauskas, M. Hofer: Appl. Phys. B**70**, S13–S23 (2000)

[4.96]   M.C. Wu, Y.K. Chen, T. Tanbun, R.A. Logan: Ultrafast Phenomena VIII, J.-L. Martin, A. Mingus, G.A. Mourou, and A.H. Ziwail Eds., Springer-Verlag, Berlin (1993), p. 212

[4.97]   M. Matsui, M.D. Pelusi, A. Suzuki: IEEE Photonics Tech. Lett. **11**, 1217 (1999)

[4.98]   A Azouz, N. Stelmakh, P. Langlois, J.M. Lourtioz, P. Gravilovic: IEEE JSTQE **1**, 577 (1995)

[4.99]   J.F. Lami and C. Hirlimann: Phys. Rev. Lett. **82**, 1032 (1999)

[4.100] C. Chudoba, J.G. Fujimoto, E P. Ippen, H.A. Haus, U. Morgner, F.X. Krtner, V. Scheuer, G. Angelow, T. Tschudi: Optics Lett. **26**, 292 (2001)

[4.101] V. Shcheslavshiy, V.V. Yakovlev, A. Ivanov: Optics Lett. **26**, 1999 (2001)

[4.102] T. Tomaru: Optics Lett. **26**, 1439 (2001)

[4.103] R. Holzwarth, M. Zimmermann, Th. Udem, T.W. Hnsch, P. Russboldt, K. Gbel, R. Poprawe, J.C. Knight, W.J. Wadsworth, P. St. J. Russell: Opt. Lett. **26**, 1376 (2001)

[4.104] F. Falcoz, F. Balembois, P. Georges, A. Brun: Optics Lett. **20**, 1874 (2001)

[4.105] C. Hönninger, R. Paschotta, M. Graf, F. Morier-Genoud, G. Zhang, M. Moser, S. Biswal, J. Nees, A. Braun, G.A. Mourou, I. Johannsen, A. Giesen, W. Seeber, U. Keller: Appl. Phys. B**69**, 3 (1999)

[4.106] C. Hönninger, F. Morier-Genoud, M. Moser, U. Keller, L.R. Brovelli, C. Harder: Opt. Lett. **23**, 126 (1998)

[4.107] F. Druon, G. Balembois, P. Georges, A. Brun, A. Courjaud, C. Hönninger, F. Salin, A. Aron, G. Aka, D. Vivien: Opt. Lett. **25**, 423 (2000)

[4.108] F. Druon, S. Chenais, P. Raybaut, F. Balembois, P. Georges, R. Gaumé, G. Aka, B. Viana, S. Mohr, D. Kopf: Opt. Lett. (2002)

[4.109] C. Hönninger, G. Zhang, U. Keller, A. Giesen: Opt. Lett. **20**, 2402 (1995)

[4.110] F. Brunner, G.J. Sphler, J. Aus der Au, L. Kranier, F. Morier-Genoud, R. Paschotta, N. Lichtenstein, S. Weiss, C. Harder, A.A. Lagatshy, A. Abdolvand, N.V. Kuleshov, U. Keller: Opt. Lett. **25**, 1119 (2000)

[4.111] H. Liu, J. Nees, G. Mourou: Opt. Lett. **26**, 1723 (2001)

[4.112] R.W. Schoenlein, W.P. Leemans, A.H. Chin, P. Volfbeyn, T.E. Glover, P. Balling, M. Zolotorev, K.J. Kim, S. Chattopadhyay and C.V. Shanh: Science **274**, 236 (1996)

[4.113] M. Hentschel, R. Kienberger, Ch. Spielmann, G.A. Reider, N. Milosevic, T. Brabec, P. Corkum, U. Heintzmann, M. Drescher, F. Krausz: Nature **414**, 509 (2001)

[4.114] A. Rousse, C. Rischel, J.C. Gauther: Rev. Mod. Phys. **73**, 17–31 (2001)

[4.115] K.W.D. Ledingham, I. Spencer, T. Mccanny, R.P. Singhal, M.I.K. Santala, E. Clark, I. Watts, F.N. Beg, M. Zepf, K. Krushelnick, M. Tatarakis, A.E. Dangor, P.A. Norreys, R. Allott, D. Neely, R.J. Clark, A.C. Machacek, J.S. Wark, A.J. Cresswell, D.C.W. Sanderson, and J. Magill: Phys. Rev. Lett. **84**, 899 (2000)

[4.116] T.E. Cowan, A.W. Hunt, T.W. Phillips, S.C. Wilks, M.D. Perry, C. Brown, W. Fountain, S. Hatchett, J. Johnson, M.H. Key, T. Parnell, D.M. Pennington, R.A. Snavely, and Y. Takahashi: Phys. Rev. Lett. **84**, 903 (2000)

[4.117] T. Ditmire, J. Zweiback, V.P. Yanovshy, T.E. Cowan, G. Hays, K.B. Wharton: Nature **398**, 489 (1999)

5

Pulsed Semiconductor Lasers

T. Amand and X. Marie

With 39 Figures

5.1 Introduction

Semiconductor (SC) lasers are widely used as coherent light sources in many applications. Their pulsed operation is particularly attractive in the fields of optical sampling, optical spectroscopy and telecommunications. SC lasers have very different characteristics compared to conventional atomic or molecular laser sources [5.1, 2]. We recall here briefly the main features of SC lasers [5.3]:

- very small size ($300 \times 300 \times 100 \, \mu m^3$ typically);
- high efficiency (up to 50 % in commercial devices); for instance, an optical output power $P_{opt.}$ of 100 mW for an electric input power of 2 V × 100 mA;
- low prices thanks to the performance of semiconductor fabrication technology.

Two other characteristics of SC lasers are very important for the pulsed operation mode:

- direct pumping by current injection, and consequently the possibility of direct optical modulation;
- the possibility to make SC modulators and of subsequent integration of the laser, the driving circuit and the modulator on the same chip.

The main drawbacks of SC lasers in the pulsed operation mode are their limited pulse energies (a few tens of pJ typically, in the picosecond regime) and their very high divergence. Their temperature sensitivity is also a problem in many applications.

We propose first to recall the basic semiconductor physics background which is essential for the understanding of the operation of SC lasers. We present then the main semiconductor laser devices which are commercially available (double-heterostructure, quantum well and vertical-cavity surface-emitting lasers). The second part of the chapter is devoted to the pulsed operation, with a detailed description of the gain-switching method, which

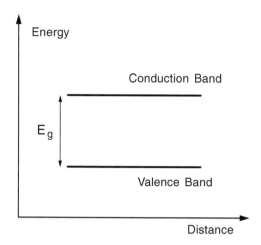

Fig. 5.1. Schematic energy-level diagram of a semiconductor

gives picosecond pulses very easily. Q-switched and mode-locked SC lasers, which are being studied in different laboratories, will also be described.

5.2 Semiconductor Lasers: Principle of Operation

5.2.1 Semiconductor Physics Background

Unlike those of other active media (such as in atomic lasers), the energy levels of the states which describe the electronic excitations in crystal semiconductors are spread into energy bands. This spreading is due to the strong interactions between the atoms which constitute these dense materials (about 10^{23} active atoms per cm^3). The two highest energy bands are determinative for laser emission: the valence band (VB) and the conduction band (CB), which are separated by an energy gap E_g (Fig. 5.1). At low temperature, the valence band is the last populated one, filled by electrons, while the conduction band, which lies at higher energies, is the first empty band. The latter corresponds to excited states of the electrons which, under the influence of an electric field, can move throughout the crystal and conduct an electrical current. Light emission occurs when an electron, in an activated state of the conduction band, makes a transition down to a valence-band state. The photon liberated by this process has an energy determined by the energy difference between the two states. Figure 5.2 shows the typical energy dispersion curves $E(K)$ of a direct semiconductor (E is the electron energy and K its wavevector). For the sake of simplicity, the band edges of the conduction and valence bands are assumed to be parabolic. The electrons can then be treated as free electrons by introducing the concept of the effective mass m_c, which takes into account the shape of the dispersion curve around $K = 0$ ($m_c^{-1} = (1/\hbar^2)(\partial^2 E/\partial K^2)$,

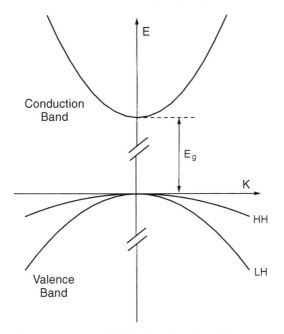

Fig. 5.2. Dispersion curve $E(K)$ of a direct-band-gap semiconductor

where $\hbar = h/2\pi$, and h is the Planck constant); the effective mass decreases when the band curvature around $K = 0$ increases.

The crystal state corresponding to a full valence band from which one electron has been removed can be conveniently represented by a pseudoparticle called a hole, for which an effective mass m_h can also be defined. The VB structure is quite complex, but we see in Fig. 5.2 that it is mainly divided into a heavy-hole (HH) and a light-hole (LH) band, which are degenerate at $K = 0$ in a bulk semiconductor. Under these conditions, the energies associated with an electron in the CB and a hole in the VB are

$$E - E_c = \frac{\hbar^2 K^2}{2m_c} \quad \text{and} \quad E - E_v = \frac{\hbar^2 K^2}{2m_h} \tag{5.1}$$

(the hole energy axis is opposite to the electron one).

The electronic properties of semiconductor materials depend also on the densities of states (DOS) of the conduction and the valence bands; it is defined as the number of electron (or hole) states per unit energy and per unit volume of the crystal. In bulk material, it can be written as:

$$\rho(E - E_c) = \frac{1}{2\pi^2} \left(\frac{2m_c}{\hbar^2} \right)^{3/2} (E - E_c)^{1/2} \quad \text{(CB)},$$

$$\rho(E - E_v) = \frac{1}{2\pi^2} \left(\frac{2m_v}{\hbar^2} \right)^{3/2} (E - E_v)^{1/2} \quad \text{(VB)}. \tag{5.2}$$

In a semiconductor material, the three mechanisms of light–matter interaction, namely absorption, spontaneous emission and stimulated emission, can be summarized as follows:

(i)   photon $\longrightarrow$ electron + hole (absorption);
(ii) electron + hole $\longrightarrow$ photon (spontaneous emission);
(iii) electron + hole + photon $\longrightarrow$ 2 photons (stimulated emission).

The distributions of electrons in the CB and holes in the VB obey Fermi–Dirac statistics, characterized by a Fermi level E_F and a temperature T. The occupation probability of a state with an energy E is given by

$$f^{c(h)}(E) = \frac{1}{1 + e^{\left(E - E_F^{c(h)}\right)/K_B T}}.$$ (5.3)

$E_F^{c(h)}$ is the quasi-Fermi level of electrons (or holes) in their band (the same origin is chosen, at the top of the valence band, for both conduction electrons and valence holes) and T is their temperature. This means, for instance, that at $T = 0$ K all the electron energy levels above E_F^c are empty, whereas below they are full (if the thermal equilibrium is disturbed, as in a pn junction under forward bias, E_F^c and E_F^h may be different).

Neglecting the correlations in the electron–hole system, one can deduce from the quantum-mechanical theory of time-dependent perturbation applied to electron–photon interaction in semiconductors, the net rate of optical transitions per unit time for a given optical mode (w_q, q):

$$W(w_q, q) = \frac{4\pi e^2 P_0^2}{n^2 \omega_q V} \sum_K \left[f_{K+q}^c f_K^h + \left(f_{K+q}^c + f_K^h - 1 \right) n_q \right] \delta\left(E_{K+q}^c + E_K^h - \hbar\omega_q \right),$$

              (ii)   (i or iii)   (5.4)

where P_0 is the matrix element which characterizes the optical transition for a given semiconductor material, V is the volume, n_q is the average number of photons in the mode (w_q, q), $\omega_q = cq/n$ with n the refractive index of the material and δ is the Dirac distribution, which ensures energy conservation. In the summation over K, the first term represents the spontaneous emission rate (ii):

$$R_{\text{spont}}(q, \hbar\omega_q) = \frac{4\pi e^2 P_0^2}{n^2 \omega_q V} \sum_K f_{K+q}^c f_K^h \delta\left(E_{K+q}^c + E_K^h - \hbar\omega_q \right),$$ (5.4a)

while the second term represents the balance between stimulated emission (iii) and absorption (i):

$$R_{\text{stim}}(q, \hbar\omega_q) = \frac{4\pi e^2 P_0^2}{n^2 \omega_q V} \sum_K \left(f_{K+q}^c + f_K^h - 1 \right) n_q \delta\left(E_{K+q}^c + E_K^h - \hbar\omega_q \right).$$

(5.4b)

The amplification/absorption coefficient of the optical mode $(q, \hbar\omega_q)$ in the material is then simply given by

$$\alpha(q, \hbar\omega_q) = \frac{n}{c}\frac{4\pi e^2 P_0^2}{n^2 \omega_q V}\sum_K \left(f_{K+q}^c + f_K^h - 1\right)\delta\left(E_{K+q}^c + E_K^h - \hbar\omega_q\right). \quad (5.5)$$

Using the energy conservation specified by the Dirac function, one can easily deduce the general relation between absorption/amplification and emission:

$$\alpha(q, \hbar\omega_q) = \frac{n}{c}\left(1 - e^{-(\hbar\omega_q - \mu)/kT}\right)R_{\text{spont}}(q, \hbar\omega_q), \quad (5.5a)$$

where $\mu = E_F^c + E_F^h$ represents the chemical potential of the electron–hole pairs.

The condition for obtaining net gain $(\alpha(q, \hbar\omega_q) > 0)$ in the material follows

$$E_g < \hbar\omega_q < E_F^c + E_F^h. \quad (5.6)$$

The left-hand inequality results from the fact that there is no interaction, in the first-order approximation, between light and the semiconductor for photons with energies below the material gap ($R_{\text{spont}} = R_{\text{stim}} = 0$ for $\hbar\omega_q < E_g$). The right-hand inequality is the so-called *Bernard and Duraffourg condition*. It means that only the photons with energy $\hbar\omega_q$ smaller than the chemical potential μ of the electron–hole pair can be amplified. This condition simply states that light amplification will occur at a spectral energy $\hbar\omega_q$ in a semiconductor if, in the subset of electron and hole states involved in the corresponding optical transitions, there are more electrons in the conduction band than in the valence band.

In direct semiconductors, in which the electron–photon coupling intensity is strong, the optical transition is vertical in K space (the optical wavelength is very large compared to the interatomic distances, so that $|q| \ll |K|$). We deduce, in the vicinity of the gap, after integrating over K, the expression for the amplification/absorption coefficient:

$$\alpha^{3D}(\omega) \propto \rho(\omega)\left[f^c\left(E_c + \frac{m_h}{m_c + m_h}(\hbar\omega - E_g)\right)\right.$$
$$\left. + f^h\left(E_v + \frac{m_c}{m_c + m_h}(\hbar\omega - E_g)\right) - 1\right],$$

where $\rho(\omega) = (1/2\pi^2)[2m_c m_h/\hbar^2(m_c + m_h)]^{3/2}(\hbar\omega - E_g/)^{1/2}$ is the joint DOS per optical transition at the energy $\hbar\omega$, per unit energy and per unit volume of the crystal. As usual the convention is $\alpha > 0$ or $\alpha < 0$ for gain or absorption respectively.

5.2.2 pn Junction – Homojunction Laser

In an intrinsic semiconductor at equilibrium, the electron population in the CB and the hole population in the VB are usually very small ($f^{c(h)} \ll 1$).

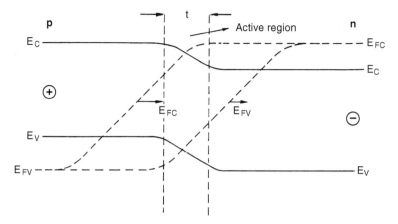

Fig. 5.3. Quasi-Fermi levels in a forward-biased pn junction

The light will thus be strongly absorbed under these conditions, according to mechanism (i) [see (5.4)]. On the other hand, light amplification may occur if there are, simultaneously and in the same region, a high density of electrons in the CB and of holes in the VB: the semiconductor must be out of thermodynamical equilibrium. These conditions can be fulfilled in a pn junction under forward bias. A junction is obtained when two types of semiconductor material are brought into contact with each other. In the first material, impurities (called donors) have been added so that there are extra electrons in the crystal. Such a semiconductor is called n-type. On the other side of the junction, other impurities (called acceptors) have been added to make the semiconductor p-type: there are in this case extra holes. If both sides of the junction are composed of the same semiconductor (for instance GaAs), it is called a homojunction. If the semiconductor materials on the two sides are different (for instance GaAs and AlGaAs), it is a heterojunction.

Most electronic and optoelectronic devices are based on the electronic properties of these junctions. If a pn junction is forward biased (direction of current p → n), the electrons and the holes diffuse and recombine radiatively, creating a depleted region, also called the active region (Fig. 5.3); such a device is called an LED (light-emitting diode).

If we want to obtain laser action, it is first necessary to find the conditions for amplification: we must have a population inversion. In atomic or molecular four-level lasers (such as the argon-ion or the Nd:YAG laser), the condition for amplification is simply given by $N_2 - N_1 > 0$, where N_1 and N_2 are the populations of the lower and higher levels involved in the optical transition. In a semiconductor, light amplification will occur if the Bernard and Duraffourg condition expressed in (5.6) is fulfilled. This condition can be reached in a highly doped pn junction under forward bias. If the pn junction is placed in a Fabry–Pérot resonator, laser light can be obtained. This resonator geometry is

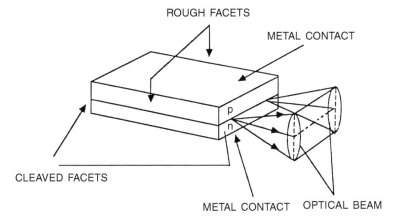

Fig. 5.4. Homojunction laser

very specific to SC lasers. The cavity length can be very short (200 to 300 μm typically) since the semiconductor can exhibit very large gain coefficients, up to 10^3 cm^{-1}. This very large gain also allows larger losses than in conventional lasers. As a matter of fact, no external mirrors are needed. The refractive-index discontinuity at the semiconductor–air interface leads to a reflectivity of about 30 %. This is enough to maintain laser operation. So the two cleaved facets of the diode perpendicular to the junction plane act simply as mirrors. When the bias current reaches a critical value, called the threshold current, a coherent light beam is emitted perpendicular to the cleaved facets (Fig. 5.4). The first SC laser operated in 1962; it was a homojunction which operated only at liquid-nitrogen temperature in a pulsed regime with a fantastically high current density to reach threshold ($J_{\text{th.}} \approx 10^5 \text{A/cm}^2$).

5.3 Semiconductor Laser Devices

Most of the commercial SC lasers use III–V compounds. In terms of applications, we can mention three important families:

• $(\text{Al}_x\text{Ga}_{1-x})_y\,\text{In}_{1-y}\text{P}$	$0.6 < \lambda < 0.8\,\mu$m;
• $\text{Al}_x\text{Ga}_{1-x}\text{As}$	$0.7 < \lambda < 0.9\,\mu$m;
• $\text{In}_{1-x}\text{Ga}_x\text{As}_y\text{P}_{1-y}$	$1 < \lambda < 1.65\,\mu$m.

We see that it is possible to combine two semiconductors, for instance GaAs and AlAs, to make $\text{Al}_x\text{Ga}_{1-x}\text{As}$. This ternary compound has a band-gap energy which varies monotonically with x. This band-gap energy control offers the possibility to obtain various wavelengths for specific applications. For example, there is a very important transmission window for silica-glass optical fibers at $1.3\,\mu$m (low attenuation, zero chromatic dispersion). A quaternary compound $In_{0.73}Ga_{0.27}As_{0.4}P_{0.6}$, lattice-matched to an InP substrate, is used to generate a laser beam at $1.3\,\mu m$ in telecommunications systems. There is a very active research activity in the field of SC-laser blue light emission; GaN or II–VI materials such as ZnSe are good candidates.

Note that commonly used semiconductor materials such as silicon or germanium are not used as laser sources. The reason is that silicon and germanium are indirect semiconductors. This means that the band edges of the CB and the VB do not coincide in K space. In other words, the optical transition must be accompanied by the emission of a phonon. The optical transition probability is thus much lower than in direct semiconductors, which leads to much lower gain.

We present now the most common laser structures.

5.3.1 Double-Heterostructure Laser

The great majority of commercial laser diodes are double-heterostrucure (DH) lasers.

5.3.1.1 Optical and Electronic Confinement.
In contrast to the homojunction of Fig. 5.4, the double heterostructure ensures the confinement of both the electronic carriers and the electromagnetic wave. A DH laser consists of a very thin layer (approximately $0.2\,\mu m$) of a semiconductor A, sandwiched between two thicker layers of a semiconductor B (Fig. 5.5; A = GaAs, B = AlGaAs).

The electronic carrier confinement is obtained thanks to the band-gap energy difference between the two semiconductors. When the structure is forward-biased, the electrons and the holes are injected from the outer p and n regions respectively into the central active layer; they become trapped between the potential-energy barriers created by the wider band gap of semiconductor B. The laser efficiency is thus much increased with respect to the homojunction laser, since the carriers do not diffuse out of the active region (where population inversion is achieved).

The optical confinement is achieved through the refractive-index discontinuity between the two semiconductors, which creates a planar waveguide (Fig. 5.5). There is a much better guiding of the electromagnetic wave in the structure and consequently a better overlap between the change carriers and the electromagnetic wave.

The combined electronic and optical confinements of a DH laser lead to high gain, a low threshold and high efficiencies. As a matter of fact, CW room-temperature operation of an SC laser became possible in the early seventies through the development of the DH laser. The threshold current density of a DH laser is of the order of $1\,kA/cm^2$.

5.3.1.2 Gain-Guided and Index-Guided Double Heterostructures.
To improve DH laser performances, lateral confinement (in the junction plane)

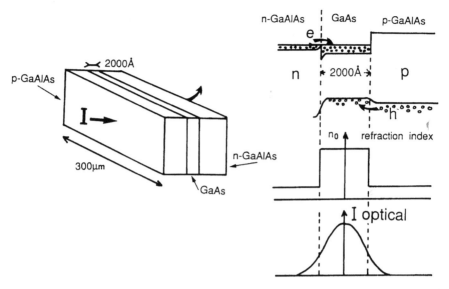

Fig. 5.5. Schematic of a double-heterostructure laser and the associated profiles of band structure, index of refraction and guided optical wave [5.2]

can also be achieved. There are two main techniques to restrict the active region laterally. The first one, called gain-guiding, consists of the deposition of a thin ribbon of metal on the structure, from which current can flow through the semiconductor layer (Fig. 5.6). In the second technique, called index-guiding, the lateral confinement is achieved through a refractive-index discontinuity in the vicinity of the active region. Because of the narrow confinement of charge carriers and light which is made possible in index-guided and gain-guided structures, laser diodes based on these architectures are very efficient. Index-guided lasers generate a tight beam of light. Gain-guided structures generate a broader one, but this allows them to reach higher powers than index-guided diodes.

5.3.1.3 Characteristics of Optical Beam. The active region of a SC laser is very small and asymmetric (it has a section of about $0.1 \times 1.5\,\mu\mathrm{m}^2$ in an index-guided DH laser). The beam waist in the structure is thus also very small and asymmetric. This leads to a highly divergent and asymmetric optical beam at the output of the device. The half-angle divergence in the direction perpendicular to the junction plane is typically 30–40°. The divergence in the other direction is two or three times smaller. As almost all the applications require collimated symmetric beams, spherical lenses with high numerical aperture are usually used to correct this divergence. The correction of the asymmetry is obtained with a pair of anamorphic prisms.

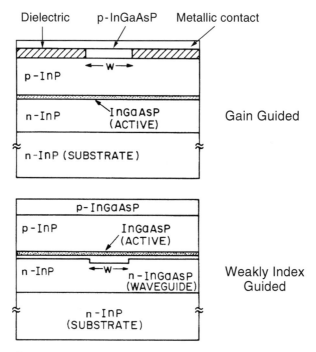

Fig. 5.6. Gain-guided and index-guided double-hetero-structure lasers [5.1]

5.3.1.4 Optical Confinement, Differential Gain and Threshold Current. In order to understand the further improvements of SC lasers, it is necessary to give the precise definitions of three parameters which are very important for the optimization of the device.

The Optical Confinement Factor Γ. Γ is the ratio between the electromagnetic energy in the active region and the total electromagnetic energy of the optical wave in the structure:

$$\Gamma = \frac{\int_{-d/2}^{d/2} |E(z)|^2 \, \mathrm{d}z}{\int_{-\infty}^{+\infty} |E(z)|^2 \, \mathrm{d}z}, \tag{5.7}$$

where d is the thickness of the active region in the z direction. It characterizes the overlap between the electromagnetic wave and the confined electronic states. Figure 5.7 shows that the optimum width for a GaInAsP laser is about $d = 1\,\mu\mathrm{m}$ ($\Gamma \approx 1$).

The Differential Gain. For a bulk semiconductor laser, the variation of the gain curve maximum as a function of the injected current density is linear (Fig. 5.8a). This variation can be written as

$$g_{\mathrm{max}} = A_J(J - J_0), \tag{5.8}$$

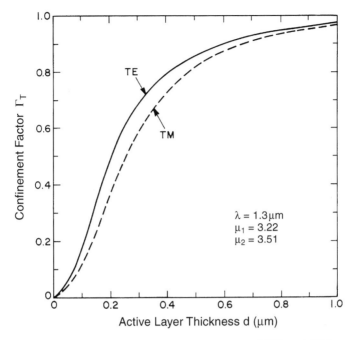

Fig. 5.7. Optical confinement factor of the fundamental TE and TM modes as a function of the active-layer thickness for a $1.3\,\mu m$ GaInAsP DH laser [5.1]

where A_J is the current differential gain, defined more generally as $A_J = \partial g/\partial J$, and J_0 is the transparency current density (for $J = J_0$, $\alpha = 0$ and $E_F^c + E_F^h = E_g$).

The Threshold Current. By equating the gain and the losses in the cavity, we obtain the threshold current density

$$J_{th.} = J_0 + \frac{\alpha_i}{A_J} + \frac{(1-\Gamma)\alpha_c}{\Gamma A_J} + \left(\frac{1}{\Gamma A_J}\right)\left(\frac{1}{2L}\right)\ln\left(\frac{1}{R_1 R_2}\right), \qquad (5.9)$$

where α_i is the internal loss coefficient of the active region (Auger effects, free-carrier absorption, carrier trapping on impurities, etc.), α_c is the loss coefficient of the optical confinement region, L is the cavity length and R_1 and R_2 are the reflectivities of the facets. It is clear from (5.9) that the threshold current decreases when the differential gain A_J and the optical confinement factor Γ increase (and obviously when the losses are reduced). An important optimization step of SC laser structures consists of finding the best (A_J, Γ) pairs which minimize $J_{th.}$ for a given wavelength.

5.3.1.5 Distributed-Feedback (DFB) Structures. In a conventional SC laser (such as a DH laser), the gain curve is much wider than the longitudinal intermode frequency spacing $c/2nL$ (where n is the refractive index and L the

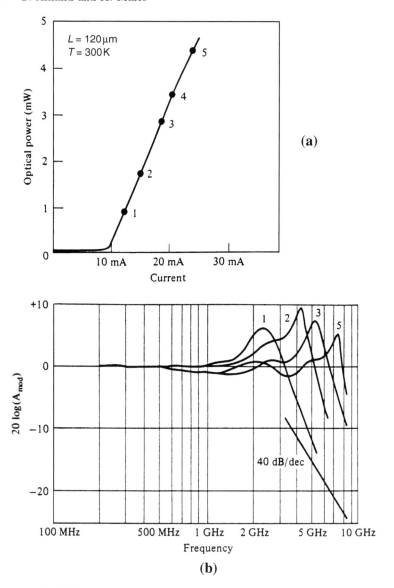

Fig. 5.8. (a) Light output power versus current for a DH laser; **(b)** modulation characteristics of this laser at the various bias points indicated in **(a)**, where A_{mod} is the current differential gain defined as A_J in (5.8) [5.3]

cavity length). So multimode operation usually occurs. In optical telecommunications systems, because of the chromatic dispersion of the fiber (mainly at $1.55\,\mu\mathrm{m}$), a monomode operation is required. This can be achieved in a distributed-feedback structure (Fig. 5.9), in which a grating is etched along

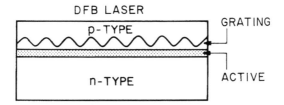

Fig. 5.9. Schematic illustration of a distributed-feedback (DFB) laser

the cavity so that the thickness of one of the cladding layers varies periodically. The wavelength selection rule results from the Bragg condition, which depends on the grating period.

5.3.2 Quantum Well Lasers

In a DH structure, the active-region thickness d varies typically between 0.1 and $1\,\mu$m. If this thickness decreases further so that

$$d < \Lambda_{\mathrm{d}}, \tag{5.10}$$

where Λ_{d} is the de Broglie wavelength, the electrons in the CB and the holes in the VB are still free to move in the plane of the layers but are confined in a well in the perpendicular direction (z). Such a structure is called a quantum well (QW). In practice, significant effects are observed if d is less than 20 nm, the Bohr radius of the excitons. With the development of semiconductor fabrication techniques, such as molecular beam epitaxy, it is now possible to grow such thin layers. The electronic and optical properties of these two-dimensional (2D) structures are drastically changed compared to bulk materials. The main features are the following.

The energy levels of the CB and the VB are quantized (Fig. 5.10a). The dispersion curves $E(K)$ (in the infinite-well approximation) become

$$E_{\mathrm{c(h)}}^{n}(K) = E_{\mathrm{c(h)}} + \frac{\pi^{2}\hbar^{2}}{2m_{\mathrm{c(h)}}d}n^{2} + \frac{\hbar^{2}K^{2}}{2m_{\mathrm{c(h)}}} \qquad (n \text{ integer}). \tag{5.11}$$

Laser operation thus occurs at photon energies fixed by the band-gap energy and the confinement energies $(\pi^{2}\hbar^{2}/2m_{\mathrm{c(h)}}d)n^{2}$. Consequently, for a given semiconductor material, the emission wavelength can be adjusted by slight variations of the well width d.

Another consequence of quantum confinement is that there is a lifting at $K = 0$ of the heavy-hole–light-hole degeneracy. This leads to a reduction of the hole density of states, since light holes do not contribute to it near the gap of the quantum well. Moreover, the confinement leads to a reduction of the heavy-hole effective mass near the band edge, which leads in turn to a further reduction of the hole DOS, which in two-dimensional structures

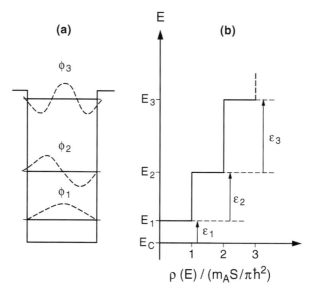

Fig. 5.10. (a) Energy levels of the conduction band in a semiconductor quantum well. (b) Density of states. $\varphi_1, \varphi_2, \varphi_3$ are the different electron wavefunctions and $\varepsilon_1, \varepsilon_2, \varepsilon_3$ are the corresponding energies; m_A is the electron effective mass in the well material; S is the QW surface

is simply equal to $m/\pi\hbar^2$ per surface unit (Fig. 5.10b). The carrier injection density current required to reach the threshold is thus also reduced. By similar arguments to those for bulk semiconductors, we find the expression for the amplification/absorption coefficient in 2D semiconductors:

$$\alpha_{n,n'}^{2D}(\omega) \propto \rho^{2D}(\omega)|\langle \phi_{cn}|\phi_{hn'}\rangle|^2$$

$$\times \left[f^c \left(E_{cn} + \frac{m_h}{m_c + m_h}(\hbar\omega - E_g - E_{cn} - E_{vn'}) \right) \right.$$

$$\left. + f^h \left(E_{vn'} + \frac{m_c}{m_c + m_h}(\hbar\omega - E_g - E_{cn} - E_{vn'}) \right) - 1 \right], (5.12)$$

where $|\langle \phi_{cn}|\phi_{hn'}\rangle|^2$ characterizes the spatial overlap between the electron and hole envelope wavefunctions in their respective subbands n and n', and $\rho^{2D}(\omega)$ is the joint 2D DOS ($\rho^{2D}(\omega) = [m_e m_h/(m_e + m_h)](1/\pi\hbar^2)$, which is energy-independent for a 2D system). Figure 5.11b shows the gain curves for increasing injected carrier densities in a DH and in a quantum well structure. We see that the carriers in quantum wells are more "efficient" in 2D structures compared to bulk structures [5.2], since the added carriers contribute to the gain at its maximum, whereas in the bulk the added carriers move the gain peak away from the band edge, so the carriers with energies lower than g_{max} are lost for the laser effect. The density differential gain A_n defined by $A_n = \partial g/\partial n$ is thus stronger in 2D structures than in the bulk in the vicinity

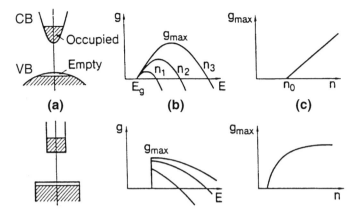

Fig. 5.11. Schematic illustrations of gain formation in DH lasers (*top*) and QW lasers (*bottom*), after [5.2]

of the transparency threshold n_0 (Fig. 5.11c). But the disadvantage of quantum well lasers is that their gain will saturate at a finite value when all the electron and hole states are inverted (Fig. 5.11c). In bulk lasers, g_{max} never saturates since the DOS increases with energy (see (5.2)).

Figure 5.12 shows that the optical confinement factor Γ in a single quantum well (SQW) is worse than in a classical DH. In order to overcome this problem of optical-wave leakage into the barrier, separate-confinement heterostructures (SCHs) or graded-index separate-confinement heterostructures (GRINSCHs) are grown, in which the confinement of the electromagnetic wave occurs in an optical-confinement layer, whereas the electronic confinement occurs in the quantum well (Fig. 5.13). So both optical and electronic confinement can be optimized separately.

5.3.3 Strained Quantum Well and Vertical-Cavity Surface-Emitting Lasers

The latest developments of SC lasers are based on the so-called strained quantum wells [5.4]. In these structures, the lattice parameters of the semiconductor materials constituting the well and the barrier are different in the relaxed materials. If one grows a thin layer of the well semiconductor on a thick layer of the barrier semiconductor, the barrier semiconductor imposes its lattice parameter in the well plane. It is thus possible to grow elastically strained quantum wells in compression or extension. The strain is then a new degree of freedom available to optimize the SC laser characteristics: the structural modifications with respect to a bulk, relaxed semiconductor, added to the size quantization effects, lead to drastic changes of the electronic properties. Figure 5.14 shows that in a compressively strained structure, the heavy-hole–light-hole splitting at $K = 0$ increases and the heavy-hole effective mass decreases.

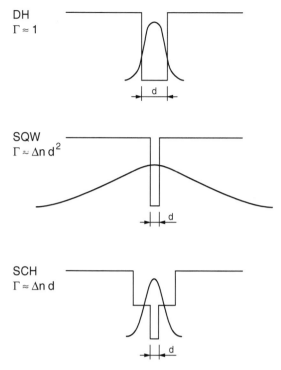

Fig. 5.12. Schematic illustrations of the confinement of an optical guided wave for a double heterostructure (DH), a single quantum well (SQW) and a separate-confinement heterostructure (SCH) [5.2]

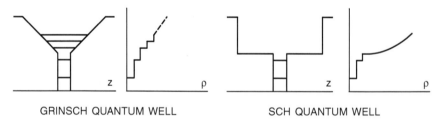

GRINSCH QUANTUM WELL   SCH QUANTUM WELL

Fig. 5.13. Energy diagram of the conduction-band edge and corresponding DOS ρ for graded-index separate-confinement heterostructure (GRINSCH) and separate-confinement heterostructure (SCH) quantum well lasers [5.2]

This leads to the reduction of the hole DOS and thus to a reduction of the threshold current density. Moreover, the light-hole states, which do not participate in the lasing transition, are further in energy from the heavy-hole states in compressively strained structures compared to the unstrained one. These states are thus less populated, which leads to an increase of the SC laser efficiency. On can also demonstrate that the differential gain A_J is higher in a

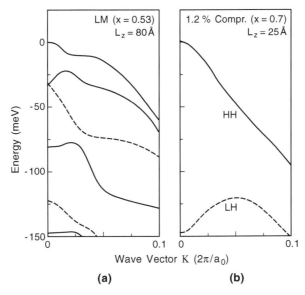

Fig. 5.14. Calculated valence band structure of (**a**) a lattice-matched (LM) and (**b**) a compressively strained $In_xGa_{1-x}As/InGaAsP$ quantum well laser designed for emission at $1.5\,\mu m$ (from [5.4] © 1994 IEEE). a_0 is the lattice parameter. L_z is the QW width

strained structure than in a lattice-matched one. The most common strained lasers are based on InGaAs/GaAs and InGaAsP/InP quantum well structures; they exhibit excellent $J_{th.}$ and A_J parameters.

We finish this very quick review of the common SC laser structures with the most recent ones, called vertical-cavity surface-emitting lasers (VCSELs) [5.5].

All of the SC laser structures discussed so far emit light from their edges. In a VCSEL, however, the laser cavity is perpendicular to the junction plane. A VCSEL structure consists of a large number of layers with a Bragg reflector at each end and an active region composed of a quantum well (Fig. 5.15). The VCSELs are very compact (the cavity length is a few microns) and the light beam is circular and less divergent than in conventional horizontal lasers. From the point of view of pulsed operation, they are characterized by a very high intracavity photon density, which leads to high relaxation oscillation frequencies (see Sect. 5.4). The main advantage for applications is that two-dimensional arrays of VCSELs can now be grown.

5.4 Semiconductor Lasers in Pulsed-Mode Operation

Generation of ultrashort optical pulses from SC lasers has been the subject of intense activity since the end of the seventies. Potential applications in the

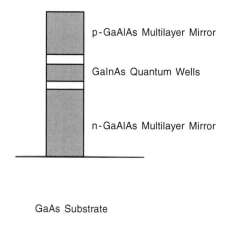

p-GaAlAs Multilayer Mirror

GaInAs Quantum Wells

n-GaAlAs Multilayer Mirror

GaAs Substrate

Fig. 5.15. Schematic illustration of a vertical-cavity surface-emitting laser (VCSEL)

fields of optical telecommunications and time-resolved spectroscopy are at the origin of such interest.

Several approaches have been investigated in order to generate short optical pulses from semiconductor lasers.

The simplest mode of operation, specific to semiconductor lasers, and which was the subject of the most intense research effort, is the gain-switching mode [5.7]. This mode of operation was investigated by *Lasher* in 1964 [5.8]. In contrast to other laser sources, where optical modulators are used, semiconductor lasers can be modulated directly through the injection current. Moreover, here one can take advantage of the possibility of monolithic integration of both the laser and the electronic modulator circuit. The gain-switching operation mode consists of driving the laser with ultrashort current pulses which are superimposed on a stationary (DC) polarization current below or at the laser threshold. We show in Sect. 5.4.1 that this allows one to generate pulses of variable temporal width, frequency and peak power, which makes this mode of operation probably the most versatile.

The second approach to the generation of ultrafast optical pulses was to apply classical Q-switching techniques to semiconductor lasers. The early attempts at this approach were performed by *Kurnosov* et al. [5.9] in 1966, using a two-section laser geometry. This will be described further in Sect. 5.4.2.

The third approach one might think of relies on conventional mode-locking, active or passive. It is based on the large gain bandwidth of the active medium, which allows, in principle, the generation of ultrashort pulses. Passive mode-locking was primarily investigated by *Van der Ziel* et al. in 1981 [5.10] and *Yokoyama* et al. in 1982 [5.11], while active mode-locking was demonstrated later by *Bowers* et al. in 1989 [5.12]. We shall see in Sect. 5.4.3 that regularly spaced picosecond or subpicosecond pulses can be obtained in such a way. However, in the active mode-locking method, the applied modulation signal must be accurately matched to the laser cavity length. In the passive mode-locking configuration, the pulsed operation depends critically on the saturable absorber used.

In the description of these different modes of operation, we will stress the specific details of the semiconductor lasers, in particular those concerning pumping and the possibilities of integration in the same device of the optical amplifier, the modulator and the waveguide.

5.4.1 Gain-Switched Operation

The output light intensity of SC lasers can be quite easily modified by changing the excitation level of the amplifying medium. This allows one to obtain an optical replica of an electrical pulse train. Owing to the small size of the semiconductor laser cavities, the build-up and disappearance of the light intensity can occur on a timescale of a few picoseconds, thus making this approach quite attractive for generating short optical pulses. Two modes of operation are currently used: the pulsed mode and the amplitude-modulation mode. Although the latter is not equivalent, strictly speaking, to the generation of pulse trains, we will present it here, because of its considerable technological applications in the fields of telecommunications and information processing. The specifications reached in both operation modes are basically determined by the oscillations of the relaxation process, which we describe now.

5.4.1.1 Relaxation Oscillation Phenomenon. This phenomenon is common to all laser sources, but the frequency of these oscillations is much higher in SC lasers than in conventional sources. Here, they can be observed when polarizing the device with a current step above the current threshold. The output power increases first and oscillates before reaching the new steady-state operation point, as displayed in Fig. 5.16a. These periodic oscillations are basically due to the time delay between the photon generation and the build-up of the gain in the active medium (a delay which may be as long as a few tens of picoseconds, typically). It can be understood qualitatively as follows: owing to the injection current increase, the carrier population in the active medium first rises, since its decay time is long (of the order of 1 ns), governed by spontaneous recombination. Net positive gain is then obtained within the medium, thus allowing the electromagnetic field to build up from the spontaneous radiation by stimulated emission. As the photon density rises, the carrier lifetime quickly decreases to the stimulated radiative time (of the order of 10 ps), thus leading to the collapse of the free-carrier density, eventually to values below the transparency threshold. The gain within the cavity then drops, leading to a decrease of the electromagnetic field intensity, and a subsequent increase of the carrier lifetime. A new cycle can then start [5.6].

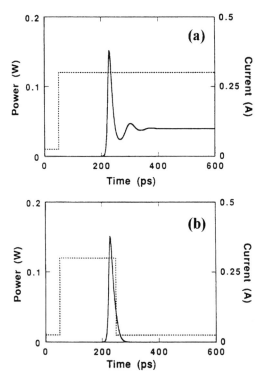

Fig. 5.16. Laser responses under strong electrical excitation. (**a**) Single-step excitation, (**b**) rectangular pulse-excitation [5.6]

The relaxation oscillation process can be described in terms of a nonlinear system of two coupled differential equations, which expresses the interplay between the injected electron (or hole) density N and the photon density P within the active medium. Neglecting the fraction of spontaneous emission entering the laser mode, it can be written as [5.13]

$$\frac{dN}{dt} = \frac{J}{ed} - \frac{N}{\tau} - \frac{c}{n}A_N(N - N_{th})P, \qquad (5.12a)$$

$$\frac{dP}{dt} = \frac{c}{n}A_N(N - N_{th})P - \frac{N}{\tau_p}, \qquad (5.12b)$$

where J/ed is the injection rate of the electrons in $cm^{-3}s^{-1}$ (J is the current density, e the electron charge and d the thickness of the active region); $(c/n)P = F$ is the photon fluence within the active region in $cm^{-2}s^{-1}$ (c is the speed of light and n the refractive index in the active region); $A_N(N - N_{th})$ is the gain within the active region in cm^{-1} (A_N is the optical differential gain and N_{th} is the threshold carrier density), so that $(c/n)A_N(N - N_{th})P$ represents the rate of stimulated recombinations per unit volume; τ is the free-carrier spontaneous lifetime within the active region; and τ_p, the photon lifetime inside the cavity, is limited primarily by the transmission of the cavity mirrors; it is given basically by the expression

$$\tau_p = -\frac{2nL}{cln(R_1 R_2)},$$ (5.13)

where L is the cavity length, n is the refractive index of the amplifying medium and R_1, R_2 are the reflectivities of the facets. The absorption in the active region (essentially the free-carrier absorption and Auger effect) and diffusion also limit the photon lifetime (we neglect these in this simple approach). Linearizing the system around a given operating point and solving the corresponding characteristic equation yields the oscillation frequency of the laser relaxation:

$$f_r = \frac{1}{2\pi}\left(\frac{A_N F_0}{\tau_p}\right)^{1/2},$$ (5.14)

where F_0 is the mean photon fluence within the cavity at the operating point.

5.4.1.2 Pulsed-Operation Mode. The basic idea here, to obtain a short optical pulse, is to switch off the current polarization step just after the first oscillation of the output power, as shown in Fig. 5.16b. So, by pumping the laser with an ultrashort electrical pulse, a single light pulse can be generated for each excitation pulse. The temporal width of the pulse is basically of the order of half the period of the relaxation oscillations. The cavity length plays an important part in determining the pulse duration, as can be inferred from (5.13) and (5.14). It can be shown that the differential gain A_N is inversely proportional to the number of inverted states for a given device, which in turn is proportional to the volume of the active region. Thus A_N is proportional to L^{-1} (where L is the cavity length). From (5.13), one can see that the photon lifetime is proportional to L. So, f_r is inversely proportional to L (5.14). As a consequence, the minimum pulse width increases linearly with the cavity length, as is observed experimentally (Fig. 5.17) [5.14]. The fundamental limitation comes indeed from the laser relaxation oscillation frequency. Possible improvements will be described in the following.

As early as 1980, this technique allowed the generation of 40 ps FWHM optical pulses from 50 ps wide current pulses, at a repetition frequency of up to 400 MHz [5.15]. This result was obtained by using a DC polarization current just below the threshold and driving the laser with an electrical pulse train. The amplitude of the optical pulse increased with that of the electrical pulses, and the spectral width was about 3 nm.

Recently, 16 ps FWHM pulses were obtained by *Luo* et al. [5.16] from a monomode DFB laser working in gain-switched mode at a repetition rate of 100 MHz. The mean power was 10 mW. More advanced techniques are presently under investigation to generate pulses of the order of 1–2 ps at repetition frequencies of the order of 10–20 GHz. These depend on shortening the pulse width using a pulse compressor made of an optical fiber and gratings [5.17].

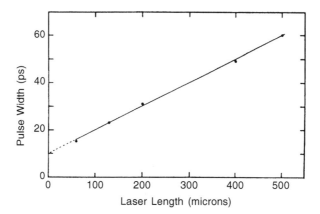

Fig. 5.17. Laser pulse width as a function of the laser cavity length (reprinted from [5.14] © 1981 American Institute of Physics)

5.4.1.3 Amplitude-Modulation Mode. The applications of SC lasers in the field of optical telecommunications stimulated research efforts on optical encoding of information and the possibility of obtaining very high modulation frequencies. The most direct way to transfer information to the output beam of the laser is to modulate the amplitude of the optical wave by varying the polarization current of the laser. The current–optical power characteristic of a 3D semiconductor laser is linear over a large range (Fig. 5.8a). Moreover, as the photon lifetime in the laser cavity (5.13) can be extremely short (typically less than 10 ps), amplitude modulation can be performed at high frequencies. This modulation is achieved by superimposing a small time-dependent signal $I(t)$ on the average pumping current I_m $(I_\mathrm{m} > I_\mathrm{th})$. We see from (5.14) and on Fig. 5.8b that the modulation bandwidth is then limited by the relaxation oscillation frequency f_r, which is due to the combined effects of the inverted population within the cavity and the electromagnetic field. Equation (5.12) suggests three simple means to increase the relaxation oscillation frequency:

(i)   increase the optical differential gain A_N,
(ii) increase the photon fluence F_0,
(iii) decrease the photon lifetime within the cavity.

The increase in the gain coefficient A_N can be obtained by optimizing the structure. In the case of a QW laser, the number of states which must be inverted to achieve the inversion condition is much lower than in a DH laser. Moreover, the injected carriers are more efficient in 2D systems than in 3D ones, since they all contribute to the gain at its peak, near the band edge. So a QW laser is better than a DH one in this respect (see Sect. 5.3.2). In 1985, it was shown theoretically that the modulation bandwidth could thus be increased by a factor 2 in multiple-quantum-well (MQW) lasers relative to the usual DH structures [5.18]. Numerous experimental verifications followed.

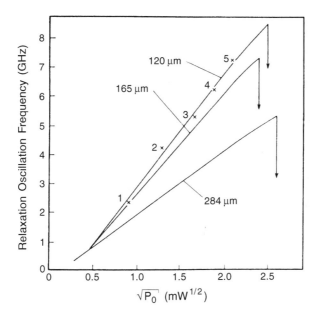

Fig. 5.18. Relaxation oscillation frequency measurements for various laser cavity lengths as a function of $\sqrt{P_0}$. P_0 is the CW optical output power (vertical arrows: catastrophic damage threshold) (from [5.13] © 1995 IEEE)

Increasing the laser polarization current allows the internal fluence to rise in the active region, which results in an increase of the output intensity (Fig. 5.8a). The problem which arises here comes from the catastrophic mirror damage which occurs for irradiances of the order of $1\,\mathrm{MW/cm^2}$ for AlGaAs SC lasers. This puts a definite limit on the photon fluence and thus on the maximum frequency bandwidth, as can be seen in Fig. 5.18.

The third means of increasing the bandwidth consists of reducing the photon lifetime by decreasing the cavity length. By this means, modulation frequencies up to $8\,\mathrm{GHz}$ were obtained with $120\,\mu\mathrm{m}$ cavities in DH lasers, as shown in Fig. 5.18.

Frequencies f_r higher than $15\,\mathrm{GHz}$ were obtained in this way in MQWs made from the GaAs/AlGaAs system.

An additional increase of the bandwidth can be achieved using a p-type doped active region. In this case, the differential gain is greater than that obtained in a similar laser with an intrinsic active region, since the electron quasi-Fermi level E_F^c changes very quickly with the injected electron density. According to the Bernard and Duraffourg condition (5.6), the population inversion condition is achieved as soon as E_F^c reaches the conduction-band minimum (or the confined-electron level in a QW laser), since the holes are already quasi-degenerated because of the doping. The current threshold is thus lowered, as seen in Fig. 5.20a. When the active region is n-doped or undoped, the

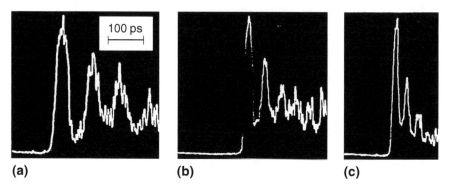

Fig. 5.19. Output pulse of a p-doped MQW laser at different output powers: (a) $P_0 = 18.5$ mW, $f_r = 11.5$ GHz; (b) $P_0 = 64$ mW, $f_r = 20.5$ GHz; (c) $P_0 = 160$ mW, $f_r = 30$ GHz (reprinted from [5.19] © 1987 American Institute of Physics)

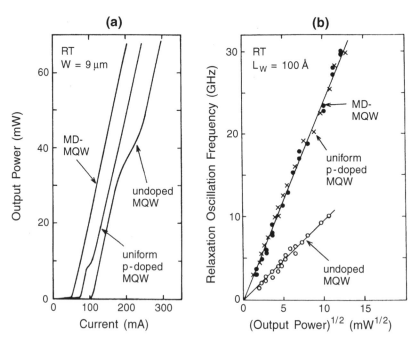

Fig. 5.20. (a) Light output power versus current for a p-type modulation-doped (MD) MQW laser, a uniformly p-doped MQW laser and an undoped MQW laser. (b) Measured relaxation oscillation frequency of a p-type MD-MQW laser (●), a uniformly p-doped MQW laser (×) and an undoped MQW laser (○) as a function of the square root of the output power (preprinted from [5.20] © 1990 American Institute of Physics). L_W and W are resepectively the QW and strip widths

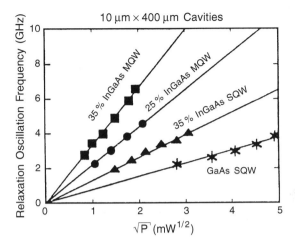

Fig. 5.21. Relaxation oscillation frequency versus the square root of the output power for unstrained GaAs SQW laser, and $x = 35\%$ SQW, $x = 25\%$ MQW and $x = 35\%$ MQW strained $In_xGa_{1-x}As$ quantum well lasers (from [5.21] © 1991 IEEE)

situation is less favorable since it is then necessary to populate hole states for which the density of states is high, and the condition (5.6) is harder to achieve. Figure 5.19 displays relaxation oscillations of the transient characteristics of a p-doped MQW laser at room temperature [5.19]. The frequency f_r is deduced from the measurement of the time delay between two successive oscillations. Using such p-doped structures, relaxation oscillation frequencies greater than 30 GHz were measured [5.20], as seen in Fig. 5.20b.

To sum up in a rule of thumb, the relaxation oscillation frequencies obtained in the various structures analyzed so far scale as follows:

$$f_r \text{ (MQW p-doped)} \approx 2.5 f_r \text{ (MQW undoped)} \approx 5 f_r \text{ (DH)}.$$

Progress in the fabrication process has allowed one to achieve compressively (or extensively) strained quantum wells (see Sect. 5.3.3). We have seen that reducing the hole transverse effective mass results in an increase of the differential gain (since the hole density of states decreases), leading to an increase of f_r in strained structures relative to those without lattice mismatch. Recent results (1991) showed a differential gain for a strained InGaAs QW three times greater than that for a lattice-matched one [5.21]. Figure 5.21 shows that f_r increases with the compressive strain, for both single-quantum-well (SQW) and multiple-quantum-well (MQW) lasers.

This method of direct intensity modulation, specific to semiconductor lasers, allows nowadays a modulation rate of 20 Gbit/s [5.22]. MQW InGaAsP DFB laser diodes operating at a wavelength of 1.55 μm and at modulation frequencies up to 4 Gbits/s are now currently available [5.23]. The serious drawback of this technique comes from the spectral broadening of the pulses due to the phase modulation induced by the injection current variations and the correlative frequency chirp which occurs [5.6, 24]. As a matter of fact, when the injection current rises suddenly, the carrier density inside the active region increases before it can be compensated by stimulated recombination, owing to the time delay necessary to build up the electromagnetic field, as seen in the previous section. A jump in the carrier density then occurs, resulting in a temporary decrease of the refractive index in the active region, as the electronic contribution to this index is a decreasing function of the electron density. The optical path within the cavity thus decreases, and consequently the laser wavelength shifts initially towards the blue. Similarly, a sudden decrease of the injection current results in a decrease of the carrier density within the active region to below its steady-state value because of strong stimulated emission, and the laser wavelength temporarily shifts towards the infrared. This effect is shown in Fig. 5.22. For structures lasing at 1.55 μm, the resulting frequency chirp is enhanced when transmitted in silicia optical fibers, for which the dispersion is negative in this spectral region, thus leading to a further temporal broadening of the pulse, and limiting the maximum modulation rate. However, long-distance communication networks are now operating with such devices, without intermediary amplifiers, over more than 100 km.

Research attempts have been made to minimize the combined effect of frequency chirp and dispersion by compressing the pulse with, for example, a grating-pair pulse compressor, with a spatial filter in its telescope to select only the frequency domain which contains a linear chirp. It is also possible to match the laser and the fiber, e.g. by operating in a spectral domain where the fiber dispersion is positive. Actually, because of this temporal broadening of pulses in the case of direct amplitude modulation of SC lasers, very-high-repetition-rate optical telecommunications systems use SC lasers operating in CW mode, coupled to an external light modulator. Amplitude modulation with a narrow spectrum and high extinction rates is thus obtained. Two types of modulators are currently used for such a purpose: the LiNbO$_3$ electro-optical modulator [5.25] and electroabsorptive modulators, which allow one to obtain information transmission rates higher than 12 Gbit/s.

5.4.2 Q-Switched Operation

5.4.2.1 Q-Switching in the Context of SC Lasers. Semiconductor lasers operating in Q-switched mode are potentially interesting in optical telecommunications systems, or more generally as optical pulse sources when high peak power, high amplitude modulation and high repetition rates are required.

The basic idea is to modulate the intracavity losses. This modulation is obtained by creating regions of saturable absorption in series with the gain medium (Fig. 5.23). High losses allow the build-up – without oscillations – of

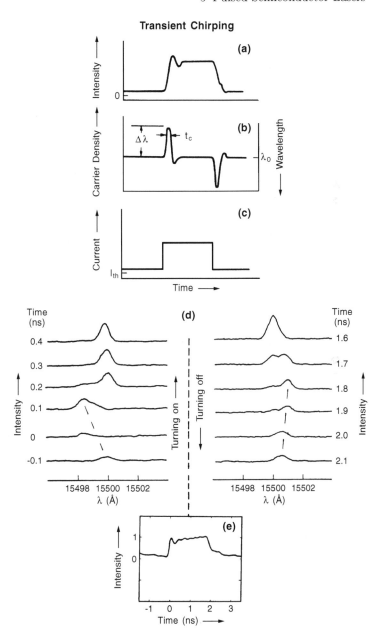

Fig. 5.22. Transient chirping phenomena in gain-switched operation: schematic representation of the time dependence of (**a**) the light intensity, (**b**) the wavelength and (**c**) the current. (**d**) Time-resolved emission spectra showing the frequency chirp on the leading and falling edges of a gain-switched SC laser; the wavelength is seen to move briefly towards the blue on start-up and towards the (infra)red on shutdown. (**e**) Total light intensity versus time (after [5.24])

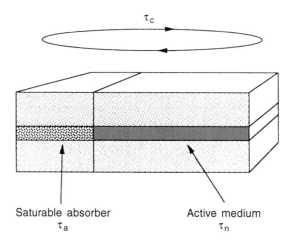

Fig. 5.23. Basic scheme of a semiconductor laser with a saturable absorber. τ_a, τ_n and τ_c are respectively the absorber relaxation time, the gain recovery time and the cavity round-trip time (see text) [5.6]

carrier densities greater than the threshold density corresponding to the minimum losses, thus increasing the energy stored within the amplifying section. Then the losses are suddenly reduced, in order to extract this energy from the active medium resulting in the formation of a short pulse. The mechanisms involved in SC laser Q-switching depend on the absorber relaxation time τ_a [5.6]. The latter must always be shorter than the gain relaxation time τ_n in order to obtain a pulsed emission. However, it may be shorter or longer than the cavity round-trip time τ_c, depending on the cavity length and the absorption process. The first situation a priori leads to mode-locking, which will be described further in Sect. 5.4.3. The laser dynamics in the second situation are similar to those of gain switching. We consider this situation now ($\tau_c < \tau_a < \tau_n$). One difference from gain switching is the possibility of generating optical pulses using a DC laser bias. In this pure Q-switching case, the pulse periodicity is fixed by τ_n. Another difference is that Q-switched pulses exhibit shorter tails than gain-switched ones, due to the absorber recovery at the end of the pulse.

As for gain switching, the Q-switching operation mode can be described by rate equations [5.6]. Equations (5.12a) and (5.12b) of Sect. 5.4.1.1 are now accompanied by a third equation for the absorber population density N_a:

$$\frac{dN_a}{dt} = -\frac{N_a - N_{eq}}{\tau_a} - \frac{c}{n} A_a N_a P, \tag{5.15}$$

where the first term on the right-hand side of this equation represents the spontaneous carrier relaxation and the second represents the induced absorption. N_{eq} is the equilibrium absorber population in the absence of light, and A_a is the absorber cross-section. The induced absorption term must be included

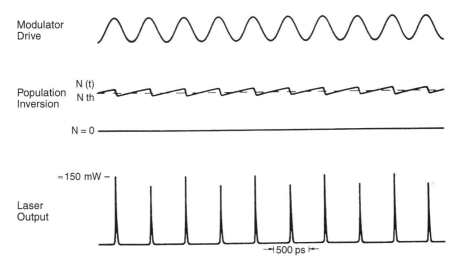

Modulator
Drive

Population
Inversion

$\frac{N(t)}{N\,th}$

$N = 0$

≈ 150 mW

Laser
Output

$\rightarrow$ 500 ps $\leftarrow$

Fig. 5.24. Typical response of a Q-switched SC laser to repetitive modulation of the intracavity absorption. The modulation frequency is 2 GHz (from [5.26] © 1983 IEEE)

as a loss term in (5.12b). Approximate solutions of these rate equations have been proposed in [5.30]. Figure 5.24 displays a simulation obtained for repetitive Q-switching [5.26] in the case of inversion-dumped Q-switching (in which the electron population fluctuates about the threshold value). As the cavity lengths of SC lasers are much shorter than in the usual laser sources, the obtained Q-switched pulses are short, with a duration of the order of 10 ps (conventional Q-switched lasers have pulse widths in the nanosecond range).

The intracavity loss modulation can be either passively or actively achieved. Passive Q-switching due to absorption saturation was observed in SC lasers with two amplification sections, nonuniformly pumped (one of the sections acts as a saturable absorber) [5.8]; more generally, it is obtained in multisection laser geometries, where some of the sections are unbiased or reverse-biased [5.26–29]. It can be also achieved by implantation of ions through the diode facets [5.10, 30, 31]. But most of the research efforts have been concentrated on active Q-switching. Before describing the different devices which have been tested, we give now a brief review of a few specific ways to modulate absorption in semiconductors, which are the most used currently and lead to technical solutions which can be integrated with the SC laser.

5.4.2.2 A Few Techniques Used to Modulate Absorption in Semiconductor Lasers. The first actively Q-switched SC lasers were built with external electro-optical or acousto-optical modulators. With the aim of integration of all the functional parts of the laser within the same device, methods to achieve light modulation specific to semiconductors were developed. Technological progress in fabrication now allows one to reach this goal, allowing the development of SC lasers with the light modulator integrated in the same device.

Most of these modulators rely on electroabsorptive effects, due to the modification of the electronic band structure under an external electric field. The most famous effect is the Franz–Keldysh effect.

Franz–Keldysh Effect in Bulk Semiconductors. This effect was predicted in 1958 independently by Franz and Keldysh, and was experimentally observed in germanium by Seraphin in 1964 [5.32]. Under an external electric field, the electronic bands are tilted in direct space, as shown in Fig. 5.25a. Owing to the spatial overlap of the evanescent tails of the electronic wave functions which develops in the band gap, some absorption appears for photon energies *below* the band gap E_g, as shown in Fig. 5.25b. Applying an electric field in the range 10^2–10^3 kV/cm results in an absorption tail below the band gap of a few 10^2 cm^{-1}.

This absorption can be directly modulated by modulating the externally applied electric field. Thus active Q-switching, as well as active mode-locking, can be achieved. As the density of states corresponding to the absorption tail is low, the absorption below E_g can be easily saturated when the density of photogenerated carriers increases, under a steady electric field, leading to self-modulation effects. In this case, the semiconductor acts as a passive saturable absorber, for photon energies around E_g, and can be integrated for passive Q-switching or mode-locking of the SC laser.

Quantum-Confined Stark Effect (QCSE) in Quantum Wells. The principle of this effect relies on the modifications of the potentials in a quantum well when applying an electric field perpendicular to the material layers. The main effect of the tilt of the electronic bands in the bottom of the quantum well is that it modifies the confinement energy of the electrons and the holes. Thus the energy of the fundamental optical transition in the quantum well is reduced, providing absorption below the position that the fundamental transition has without an electric field, as shown in Fig. 5.26 [5.33]. Although the electric field draws apart the occupation probabilities of the electrons and the holes to opposite sides of the QW, the overlap of the electron and hole wave functions is better than in the case of the Franz–Keldysh effect, owing to their confinement within the QW along the growth axis of the device. So the oscillator strength remains relatively large under an electric field, leading to higher absorption coefficients below the band gap of the structure than in the case of the bulk electro-optic effect.

As before, active Q-switching (or mode-locking) can be achieved. In this case direct modulation of the transmission of the QW absorber is obtained by modulating its electrical polarization. On the other hand, under constant electrical bias, the device acts as a saturable absorber, owing to the bleaching of the lower optical transition in the QW by the photogenerated carriers. Passive Q-switching (or mode-locking) results. Such a device can be integrated

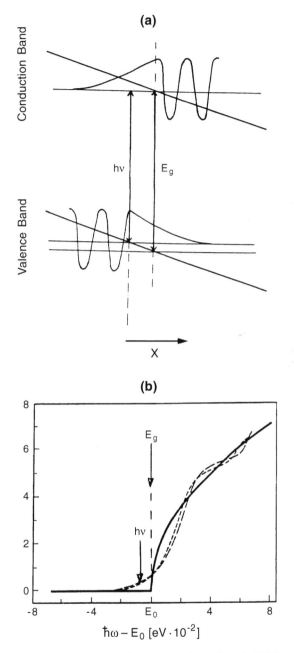

Fig. 5.25. Schematic illustration of the action of an electric field on a bulk semi-conductor (Franz–Keldysh effect) (**a**) Tilt of the band structure, and the conduction and valence band wave functions, and (**b**) the resulting absorption below the band edge [5.2]. *Dashed lines*, band-edge absorption with applied electric field; *solid line*, without electric field

QW-NO APPLIED E-FIELD   QW-WITH APPLIED E-FIELD

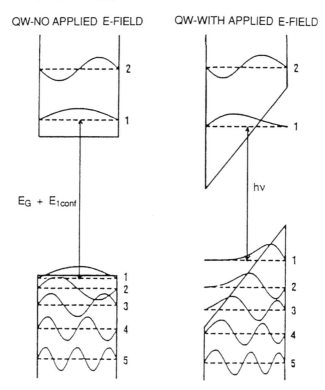

$E_G + E_{1conf}$

Fig. 5.26. Action on the conduction- and valence-band states of an electric field applied perpendicular to the plane of a quantum well (quantum-confined Stark effect) (reprinted from [5.33] © 1988 American Institute of Physics)

in AlGaAs/GaAs or GaInAsP/GaInAs QW lasers, GRINSCH structures and semiconductor colliding-pulse mode-locked lasers, as will be demonstrated below.

Semiconductor Saturable Absorber Created by Ion Implantation. This idea consists of creating localized defect states below the band gap in a limited zone of the SC laser. This can be achieved by proton irradiation in the SC laser facets, but heavy ion implantation (O^{5+}, for instance) gives better results. As shown in Fig. 5.27a, spatial fluctuations of the conduction- and valence-band energies result, leading to optical transitions between states which are eventually localized on different sites. Density-of-states tails appear below and above the conduction and valence bands respectively (Fig. 5.27b), allowing optical absorption for energies *below* the band gap of the material. These transitions can be bleached at the SC laser wavelength: the photogenerated carriers saturate the localized states, thus providing a self-modulation of the absorption. As the cross-section σ_A of the saturable absorber is greater than the cross-section of the amplifying medium σ_G, and the recovery time of the absorber

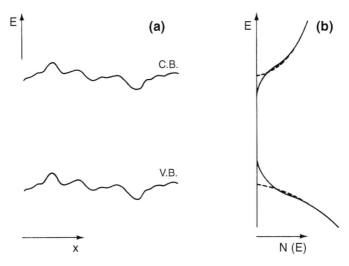

Fig. 5.27. Schematic illustration of (**a**) the perturbation of the band edges by Coulomb interaction with inhomogeneously distributed impurities, and of (**b**) the resulting formation of density-of-states tails. The *dashed lines* show the distribution of states in the unperturbed case

is of the order of a few picoseconds, such processes can be efficiently used for passive Q-switching or mode-locking in SC lasers.

5.4.2.3 Active Q-Switching. Figure 5.28 shows a quaternary-compound InGaAsP DH diode operating in the Q-switched mode [5.26]. In this schematic drawing, we can clearly identify the following three sections:

- the amplifying section, the length of which is about $200\,\mu$m,
- the passive optical waveguide,
- the electroabsorption modulator section.

The electroabsorption modulator consists of a pn junction, built close to the quaternary layer, which provides the controlled intracavity loss. When this junction is reverse biased, the optical absorption of the light emitted in the active region, increases in the modulator section, because of the Franz–Keldysh effect. In InGaAsP material under an electric field of about $400\,\mathrm{kV/cm}$, the optical absorption can reach $1000\,\mathrm{cm}^{-1}$ for photon energies down to $60\,\mathrm{meV}$ below the gap of the unbiased material.

The waveguide section has two simultaneous actions: it optically couples the amplifier and the modulator, but isolates them electrically.

Note that the amplifying GaInAsP quaternary region is p-doped. This device generates $100\,\mathrm{ps}$ pulses, with $50\,\mathrm{pJ}$ energy per pulse (corresponding to a peak power of $500\,\mathrm{mW}$), at a repetition frequency of $8\,\mathrm{GHz}$ [5.26]. More

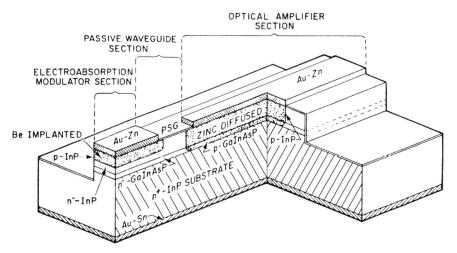

Fig. 5.28. Schematic drawing of a Q-switched SC laser. The device consists of a zinc-diffused amplifier section, a beryllium-implanted modulator section and a passive waveguide section (from [5.26] © 1983 IEEE)

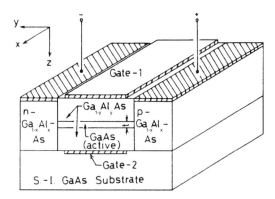

Fig. 5.29. Schematic drawing of a quantum-mechanical size-effect modulation SC laser (from [5.34] © 1986 IEEE)

recently, a similar device allowed one to obtain 40 ps pulses, with 300 pJ per pulse at a repetition frequency of 400 MHz [5.29] .

Another technique for the generation of short optical pulses is obtained from a three-terminal "quantum-mechanical size-effect modulation" laser [5.34]. This quantum well laser is sketched in Fig. 5.29. The gain modulation on which it relies is based on a quantum size effect. Although it could be classified among the gain-switched SC lasers, it leads to similar carrier density dynamics in the active region to those in Q-switched SC lasers. The modulation is obtained in the following way.

The electrons and holes are injected into the GaAs quantum-well active region from the lateral n-doped and p-doped AlGaAs regions. Applying a bias to the top and bottom gate electrodes creates an electric field in the GaAs quantum well through the AlGaAs layers, leading to the spatial separation of the electron and hole wave functions (as in Fig. 5.26). A decrease of the optical transition matrix element, and consequently a decrease of the optical recombination probability results (see (5.12)). This allows the building of an electron–hole population above the threshold density, corresponding to a zero bias on the top and bottom electrodes. When the latter is suppressed, strong gain sets in and the energy stored within the amplifier is converted into a short optical pulse. Owing to the applied potentials on the gates, the transit time of the carriers through the structure is less than one picosecond; a very fast switching of the emission results. Repetition frequencies up to 12 GHz can thus be obtained.

5.4.2.4 Passive Q-Switching. Passive Q-switching was obtained in 1981 by Van der Ziel [5.10] from an SC laser which was ion-implanted, thus creating a saturable absorber within the laser device near one end facet. The precise description of phase modulation effects in the Q-switching regime is somewhat difficult, owing to the presence of the saturable absorber. The most simple approach is to assume that the chirp amplitude during the pulse is proportional to the gain variations as well as to the electron density. This is due to the linear relation existing between the gain and the electron density. Since the number of emitted photons is approximately equal to the number of lost electrons, the total chirp amplitude is expected to be proportional to the pulse energy [5.6]. Figure 5.30 displays the autocorrelation traces of Q-switched pulses emitted by a gain-guided AlGaAs laser implanted with 20 MeV oxygen. The pulse width is ~ 22 ps before compression; after compression, it reduces to ~ 1.3 ps (a factor of 0.65 is used to deduce the laser pulse widths from the autocorrelation traces), one of the best reported results to date for a Q-switched SC laser with electrical pumping. Note the extremely high value of the compression ratio. This illustrates the possibility of very large phase modulation in single-mode Q-switched SC lasers. Owing to high power levels, resulting in strong laser nonlinearities, the ultimate performances of Q-switching may be different from those previously established for gain-switched lasers. In principle, the ultimate limit of chirp amplitude is given by the gain bandwidth. More realistically, a chirp amplitude of about 4 nm can be expected at 0.8 μm for a pulse of ~ 30 pJ, which should correspond to ~ 240 fs pulses after compression.

5.4.3 Mode-Locked Operation

As for all laser systems, the technique which leads to the generation of the shortest optical pulses in SC lasers is that of mode-locked operation. It is significantly superior to the gain-switching and Q-switching techniques with

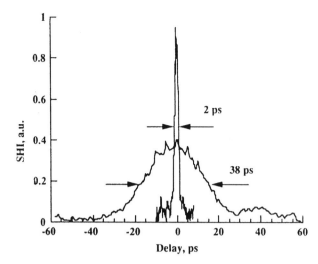

Fig. 5.30. Second-harmonic autocorrelation traces of Q-switched pulses emitted by a gain-guided AlGaAs laser implanted with 20 MeV oxygen. The energy per pulse is about 33 pJ. The trace width is 38 ps without external compression. It reduces to 2 ps after compression by a $\sim$ 200 m long fiber with a total dispersion of $\sim$ 24 ps/nm [5.6]

respect to the amplitude and the timing jitter [5.12], as well as the minimum width of the obtainable pulses [5.10, 11, 35, 36]. It should lead, in principle, to ultrashort pulses, owing to the large bandwidth of semiconductor spectral gain: for instance, a usual spectral gain spanning 30 nm would lead, in the case of Fourier-transform-limited pulses, to durations of about 30 fs. It is, however, more complex. Mode-locking is achieved by introducing a mechanism inside the laser cavity to cause the longitudinal modes to interact with one another, thereby locking them in phase. As we shall see now, this mode-locking can be active or passive.

5.4.4 Mode-Locking by Gain Modulation

5.4.4.1 Synchronous Pumping by Current Injection. Active mode-locking is based on a direct modulation of the injection current at a frequency $f = c/2nL$ matched to the cavity length. It corresponds to synchronous pumping, but here it is obtained directly through the injection current. The usual length of SC lasers being of the order of 300 μm, this leads to modulation frequencies about 150 GHz, much higher than the relaxation oscillation frequencies f_r. It is thus necessary to extend the laser cavity, to lower the frequency of operation below f_r. The first solution consists of coupling the SC laser to an external cavity of the order of 1 cm length, as represented in Fig. 5.31a, thus reducing the longitudinal intermode spacing to a few GHz, i.e. within the electrical modulation bandwidth of the device, which is typically about

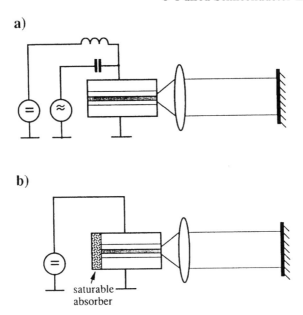

Fig. 5.31. Basic schemes of a mode-locked SC laser with extended cavity. (**a**) Active mode-locking by injection current modulation; (**b**) passive mode-locking [5.6]

20 GHz (Sect. 5.3.1.1). The SC laser facet corresponding to the extended part of the cavity is covered with an antireflection coating.

Optical pulses of 5 ps duration and 0.1 pJ energy, corresponding to a peak output power of 18 mW, were obtained at a repetition rate of 20 GHz, and thus a mean power of 2 mW, from a 1.55 μm InGaAsP laser coupled to a single-mode output fiber [5.37]. However, this technique suffers from two drawbacks: first, the external cavity length must correspond accurately to the applied current modulation frequency. The second drawback comes from the residual reflectivity of the coating deposited on the internal facet of the laser diode: a reflectivity as low as 10^{-3} creates a modulation of the modal structure of the external resonator of the order of the intermode spacing of the subcavity constituted by the semiconductor (about 150 GHz). Owing to reflection from the internal facet, the laser emission thus consists of trains of pulses separated by the round-trip time of the diode cavity. In fact, the mode-locked laser must be treated as a two-cavity system, as first analyzed by *Haus* [5.38, 39]. Figure 5.32 gives a schematic representation in the frequency domain [5.6, 38]. Owing to the presence of the diode cavity, the longitudinal modes of the extended cavity can be gathered into different cluster, each clusters being centered around one of the diode-cavity resonances. The different modes of a given cluster can easily lock together. In contrast, modes from neighboring clusters are not necessarily in phase with one another. In the theoretical description

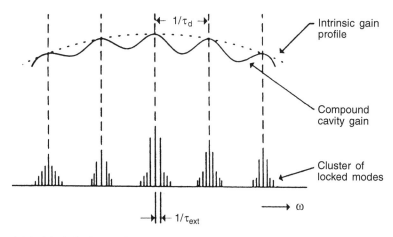

Fig. 5.32. Mode-locking of a nonideally antireflection-coated laser diode in an external cavity. The influence of the diode-cavity resonances is simulated by a periodic gain profile (compound cavity gain) [5.38]; τ_{ext} and τ_d are the round-trip times in the external cavity and in the diode subcavity respectively. Modes from neighboring clusters are not necessarily in phase with one another [5.6]

given by *Haus* [5.38, 39], the diode-cavity resonances were simulated using a periodic profile for the laser gain, as shown in Fig. 5.32.

So mode-locked operation with an external cavity composed of optical components suffers from both bad mechanical stability and multiple pulse emission. One solution consists of lowering the reflectivity of the internal facet to 10^{-4} with an antireflection coating. A better solution consists of integrating the passive resonator with the laser waveguide, which efficiently remedies both drawbacks. This is shown in Fig. 5.33, which displays a schematic of an extended-cavity laser with an integrated Bragg reflector for active mode-locking [5.40]. The device generates 20 ps wide transform-limited pulses, at a repetition frequency of 8.1 GHz.

Recent improvements in the design of extended cavities have led to better performances with respect to the peak power and pulse width. For instance, mode-locking of monolithic GaAs/AlGaAs single-quantum-well lasers with active waveguide cavities was achieved, leading to 110 mW peak power (1.5 pJ) and 10 ps pulses at a repetition rate of 5.9 GHz [5.41].

5.4.4.2 Synchronous Pumping by Optical Injection.
Mode-locking can also be achieved by optical pumping of the amplifying medium of the semiconductor laser, a technique widely used formerly for dye lasers. The SC gain is modulated by an external laser source, the repetition rate of which corresponds to the round-trip time within the semiconductor laser resonator. As the stimulated lifetime of the photogenerated excitations in the active medium is much shorter than the round-trip time, the locking of the longitu-

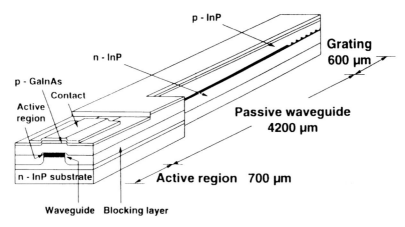

Fig. 5.33. Schematic of a monolithic extended-cavity laser with integrated Bragg reflector for active mode-locking (from [5.40] © 1992 IEEE)

dinal modes of the resonator occurs. We describe this technique here in the context of vertical-cavity surface-emitting lasers [5.42], a device described in Sect. 5.3.3. The aim here is to characterize the potential applications of such devices in pulsed-operation mode. The laser cavity is extended as indicated in Fig. 5.34a. The internal facet of the VCSEL is covered with an antireflection coating (Fig. 5.34b). The device is colinearly pumped by an external pulsed-laser source. The length of the extended resonator is tuned so that the overall round-trip time in the laser cavity corresponds to the time between two successive pulses of the external pump. The output pulses are temporally shortened using an external two-grating pulse compressor. The characteristics of the compressed pulses are displayed in Fig. 5.34c,d. The pulse duration deduced from the autocorrelation curve is about 320 fs. A high mode quality (TEM$_{00}$), reduction of all the losses due to the small active volume (about 2 μm diameter and less than 10 μm thickness, including the Bragg mirror), the high quantum efficiency, the optimal coupling to optical fibers and the possibility of arranging VCSELs in two-dimentional arrays motivate the research effort in this direction.

5.4.5 Mode-Locking by Loss Modulation: Passive Mode-Locking by Absorption Saturation

This method of laser operation is based on the self-modulation of the absorber transmission, which is bleached by the absorption of the stimulated fluorescence emitted by the active zone. As in Q-switching, the saturable absorber sharpens up the leading edge of the pulse, while gain saturation effects combined with fast absorber recovery provide an abrupt termination of the pulse [5.43, 44]. The difference from Q-switching lies in the supplementary conditions imposed on the absorber relaxation time τ_a and the cavity round-trip

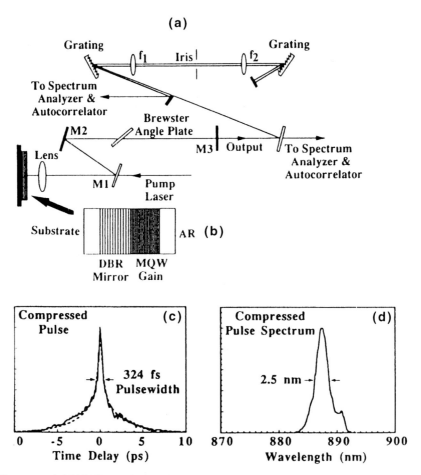

Fig. 5.34. (a) VCSEL mode-locked by synchronous optical pumping, and a grating–telescope compression system. The laser gain is provided by $120 \, \text{GaAs}/\text{Al}_{0.3}\text{Ga}_{0.7}\text{As}$ quantum wells (b). (c) Autocorrelation and (d) spectrum of the compressed pulse (reprinted from [5.42] © 1992 American Institute of Physics)

time τ_c. The most favorable situation for passive mode-locking occurs when the condition $\tau_a < \tau_n < \tau_c$ is fulfilled, τ_n being the recovery time of the gain medium. Since τ_n is of the order of 1 ns in SC lasers, this condition can be fulfilled only if the cavity length exceeds a few centimeters. As in active mode-locking, the technical solution is to use an external cavity (Fig. 5.31b). However, passive mode-locking may occur when the condition $\tau_a < \tau_c \leq \tau_n$ is satisfied. In principle, no external cavity is needed in this case, but a very fast absorber must be used. Interest in such a configuration stems from the achievement of very high repetition rates, as will be shown below. The major drawback is the weak level of peak emitted power. Since the gain medium cannot recover between two successive pulses, the population inversion rate always remains at a low value.

The main difficulty for passive mode-locking was the identification and characterization of a suitable saturable absorber. In the first work in this field, the absorption due to optically created defects in the laser was used. So, by aging the device, pulses as short as 0.5 ps (at 6 GHz) were obtained [5.11]. This mode of operation is associated with an absorption saturation mechanism induced inside the laser diode by aging. More generally, the absorption may be due to interface inhomogeneities, defects of the black-line type or regions with a very high nonradiative recombination rate.

Progress in the control of the saturable absorber properties may be made by creating the defects by heavy-ion implantation in some definite zones of the laser. The first attempts were made by Van der Ziel in 1981 [5.45] and Yokoyama in 1982 [5.11]. An SC laser ion-implanted on one side is used as in Sect. 5.4.2.4, but now the device is put into an extended external resonator. The laser facet coupled to the resonator is antireflection-coated. More recently, pulses of 5–10 ps duration and 25–50 pJ energy, leading to a peak power of about 5 W, were obtained [5.46, 47], by implanting 20 MeV O^{5+} ions to a depth L_i of about 10 μm from the cleaved laser-diode facets. Figure 5.35 displays results obtained in [5.6] with a passively mode-locked AlGaAs laser. The experimental scheme is close to that shown in Fig. 5.31b. The saturable absorber is created by a 10 MeV implantation of oxygen ions through the rear facet of the diode. The external cavity is about 15 cm long. From the autocorrelation traces shown in Fig. 5.34b, it is clear that the output of the mode-locked laser is composed of pulse trains, the pulse spacing (∼ 8 ps) being the round-trip time in the semiconductor amplifier. For most of the experimental conditions, the contrast between the pulses is moderate: the modulation ratio is less than 60 % (dotted curve in Fig. 5.35b). However, an abrupt change is observed for certain tunings of the external cavity, and the modulation ratio increases to 90 % (solid curves in Fig. 5.35). Simultaneously, the emitted spectrum is shifted towards shorter wavelengths and the spectral width is increased from 2.5 to 4.0 nm (Fig. 5.34a). The minimum pulse width deduced under these conditions is about 0.7 ps, which corresponds approximately to the phase-locking of a cluster of five modes. The average output power is 4 mW, thereby indicating a peak power of around 1 W.

Another way to mode-lock a laser diode was obtained using a multiple quantum well as a saturable absorber in an external cavity coupled to the SC laser [5.48, 49], as shown in Fig. 5.36. Note that the use of an external coupled cavity brings some drawbacks, as described previously in Sect. 5.4.3.1, but the interest here lies in the fact that, in this configuration, the absorber is completely independent of the laser structure, allowing one to study the impact of its physical characteristics on the laser operation. The physical effect used here is the saturation of excitonic absorption in the quantum well. Owing to Coulomb interaction, an electron and a hole can bind together, so forming a small hydrogenoid system, which is called an exciton. A strong absorption

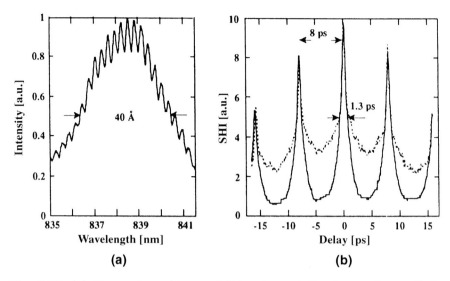

Fig. 5.35. (a) Measured spectrum and (b) autocorrelation traces of a passively mode-locked SC laser with a 15 cm long external cavity. *Dotted* and *solid curves* in (b) correspond to two different tunings of the the external cavity (see text) [5.6]

thus takes place in the crystal, at an energy corresponding to the energy of the interacting electron–hole pair; this corresponds to the band gap of the material minus the binding energy of the exciton (of the order of 10 meV in quantum wells). When the photon irradiance I on the semiconductor increases, the optical absorption α saturates according to

$$\alpha = \frac{\alpha_0}{1 + I/I_0}, \tag{5.16}$$

where α_0 is the absorption at low irradiance and I_0 the saturation irradiance. The latter corresponds, roughly speaking, to the case where all possible excitonic transitions have been made in the semiconductor. The use of a multiple-quantum-well system is more favorable since it is possible to design the quantum well in such a way that the exciton resonance corresponds to the energy of the photons emitted by the laser diode. Moreover, the nonlinear behavior is enhanced in quantum wells compared to the bulk situation. Self-modulation occurs, due to photocreated excitons in the quantum wells. Haus has analyzed the conditions necessary for mode-locking a homogeneously broadened laser with a saturable absorber having a relaxation time longer than the pulse width [5.43]. He showed that the relaxation time of the absorber must be faster than that of the gain and that the condition

$$\frac{\sigma_A}{S_A} > \frac{\sigma_G}{S_G} \tag{5.17}$$

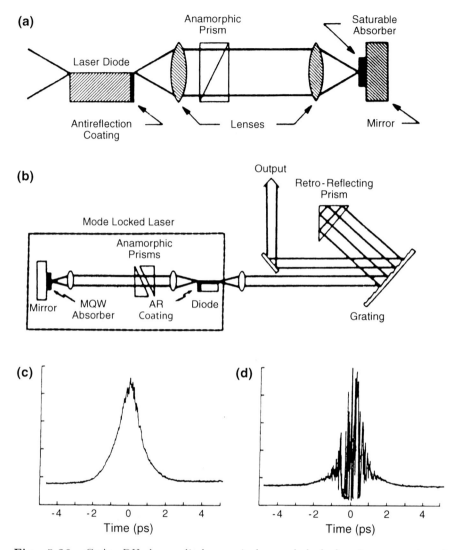

Fig. 5.36. GaAs DH laser diode passively mode-locked using an external GaAs/AlGaAs multiple-quantum-well saturable absorber. (**a**) Experimental setup [5.48]. (**b**) Setup with external grating compressor. Intensity (**c**) and interferometric (**d**) autocorrelation traces for the compressed output pulses [5.49]

must be fulfilled, where σ_A and σ_G are the cross-sections of the absorber and gain medium, respectively, and S_A and S_G are the cross-sectional areas of the laser beam in the absorber and in the gain medium.

It is thus possible to generate pulses of 1–2 ps duration at a repetition frequency of 1 GHz. The temporal width of the pulses in such a mode-locked laser is determined by the balance between two opposing effects: the compression

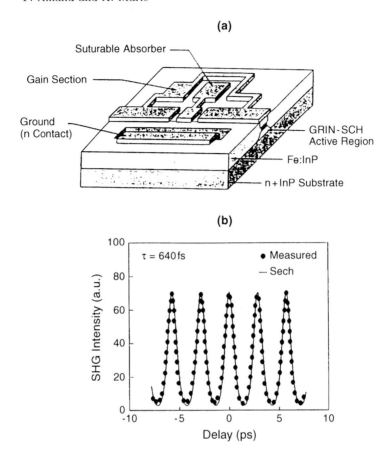

Fig. 5.37. Passive monolithic colliding-pulse mode-locked (CPM) quantum well laser. (**a**) Schematic diagram. (**b**) Autocorrelation trace of the CPM laser output with 350 GHz repetition rate. The pulse width is 640 fs [5.50, 5.51]

due to the combined effect of the gain and losses, and the broadening due to the positive group velocity dispersion (GVD) resulting from wave propagation in the amplifying medium, the saturable absorber and the optical components of the cavity. To compensate this temporal broadening, the pulse can be compressed by passing it through in an optical device providing a GVD of the same amplitude, but with opposite sign. A grating pair (or a single grating used twice, as in Fig. 5.36b) placed outside the cavity provides the required negative GVD. Pulses as short as about 500 fs have been generated in such a way.

Recently, progress in semiconductor integration technology has allowed one to apply the technique of colliding-pulse mode-locking (CPM), already used in dye lasers (see Chap. 4), to semiconductor lasers [5.50, 51]. Figure 5.37a shows a schematic of a monolithic multiple-quantum-well CPM laser. The linear cavity consists of a GRINSCH including InGaAsP/InGaAs strained multiple quantum wells. The cleaved end facets are used as mirrors. The optical cavity is divided into three sections, corresponding to the ribbon metallic contacts on the top. The quantum wells in the central region are polarized with a reverse bias, so that this section acts as a saturable absorber working on the principle of the QCSE (see Sect. 5.4.2.2). The two neighboring regions are polarized with a forward bias, so that they constitute the gain region of the laser. As in dye CPM lasers, two optical pulses travel in opposite directions within the laser resonator (see Chap. 4). These two pulses interfere constructively in the saturable absorber to achieve its bleaching. The synchronization of the two pulses results automatically. The transient grating produced in the saturable absorber by the coherent superposition of these two pulses stabilizes and shortens the pulses. The repetition frequency corresponds then to half the round-trip time of the resonator, i.e. the travel time between the two facets. Figure 5.37b displays the autocorrelation trace of this monolithic CPM laser, with a repetition frequency of 350 GHz and a temporal width of 640 fs.

This particular mode-locked configuration presents numerous advantages. The use of a cavity with an integrated waveguide eliminates the problems encountered by SC lasers with external extended cavities. On the other hand, the repetition frequency is not limited by the relaxation oscillation frequency, as in active mode-locking by current injection (Sect. 5.4.3.1). So, very high repetition frequencies can be achieved, by decreasing the cavity length (up to 350 GHz, as in the example here, which corresponds to a cavity length of 250 μm). This system is thus a very promising one for the optical generation of waves in the millimetric wavelength range. Finally, this CPM laser allows one to generate subpicosecond pulses without external pulse compression. A hybrid version of this monolithic CPM laser was developed more recently [5.51]. An electro-optical modulator based on the QCSE was integrated within the laser structure, which allows one to modify the repetition frequency from 30 to 350 GHz. Cavities shorter than 100 μm are expected to produce pulses at repetition rates in the THz range in CPM laser diodes, but such short-cavity devices will inevitably impose a limitation on the output power.

Harmonic mode-locking in monolithic laser diodes provides an alternative for generating ultrafast pulses without using a short-cavity laser [5.52]. A multicontact, monolithic, passively mode-locked distributed Bragg reflector (DBR) is achieved according to the scheme shown in Fig. 5.38. The device includes an intracavity electro-absorptive saturable absorber. Mode-locking is achieved at very high repetition rates of more than a terahertz, with high output power. Operating this laser at a high gain level and using the selectivity of the harmonic numbers related to the spectrum-filtering properties of the intracavity Bragg reflector allows one to generate transform-limited pulses at repetition rates from 500 GHz to 1.54 THz, with an output power exceeding 15 mW. The device is tunable in terms of wavelength and pulse repetition frequency: the central wavelength of the emitted pulses ($\sim 1.55\,\mu$m) can be tuned over a wide range (~ 4 nm) by changing the refractive index of the

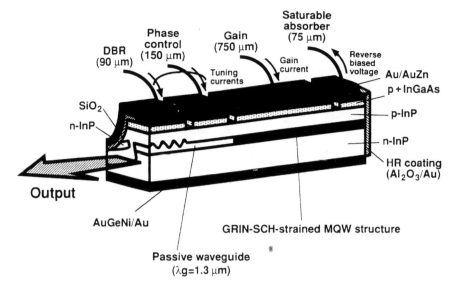

Fig. 5.38. Monolithic mode-locked DBR laser for passive harmonic mode-locking (from [5.52] © 1996 IEEE)

Bragg reflector section, either by carrier injection or by thermal effects, while maintaining the pulses transform-limited. Tuning the repetition rate can be achieved over a total range of more than 1GHz by combining different tuning schemes, namely current injection in the passive phase-control section, varying the reverse bias applied to the absorber, varying the injection current in the gain region and using thermal effects.

5.4.6 Prospects for Further Developments

As is clear from the previous sections, the main desirable aims in improving the characteristics of pulsed SC lasers concern the lowering of the threshold current, the reduction of its temperature dependence, the increase of the differential gain, the shortening of the generated pulses down to 100 fs (or less) and the increase of the energy of short laser pulses from SC lasers.

On the basis of the considerable progress achieved since quantum well SC lasers appeared, the route towards further reduced dimensionality (quantum wires, 1D, and quantum boxes, 0D) offers encouraging prospects for the next few years. The main characteristic of low-dimensional structures consists of a reduced density of allowed states $E(K)$. Calculations show that quantum dot lasers should have threshold currents 10 to 100 times lower than in quantum well lasers, and a reduced temperature dependence of this threshold current. An increase of the modulation bandwidth is also predicted, related to potentially higher differential gain, due to the reduced density of states which need to be inverted, and to the fact that the excited states contribute more

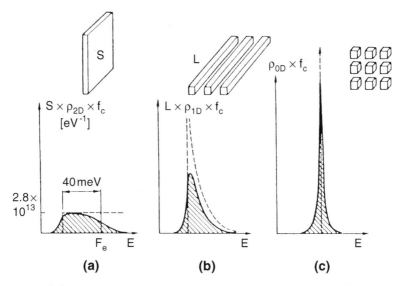

Fig. 5.39. Schematic illustrations of gain curves in (**a**) 2D, (**b**) 1D and (**c**) 0D structures. Here, similar numbers of electrons and holes are being injected above transparency, yielding equal integrated gain. $\rho_{2D}, \rho_{1D}, \rho_{0D}$ are the 2D, 1D, 0D DOS, respectively. S is the QW surface; L is the length of the quantum wires; f_c is the conduction Fermi function (after [5.2, 53])

efficiently to the gain when the dimensionality of the active region decreases, as shown in Fig. 5.39. Up to now, the technical difficulties met in the process of fabrication of 1D devices have resulted in specifications which are merely similar to those of quantum well SC lasers. A promising way for the fabrication of quantum dot lasers was opened recently by using 3D self-organized growth of strained InAs 0D islands on a GaAs substrate [5.54].

From the discussion in the previous sections, it is also clear that pulse durations shorter than 300 fs will soon be feasible. Mode-locking remains the most efficient technique for the generation of short pulses. However, integration of the laser diode and the passive waveguide in the same device is preferable to external extended cavities in order to avoid parasitic subcavity effects (Sect. 5.4.3). On the other hand, gain-switched and Q-switched in VCSELs are promising devices. The main advantages of the VCSEL are (i) low beam divergence, (ii) dynamical single-mode operation due to a very short cavity length ($< 10\,\mu$m), (iii) the ease of forming a two-dimensional laser array and (iv) the capability for high-speed modulation [5.55, 56].

Finally, it is desirable to increase the energy of short pulses from SC lasers, for several applications such as long-haul soliton transmission in optical fibers and the triggering of fast electronic circuits. Up to now, the output energy of semiconductor lasers has seldom exceeded 30 pJ per pulse in the short-pulse regime, and further improvements are still possible. However, one must be aware of the fundamental limitation which arises in semiconductors, dictated by two-photon absorption (TPA). A typical value for TPA in III–V compound semiconductors is about $0.03\,\mathrm{cmMW^{-1}}$, which leads to a threshold power of the order $10^9\,\mathrm{W\,cm^{-2}}$. Thus for a laser stripe of standard dimensions ($\sim 0.2 \times 2.5\,\mu\mathrm{m}^2$), the maximum peak power is then about 5 W, and the maximum energy emitted in a 10 ps pulse is 50 pJ [5.6]. Further improvement requires the use of SC lasers with larger emitting surfaces. Various solutions are under investigation, such as broad stripe lasers, laser diode arrays (both commercially available now) and thick stripe lasers, for which peak powers of about 100 W should be obtained. However, VCSELs, in which all losses are reduced owing to the small size of the device, here again present the most favorable geometry, since they provide the largest emitting surface. One should then expect to increase the peak emitted power by up to one order of magnitude. Optical pumping of external-cavity surface-emitting InGaAs/InP MQW lasers demonstrates the potential in this respect, since 36 ps pulses with a high average power of 260 mW were obtained, leading to 150 fs pulses with peak powers exceeding 1 kW by means of pulse compression [5.57].

References

[5.1]   G.P. Agrawal, N.K. Dutta: Long Wavelength Semiconductor Lasers (Chapman and Hall, New York 1986)

[5.2]   C. Weisbuch, B. Vinter (ed.): *Quantum Semiconductor Structures* (Academic Press, San Diego 1991)

[5.3]   A. Yariv: *Optical Electronics*, 3rd edn. (Wiley, New York 1989)

[5.4]   P.J.A. This, L.F. Tiemeijer, J.J.M. Binsma, T.V. Dongen: IEEE J. Quant. Electr. **30**, 2 (1994); E. Yablonovitch, E.O. Kane: J. Lightwave Technology **6**, 8 (1988)

[5.5]   J.L. Jewel, A. Scherer, S.L. McCall, Y.H. Lee, S. Walker, J.P. Harbison, L.T. Florez: Electron. Lett. **25**, 1124 (1989); A. Scherer, J.L. Jewel, Y.H. Lee, J.P. Harbison, L.T. Florez: Appl. Phys. Lett. **55**, 2724 (1989)

[5.6]   J.-M. Lourtioz, L. Chusseau, N. Stelmakh: J. Phys. III (France) **2**, 1673 (1992)

[5.7]   P.L. Liu, Chinlon Lin, T.C. Damen, D.J. Eilenberger: *Picosecond Phenomena II*, ed. R.M. Hochstrasser, W. Kaiser, C.V. Shank, Springer Ser. in Chem. Phys., Vol. 14, (Springer, Berlin, Heidelberg 1980) p. 30

[5.8]   G.J. Lasher: Solid State Electr. **7**, 707 (1964)

[5.9]   V.D. Kurnosov, V.I. Magalyas, A.A. Pleshkov, L.A. Rivlin, V.G. Trukhan, V.V. Tsvetkov: JETP Lett. **4**, 303 (1966)

[5.10]   J.P. Van der Ziel, W.T. Tsang, R.A. Logan, R.M. Mikulyak, W.M. Augustyniak: Appl. Phys. Lett. **39**, 525 (1981)

[5.11]   H. Yokoyama, H. Ito, H. Inaba: Appl. Phys. Lett. **40**, 105 (1982)

[5.12]   J.E. Bowers, P.A. Morton, A. Mar, S.W. Corzine: IEEE J. Quant. Electr. **QE 25**, 1426 (1989)

[5.13]   K.Y. Lau, A. Yariv: IEEE J. Quant. Electr. **QE 21**, 121 (1985)

[5.14]   J. Aspin, J.E. Caroll, R.G. Plumb: Appl. Phys. Lett. **39**, 11 (1981)

[5.15] P.L. Liu, Chinlon Lin, T.C. Damen, D.J. Eilenberger: in *Picosecond Phenomena II*, ed. R. M. Hochstrasser, W. Kaiser, C.V. Shank. Springer Ser. Chem. Phys., Vol. 14 (Springer, Berlin, Heidelberg 1980) p. 30

[5.16] Y. Luo, R. Takahashi, Y. Nakano, K. Tada, H. Hosomatsu: Appl. Phys. Lett. **59**, 1 (1991)

[5.17] M. Cavelier, N. Stelmakh, J.M. Xie, L. Chusseau, J.M. Lourtioz, C. Kazmierski, N. Bouadma: Electr. Lett. **28**, 3 (1992)

[5.18] K. Uomi, N. Chinone, T. Ohtoshi, T. Kajimura: Japan J. Appl. Phys. **24**, L539 (1985)

[5.19] K. Uomi, T. Mishima, N. Chinone: Appl. Phys. Lett. **51**, 78 (1987)

[5.20] K. Uomi, T. Mishima, N. Chinone: Japan J. Appl. Phys. **29**, 88 (1990)

[5.21] D. Offsey, W.J. Schaff, L.F. Lester, L.F. Eastman, S.K. McKernan: IEEE J. Quant. Electr. **QE 27**, 1455 (1991)

[5.22] M. Shirasaki, I. Yokota, T. Touge: Electr. Lett. **26**, 33 (1990)

[5.23] T.L. Koch, T.J. Bridges, E.G. Burkhardt, P.J. Corvini, L.A. Coldren, R.A. Linke, W.T. Tsang: Appl. Phys. Lett. **47**, 12 (1985)

[5.24] R.A. Linke: IEEE J. Quant. Electr. **QE 21**, 593 (1985)

[5.25] C.M. Gee, G.D. Thurmond, H.W. Yen: Appl. Phys. Lett. **43**, 998 (1983)

[5.26] D.Z. Tsang, J.N. Walpole: IEEE J. Quant. Electr. **QE 19**, 145 (1983)

[5.27] H. Ito, N. Onodera, K. Gen-Ei, H. Inaba: Electr. Lett. **17**, 15 (1981)

[5.28] K.A. Williams, D. Burns, I.H. White, W. Sibbett, M.J. Fice: IEEE Photonics Technology Lett. **5**, 867 (1993)

[5.29] P. Gavrilovic, N. Stelmakh, J.H. Zarrabi, D.M. Beyea: Electr. Lett. **31**, 1154 (1995)

[5.30] M. Kuznetsov: IEEE J. Quant. Electr. **QE 21**, 587 (1985)

[5.31] Zh.I. Alferov, A.B. Zhularev, E.L. Portnoi, N.M. Stelmakh: Sov. Tech. Phys. Lett. **12**, 452 (1986)

[5.32] L.V. Keldysh: Sov. Phys. JETP **7**, 788 (1958); W. Franz: Z. Naturforsch. **13a**, 484 (1958); B.O. Seraphin, R.B. Hess, N. Bottka: Bull. Am. Phys. Soc., Ser. II **9**, 714 (1964)

[5.33] S. Schmitt-Rink, D.S. Chemla, D.A.B. Miller: Adv. Phys. **38**, 89 (1988); D.A.B. Miller, D.S. Chemla, S. Schmitt-Rink: Phys. Rev. B **33**, 6976 (1986)

[5.34] I. Suemene, T. Takeoka, M. Yamanishi, Y. Lee: IEEE J. Quant. Elect. **QE 22**, 1900 (1986)

[5.35] S.W. Corzine, J.E. Bowers, G. Przybylek, U. Koren, B.I. Miller, C.E. Soccholich: Appl. Phys. Lett. **52**, 348 (1987)

[5.36] P.J. Delfyett, C.H. Lee, L.T. Florez, N.G. Stoffel, T.J. Gmitter, N.C. Andreakis, G.A. Alphonse, J.C. Conolly: Opt. Lett. **15**, 1371 (1990)

[5.37] R.S. Tucker, S.K. Korotky, G. Eisenstein, U. Koren, L.W. Stulz, J.J. Veselka: Electr. Lett. **21**, 239 (1985)

[5.38] H.A. Haus: J. Appl. Phys. **51**, 4042 (1980)

[5.39] H.A. Haus: Japan. J. Appl. Phys. **20**, 1007 (1981)

[5.40] P.B. Hansen, G. Raybon, U. Koren, B.I. Miller, M.G. Young, M. Chien, C.A. Burrus, R.C. Alferness: IEEE Photonics Technology Lett. **4**, 1041 (1992)

[5.41] L.R. Brovelli, H. Jäckel: Electr. Lett. **27**, 1104 (1991)

[5.42] W.B. Jiang, R. Mirin, J.E. Bowers: Appl. Phys. Lett. **60**, 677 (1992)

[5.43] H.A. Haus: IEEE J. Quant. Electr. **QE 11**, 736 (1975)

[5.44] H.A. Haus: J. Appl. Phys. **46**, 3049 (1975)

[5.45] J.P. Van der Ziel: J. Appl. Phys. **52**, 4435 (1981)

[5.46] E.L. Portnoy, N.M. Stel'makh, A.V. Chelnokov: Sov. Tech. Phys. Lett. **15**, 432 (1989)

[5.47] A.G. Deryagin, D.V. Kuksenkov, V.I. Kuchinskii, E.L. Portnoi, I.Yu. Khrushchev: Electr. Lett. **30**, 309 (1994)

[5.48] Y. Silberberg, P.W. Smith, D.J. Eilenberger, D.A.B. Miller, A.C. Gossard, W. Woiegman: Opt. Lett. **9**, 507 (1984)

[5.49] Y. Silberberg, P.W. Smith: IEEE J. Quant. Electr. **QE 22**, 759 (1986)

[5.50] Y.K. Chen, M.C. Wu, T. Tanbun-Ek, R.A. Logan, M.A. Chin: Appl. Phys. Lett. **58**, 12 (1991)

[5.51] M.C. Wu, Y.K. Chen, T. Tanbun-Ek, R.A. Logan: in *Ultrafast Phenomena VIII*, ed. J.-L. Martin, A. Migus, G.A. Mourou, A.H. Zewail, Springer Ser. Chem. Phys., Vol. 55, (Springer, Berlin, Heidelberg 1993) p. 211

[5.52] S. Arahira, Y. Matsui, Y. Ogawa: IEEE J. Quant. Electr. **QE 32**, 1211 (1996)

[5.53] J. Nagle, C. Weisbuch: in *Science and Engineering of 1 and 0 Dimensional Semiconductor Systems*, ed. C.M. Sotomayor-Torres, S.P. Beaumont (Plenum, New York 1990)

[5.54] N. Kirstaedter, N.N. Ledentsov, M. Grundmann, D. Bimberg, V.M. Ustinov, S.S. Ruvimov, M.V. Maximov, P.S. Kop'ev, Zh.I. Alferov, U. Richter, P. Werner, U. Gösele, J. Heydenreich: Electr. Lett. **30**, 1416 (1994)

[5.55] J.-P. Zhang, K. Peterman: IEEE J. Quant. Electr. **30**, 1529 (1994)

[5.56] K.L. Lear, A. Mar, K.D. Choquette, S.P. Kilcoyne, R.P. Schneider Jr., K.M. Geib: Electr. Lett. **32**, 457 (1996)

[5.57] W.-H. Xiang, S.R. Friberg, K. Wanatabe, S. Machida, Y. Sakai, H. Iwamura, Y. Yamamoto: Appl. Phys. Lett. **59**, 2076 (1991); W.H. Xiang, S.R. Friberg, K. Wanatabe, S. Machida, W. Jiang, H. Iwamura, Y. Yamamoto: Opt. Lett. **16**, 1394 (1991)

6

How to Manipulate and Change the Characteristics of Laser Pulses

F. Salin

With 15 Figures

6.1 Introduction

Up to this point in this book, we have dealt successively with the basics of producing ultrashort laser pulses and the main types of laser sources that make it possible to generate these pulses. However, for a given laser system the pulse duration, wavelength and energy are, in practice, fixed. But specific applications may require one to change the pulse duration, or to get more energy or even generate pulse replicas at different wavelengths. This chapter will present various ways to compress or stretch femtosecond optical pulses, amplify them or change their wavelength. This description is not aimed at an exhaustive description of all the various techniques or technologies used in these fields, but more as an introduction to these techniques based on particular examples.

6.2 Pulse Compression

We have seen in previous chapters that the propagation of an intense pulse in a nonlinear medium is affected mainly by two phenomena: group velocity dispersion (GVD) and self-phase-modulation (SPM). GVD introduces a spectral dephasing and does not affect the spectral amplitude of the pulse. Conversely, SPM induces a temporal dephasing and hence does not change the temporal profile. Nevertheless, SPM creates new frequencies, which are redder than the pulse central frequency in the leading edge of the pulse and bluer in the trailing edge. This larger spectrum provides an opportunity to reduce the pulse duration but does not compress the pulse by itself. In order to get a shorter pulse, SPM must be followed by a dispersive stage in which the new frequencies are rephased (i.e. the redder are slowed down compared to the bluer). The dispersion introduced by this stage must have a spectral variation which follows as much as possible the dispersion introduced by the SPM.

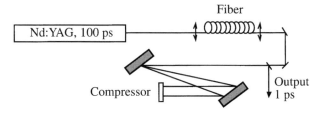

Fig. 6.1. Typical setup for pulse compression

In practice, femtosecond pulses can be compressed using two stages, one for SPM and the other using GVD to compensate for the spectral dephasing. One must remember that these two stages are absolutely necessary in order to reduce the duration of an initially transform-limited pulse. If only SPM is used, the pulse spectrum will broaden, but with no change or sometimes an increase in the pulse duration. A broad spectrum is not automatically a sign of short pulses, as the light bulb shows. It has a very broad spectrum but is a very poor short-pulse source. On the other hand, GVD can rephase an initially chirped pulse but will never be able to reduce its duration below the limit imposed by the Fourier relation. In order to get an uniform broadening of the pulse spectrum over the whole beam, one must use a nonlinear waveguide. Without this precaution, the broadening would be much smaller on the edge of the beam, where the intensity is lower than at the beam center. The most common optical waveguide is a single-mode optical fiber [6.1]. The very high peak power achieved with ultrashort pulses makes a normal fiber a very nonlinear medium. Furthermore, the nonlinear effect can be accumulated over a long distance with almost no loss. A typical compression experiment is shown in Fig. 6.1 [6.2].

An actively mode-locked Nd:YAG laser is used as a source of transform-limited 100 ps pulses around 1064 nm. This laser produces a typical average power of 1–10 W at a 100 MHz repetition rate. Its TEM_{00} beam can be efficiently coupled into a single-mode fiber. During its propagation in the fiber the pulse suffers the effect of both SPM and GVD. For a typical length of 1 km, the pulse can accumulate a nonlinear dephasing of over 200π, which implies a broadening of its spectrum by a factor 100 [6.3]. Owing to the combined effects of SPM and GVD in the fiber, the pulse becomes strongly but almost linearly chirped. After the fiber, the pulse duration is slightly larger than the initial 100 ps. The beam is then sent through a dispersive delay line formed by two diffraction gratings [6.4]. The first grating spatially disperses the different wavelengths and the second grating recollimates the beam.

As seen in Fig. 6.2, the optical path through the grating pair is longer for the long wavelengths than for the short ones. This system introduces a negative group velocity dispersion, which can be used to compensate for the positive chirp of the self-phase-modulated pulse. A pair of mirrors acting as a retroreflector is used to reflect the beam back into the grating pair at a

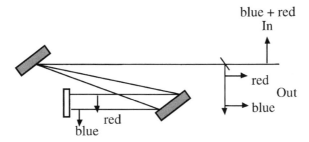

Fig. 6.2. Group velocity dispersion of a pair of gratings

different height. This second pass cancels the spatial chirp on the beam while doubling the dispersion of the system. When a transform-limited pulse is sent through such a compressor, the output pulse is negatively chirped (the front of the pulse is bluer than the tail) and its duration is increased. Conversely, when the input pulse is positively chirped, the grating pair can cancel part of this chirp and the output pulse becomes shorter than the input one. The maximum compression occurs when the different wavelengths are brought back together, which corresponds to the minimum chirp or the flattest spectral phase. Since the negative chirp introduced by the compressor may not have the very same spectral variations as the positive chirp carried by the input pulse, the compensation may not be perfect and the residual phase may not be identically zero.

Figure 6.3 shows a typical example of a calculation for a Gaussian pulse exhibiting a nonlinear chirp which can be fitted by a third order polynomial (Fig. 6.3a, c). This pulse was sent through a dispersive delay line (a pair of gratings, for instance), for which the spectral phase was essentially parabolic. The final pulse is shown in (Fig. 6.3b) (phase) and (Fig. 6.3d) (amplitude). Several observations can be made on this figure. First, the spectrum amplitude is not affected by the dispersive delay line; only the spectral phase changes. As the pulse goes through the delay line the curvature of the spectral phase is canceled, leaving a residual phase with a complicated shape. At the same time the temporal profile of the pulse is compressed (i.e. the FWHM is reduced) but rebounds appear on the pulse profile. The compression has effectively decreased the pulse duration, but the imperfect match of the phase function leaves a pulse of poorer quality [6.5, 6.6]. This is exactly what happens experimentally. Figure 6.4 shows the autocorrelation trace (see Chap. 7 for more information on the measurement of pulse durations) of a 100 ps pulse which has been compressed, in a setup similar to that of Fig. 6.1, to 1 ps [6.2, 6.7–10].

The pulse duration has been reduced by a factor of 100 but the compression has introduced wings on the pulse profile. The same kind of setup has been used to compress 100 fs pulses to 10 fs [6.11]. In that case, the optical fiber was replaced by a hollow waveguide, providing over 500 μJ energy per pulse.

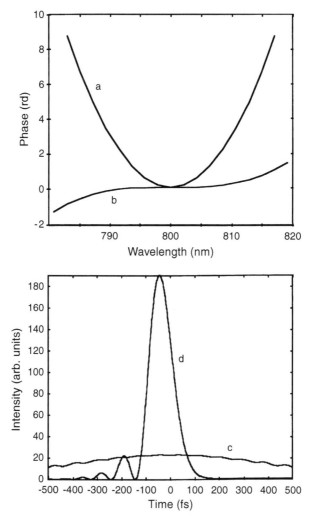

Fig. 6.3. Phase (**a**) and amplitude (**c**) of a chirped pulse. Same pulse after passing through a quadratic dispersive line: residual phase (**b**) and compressed pulse (**d**)

6.3 Amplification

Amplification of short pulses is similar to that of any optical radiation. It is based on the population inversion introduced in Chap. 1. The main difference is that hereafter we will focus only on the amplification of pulses. The parameters used are slightly different because all the calculations are done for a single pulse regardless of the repetition rate, average power, and so on. The parameters of practical interest, then, are the energy, the fluence (energy per unit area), and the total amplification through the amplifier. The relations

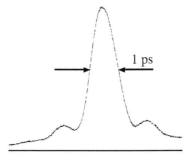

Fig. 6.4. Typical autocorrelation trace of a compressed pulse. The bottom line represents the real zero intensity level

driving the physics of an amplifier are similar to those of a cw laser, although they are expressed in different units.

A laser medium is pumped by an external source of radiation, which in most cases is a laser. If the excited state lifetime of the amplifier medium is assumed to be longer than any other time involved in the process, one can easily calculate the small signal gain of the amplifier [6.12]:

$$g_0 = \frac{J_{\text{sto}}}{J_{\text{sat}}}, \tag{6.1}$$

where J_{sto} is the pump fluence (energy density in J/cm^2) stored in the medium and J_{sat} is an intrinsic parameter called saturation fluence (J/cm^2). J_{sat} is given by the inverse of the gain cross section σ_e:

$$J_{\text{sat}} = \frac{h\upsilon}{\sigma_e},$$

where h is the Plank constant and υ the frequency of the transition. J_{sat} is related to the saturation intensity by:

$$I_{\text{sat}} = \frac{J_{\text{sat}}}{\tau_f},$$

where τ_f is the fluorescence lifetime of the laser medium.

Note that the physical length of the amplifier is meaningless; only the amount of energy stored in the amplification medium is of interest. The fluence stored in the laser medium J_{sto} can be deduced from experimental values:

$$J_{\text{sto}} = \frac{E_p \alpha}{S} \cdot \frac{\lambda_p}{\lambda_L}, \tag{6.2}$$

where E_p is the pump energy incident on the amplifier, α is the fraction of the pump radiation absorbed in the crystal, and S is the pump beam cross section. λ_{P} and λ_{L} are, respectively, the pump and laser wavelengths. Note that the stored fluence differs from the absorbed fluence E_p/S by a factor λ_p/λ_L called the quantum defect. Because one pump photon can only give rise to one excited atom, which can in turn only give one signal photon, the maximum efficiency (extracted energy / pump energy) is given by the ration of the pump to signal wavelengths: $\eta_{max} = (\lambda_p/\lambda_s)$. A pulse with a small energy E_{in} will be amplified by passing through the amplifier and its output energy will be:

$$E_{out} = E_{in}\exp(g_0). \tag{6.3}$$

This very simple formula is only valid for small input fluences. As soon as the input fluence reaches a level comparable to that of the saturation fluence J_{sat} of the amplifier, one must use a more complicated expression derived by Frantz and Nodvick [6.13]:

$$J_{out} = E_{out}/S = J_{sat}\text{Log}[g_0(\exp(J_{in}/J_{sat}) - 1 + 1], \tag{6.4}$$

where J_{out} is the output fluence of the amplified pulse. It is recommended to spend some time on this formula, which is the basis of pulse amplification. The role of the parameter J_{sat} is obvious and will dictate the precise technology that one can use with a particular laser medium. Examples of typical saturation fluences are given in Table 6.1.

Table 6.1. Saturation fluences of typical amplifier media used in short pulses amplification

Amplifier medium	J_{sat}	
Dyes	$\sim 1\,\text{mJ/cm}^2$	
Excimers	$\sim 1\,\text{mJ/cm}^2$	
Nd:YAG	$0.5\,\text{J/cm}^2$	
Ti:Al$_2$O$_3$	$1\,\text{J/cm}^2$	At 800 nm
Cr:LiSAF	$5\,\text{J/cm}^2$	At 830 nm
Nd:glass	$5\,\text{J/cm}^2$	
Yb:YAG	$5\,\text{J/cm}^2$	
Yb:glass	$100\,\text{J/cm}^2$	

One can see in this table that two groups can be found: on the one hand liquids and gases with low J_{sat}, and on the other hand solids with large J_{sat}. This distinction leads naturally to two different types of amplifier.

Dye amplifiers were the first to be used for femtosecond pulses [6.14–16]. They have now been surpassed by solid-state materials and we will not describe their technology in detail. The typical structure of a dye amplifier is a series of dye cells, in general four or five, which are pumped transversely by a nanosecond blue or green laser (excimer, second harmonic of Nd:YAG or copper-vapor laser). The optical gain per stage can be higher than a thousand, with the gain after four stages reaching 10^6. The major drawback of these amplifiers is the high optical gain. It is very difficult to avoid photons spontaneously emitted by fluorescence in the first stage being amplified in the chain. When nanosecond pulses are used to pump the amplifier, the output of the amplifier consists of a short amplified pulse on top of a very broad pedestal. This nanosecond pedestal, the so-called ASE, explained in Chaps. 3 and 4, corresponds to noise photons following the same optical path as the femtosecond pulse and hence undergoing the same amplification. Several techniques are used to reduce the amount of ASE, such as saturable absorbers or spatial filtering, but the ASE generally amounts to 5–10 % of the total output energy. The efficiency of a dye amplifier pumped by the second harmonic of a Nd:YAG laser is low, typically 0.1 to 0.3 %. This kind of amplifier has been adapted to several pump lasers with various characteristics. High-repetition-rate, low-energy pulses have been obtained using copper-vapor lasers [6.16]. Picosecond lasers running at 1 kHz have been used to produce high-quality, low-ASE pulses at the expense of considerable complication in the pump system [6.17, 18]. Direct amplification in excimer amplifiers has been the subject of several studies [6.19]. Excimers, owing to their low saturation fluence, have the same advantages (high optical gain) and drawbacks (ASE) as dyes, with the complication of working in the UV. They are still used in some high-energy systems, since they can be relatively easily scaled up to large amplifier diameters and hence large energies. However, except for very particular applications, dye amplifiers have now been abandoned and are replaced by their solid-state counterpart.

As seen in Table 6.1, solid state materials have in general a high saturation fluence, leading to a much smaller optical gain per pass in the amplifier than in dyes or excimers. In order to extract the energy stored in the medium, one must use multiple passes in the amplifier. Two techniques have been developed, namely multipass [6.20] and regenerative amplification [6.21].

Multipass amplification is based on a bow-tie type of amplifier, in which the different passes in the amplifier are separated geometrically (Fig. 6.5). The number of passes is limited by the geometrical complexity of the design and by the increasing difficulty of focusing all the passes on a single spot of the crystal. The typical number of passes is four to eight. For higher gains, several multipass amplifiers can be cascaded [6.22] or alternative schemes can be used [6.23]. This technique is simple and cheap but requires long and tedious adjustments, and the crystal is used close to the damage threshold in order to keep the gain high enough to be compatible with the small number of passes. It is nevertheless the only technique usable for producing energies above 50 mJ.

Regenerative amplification (Fig. 6.6) is based on a pulse trapped in a laser resonator. Regardless of the gain per pass, the pulse is kept in the resonator until it has extracted all the energy stored in the amplification medium.

Trapping and dumping the pulse is done using a Pockels cell and a broad-band polarizer. The Pockels cell is initially set to be equivalent to a quarter-wave plate. When a pulse is in the resonator, the voltage on the Pockels-cell crystal is switched so that it becomes equivalent to a half-wave plate. The

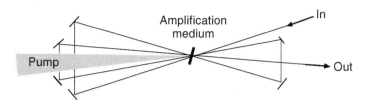

Fig. 6.5. Setup of multipass amplifier

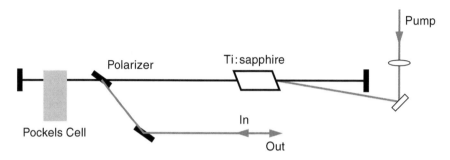

Fig. 6.6. Setup of regenerative amplifier

pulse stays in the cavity until it reaches saturation, and a second voltage step is applied to extract the pulse. From a physical point of view, the only difference between multipass and regenerative amplification is the way input and output beams are distinguished: in a multipass amplifier this is done geometrically and in a regenerative amplifier polarization is used. The propagation in a regenerative or multipass amplifier can be easily simulated using a Frantz–Nodvick model [6.13]. Equation (6.4) is applied at each pass, taking into account the depletion of the excited-state population due the gain process. One typically obtains the picture shown in Fig. 6.7. From this figure it is obvious that the pulse should be extracted from the amplifier after 20 passes, in this particular example. These calculations also show that quite good efficiencies can be expected from these amplifiers.

The amplifiers presented here can be used with any kind of pulses and any kind of laser media, and are not specific to ultrashort pulses. When femtosecond pulses are to be amplified, new problems arise. First of all, the amplifier bandwidth must be broad enough to support the pulse spectrum. As a matter of fact, the effective gain bandwidth of an amplifier is much narrower than its fluorescence spectrum. This is due to the very high gain seen by the pulse and can be explained by looking at the amplifier as a spectral filter. At each pass, wavelengths at the peak of the gain are amplified more than those on the edge of the gain curve. Pass after pass, this leads to a relative reduction of the wings of the pulse spectrum compared to its central part. This phenomenon is called "gain narrowing" and plays an important role in the amplification

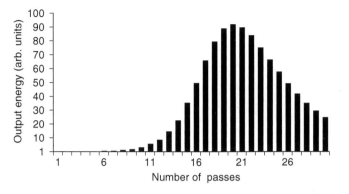

Fig. 6.7. Theoretical evolution of the pulse energy versus the number of passes in a regenerative amplifier

of short pulses [6.23]. Among the various materials which have been used in lasers, only a few can potentially be used in femtosecond amplifiers:

- Ti:sapphire (Ti:Al$_2$O$_3$)   650–1100 nm
- alexandrite (Cr:Be$_2$O$_3$)   700–820 nm
- colquirites (Cr:LiSAF, Cr:LiCAF, etc.)   800–1000 nm
- Fosterite   1250–1300 nm
- Yb doped materials   1030–1080 nm
- glasses (Nd:glass)   1040–1070 nm.

A second problem arises from the high fluence used in these amplifiers, which for short pulses leads to intensities above the damage threshold of the amplifier. The technique now widely used to avoid this problem is called chirped pulse amplification (CPA) [6.24]. In this technique a pulse is first stretched by a large factor (typically 10 000) in order to reduce its peak power. The pulse can then be safely amplified, and is finally recompressed to its initial duration. The stretcher uses a pair of gratings separated by an afocal system of magnification −1 [6.25]. This setup works in a similar way to the compressor described at the beginning of this chapter. The main difference is that the afocal system introduces an image inversion which makes the red optical path shorter than the blue one (Fig. 6.8). It can be shown that if the optical system included in the stretcher is free of chromatic and geometrical aberrations, the dispersion it introduces can be exactly compensated by a pair of parallel gratings used at the same incidence angle. After passing through the stretcher, the pulse duration is increased by up to 400 ps and can be amplified. Depending on the pump laser, a Ti:sapphire amplifier can work at repetition rates from 10 Hz [6.26] to 1 kHz [6.27] and is now commercially available at 250 kHz. Similar techniques can be used with other wavelengths and laser media. A typical amplifier setup is shown in Fig. 6.9. The gain medium is a Ti:sapphire crystal pumped by the second harmonic of a Nd:YAG laser

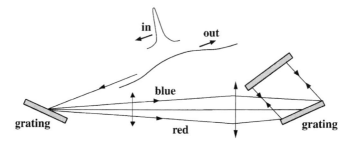

Fig. 6.8. Typical stretcher setup. The lenses located between the two gratings can be replaced by mirrors to reduce the chromatic aberrations

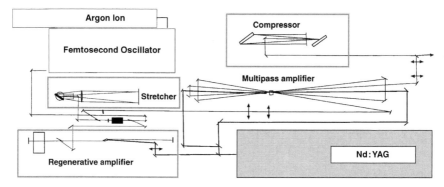

Fig. 6.9. Typical setup of a CPA amplifier, including a stretcher, a regenerative amplifier, a multipass amplifier and a compressor

running at 10 Hz. The very same design can be used at 1000 Hz by replacing the Nd:YAG laser by a Q-switched Nd:YLF laser. Higher repetition rates would require a continuously pumped laser.

Once trapped in the cavity, the pulse is amplified at each pass in the crystal, with a typical gain per pass of 2. After 15–20 round trips the pulse saturates the gain and reaches its maximum energy. It is extracted from the amplifier, and sent into a multipass power amplifier and finally into the compressor. It recovers its initial duration there, and pulses as short as 30 fs with energies in the joule range can be obtained [6.24]. The peak power at that point is over 10 TW, which is more than the total electrical power produced on earth (but of course only during 30 fs). The efficiency of this kind of amplifier can be in excess of 10 %, over 100 times better than a dye amplifier. At 1 kHz, energies up to 3 mJ are obtained. When focused, these amplified pulses lead to intensities up to 10^{19} W/cm^2.

The major drawback of chirped pulse amplification is the difficulty in recovering the initial pulse duration and pulse quality. Besides the dispersion of the stretcher, the compressor must compensate for the dispersion introduced by the amplifier itself. This means that the distance between the gratings must be set to a larger value in the compressor than in the stretcher. This difference in length makes it possible to cancel the overall second-order dispersion and hence get relatively short pulses, but also brings in higher-order dispersion terms which translate into wings and prepulses. These prepulses make the system unsuitable for high-intensity applications, and several designs have been introduced to keep the final pulse quality as good as possible [6.29–31].

6.4 Wavelength Tunability

Most of the femtosecond systems built nowadays are based on Ti:sapphire and can only be tuned in a small spectral range around 800 nm. Few applications need only a single wavelength in this spectral region, and pulses tunable from the UV to the IR are highly desirable.

The easiest way to get broadband tunability is to focus a high-energy ($E > 1 \, \mu J$) pulse in a transparent material such as glass, water or sapphire. When the intensity reaches $10^{13} \, W/cm^2$, the beam undergoes self-focusing and self-phase-modulation. The beam breaks up into small filaments, while the spectrum of each filament increases until the radiation becomes perfectly white [6.31]. A continuous spectrum ranging from 200 nm to 1500 nm can then be produced. The spectral density decreases rapidly when the wavelength is tuned away from the fundamental wavelength (800 nm). Moreover, the initial pulse energy of the input pulse is spread over the whole bandwidth, leading to small intensities in wavelength ranges far from the fundamental wavelength. Although continuum generation is widely used and very versatile for probe pulses (see Chap. 8), it cannot in general provide a high enough energy per pulse in a reasonable bandwidth to be used to pump or trigger the phenomenon under study. In order to get higher energies while keeping or even extending the tuning range, one must rely on parametric interactions in nonlinear crystals.

Parametric devices are based on the use of nonlinear quadratic crystals having a large $\chi^{(2)}$, such as BBO (β-barium borate), LBO (lithium triborate) or KTP (KTiOPO$_4$). In such a medium two incident fields at frequencies ω_1 and ω_2 can produce a new radiation at $\omega_3 = \omega_1 + \omega_2$ or $\omega_3 = \omega_1 - \omega_2$ (see Chap. 2). This is also true for femtosecond pulses, and all the usual second-order nonlinear phenomena such as second-harmonic generation, frequency mixing and parametric amplification can be obtained. For the mathematical background to these effects, we recommend [6.33–36]. It is enough to know that three-wave mixing can exist only if the following two conditions are fulfilled:

$$\omega_1 + \omega_2 = \omega_3, \tag{6.5}$$
$$\boldsymbol{k}_1 + \boldsymbol{k}_2 = \boldsymbol{k}_3. \tag{6.6}$$

Here, the $\boldsymbol{k}_j$ are the wave vectors of the three waves. When the three waves propagate colinearly, the second relation, known as the phase-matching condition, can be rewritten as

$$\frac{n_1}{\lambda_1} + \frac{n_2}{\lambda_2} = \frac{n_3}{\lambda_3}. \tag{6.7}$$

In normal materials, where the index varies as $1/\lambda$, this relation can never be fulfilled; one must use birefringent materials, in which the waves can propagate with different polarizations. By adjusting the angle between the optical axis of the birefringent crystal and the direction of propagation, one can change the wavelength for which (6.7) is fulfilled and hence tune the parametric interaction. General formulae giving the phase-matching angle as a function of the type of interaction and the crystal can be found in [6.35]. When ultrashort pulses are used for parametric interactions, one must pay attention to higher-order phenomena. First, the phase-matching condition must be fulfilled over the whole spectrum of the input pulses. Mathematically this translates into so-called "group velocity dispersion matching", which means that the first derivatives of the wave vector with respect to frequency for the different waves must be as close as possible. In practice, this condition means that only thin crystals can be used in three-wave interactions with short pulses. A final condition is related to the group velocity dispersion of each wave. GVD tends to chirp the pulses and hence reduces their peak power. This eventually reduces the efficiency of the process and leads also to the use of thin crystals. In a first approximation, the efficiency of a parametric process is proportional to the product of the intensity and the interaction length. With short pulses the use of thin crystals can be easily compensated by the very high peak power obtained, without damage to the crystals.

6.4.1 Second- and Third-Harmonic Generation

UV spectral range is interesting for several applications. The easiest way to produce UV femtosecond pulses is to start with a near-infrared laser source and use second- or third-harmonic generation. The ideal crystal for this application must have a small group velocity difference between the fundamental and harmonic waves and a large nonlinearity, and must exhibit no absorption at both the fundamental and the harmonic wavelengths. The crystal should also be transparent at twice the harmonic frequency to avoid two-photon absorption. A typical choice is BBO because of its high nonlinearity, which makes the use of very thin crystals possible [6.36]. The crystal thickness should be around 0.5 mm for 80–100 fs pulses and below 200 μm for 30 fs pulses. Even with such thin crystals, 40–50 % conversion efficiency from 800 nm to 400 nm is typically observed. Further tuning in the UV is obtained by mixing the 400 nm and 800 nm pulses to produce radiation at the third-harmonic wavelength (266 nm). However, the group velocity difference between the fundamental and second-harmonic pulses in the first crystal (about 200 fs/mm) shifts these

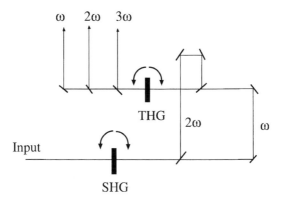

Fig. 6.10. Experimental setup for second- and third-harmonic generation (SHG, second-harmonic generator; THG, third-harmonic generator)

two pulses in time and must be compensated prior to any mixing. A typical setup for second- and third-harmonic generation is shown in Fig. 6.10.

Using a second BBO crystal of about the same thickness as the previous one, one can typically obtain a tripling efficiency $I(3\omega)/I(\omega)$ around 10 %. Owing to the large dispersion between the fundamental and third-harmonic group velocities, the third-harmonic pulses are longer than the fundamental ones. Shorter UV pulses can be obtained at the expense of efficiency by using thinner crystals. It is important to note that the peak power of the pulses produced at 266 nm exceeds 1 GW, which leads to unexpected effects. For instance, most of the UV materials exhibit two-photon absorption at 266 nm and, although the absorption cross-section is low, the very high peak power leads to very important absorption. One must be careful when using powerful UV pulses and check for any parasitic effect due to the short duration or the high peak power of these pulses. Radiation around 200 nm can subsequently be obtained by mixing the fundamental and the third-harmonic radiations with efficiencies around 1 % [6.38].

6.4.2 Optical Parametric Generators (OPGs) and Amplifiers (OPAs)

The same phenomenon as that which produces harmonic generation can be also used to produce tunable infrared pulses: in second-harmonic generation, two infrared photons add to produce a visible photon; in a parametric generator, one visible photon splits to produce two infrared photons. The main difference is that, although the total energy must be conserved, the two infrared photons can have different frequencies. Equations (6.5) and (6.6) are still valid, and the wavelengths of the photon pair are fixed by the same phase-matching condition as before. Owing to this phase matching condition, only a narrow range of frequencies can be produced for a given crystal orientation.

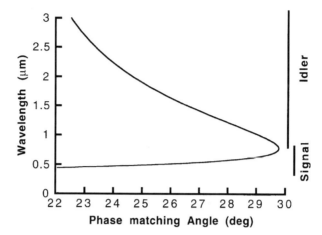

Fig. 6.11. Phase-matching curve for parametric generation in type I BBO with a pump at 400 nm

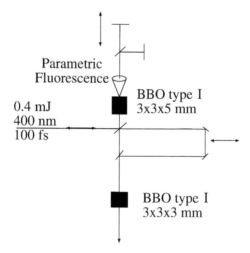

Fig. 6.12. Experimental setup for femtosecond parametric generation

A much broader range can be covered step by step, by rotating the crystal and hence changing the pair of wavelengths for which phase-matching exists. A typical phase-matching curve for an optical parametric generator (OPG) pumped at 400 nm is shown in Fig. 6.11.

One can see that each angle corresponds to a pair of wavelengths. The short-wavelength radiation is called the signal and the long wavelength radiation the idler. A particular situation is obtained when the signal and idler frequencies merge. This situation is called degeneracy and corresponds to the production of two fundamental photons by one second-harmonic photon. A typical setup for optical parametric generation is shown in Fig. 6.12 [6.39–42].

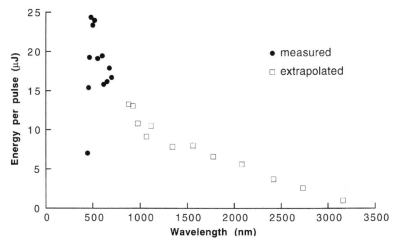

Fig. 6.13. Tuning curve of a two-stage parametric generator pumped by 100 fs, 400 nm pulses at a 1 kHz repetition rate

An input beam at 400 nm is incident on a nonlinear crystal cut at the correct angle to ensure phase matching. The small-signal gain of a femtosecond parametric amplifier can be as high as 10^6 for a single pass in the crystal. A single input photon gives 10^6 photons after one pass and about 10^{12} photons after two passes. Starting with a photon energy of a few times 10^{-19} J, one obtains after two passes a typical energy of 1 μJ. This means that even the photons coming from the thermal agitation in the crystal are sufficient to produce a measurable output and that there is no need to inject an external signal into an OPG to produce tunable femtosecond pulses. Of course, only the wavelengths fulfilling the phase-matching condition can be amplified, and the output is not a white noise but a well-defined spectral band, the central wavelength of which is tuned by rotating the crystal. As seen in Fig. 6.12, the pump beam and the parametrically amplified light coming out of the crystal after the first pass are redirected toward the crystal for a second pass, after being retimed together. A typical OPG uses a 5 mm BBO crystal cut for type I interaction and produces a few microjoules tunable from 450 nm to 2500 nm. The pump intensity incident on the crystal is around 50 GW/cm². When more energy is needed, a second crystal is used as a power amplifier. The gain is much lower because of saturation but the output energies can reach 50 μJ per pulse at 1 kHz and 500 μJ at 10 Hz. A typical tuning curve obtained with a two-stage BBO OPG pumped at 1 kHz by the second harmonic of a Ti:sapphire-amplified system is shown in Fig. 6.13. Although this type of OPG, when pumped by UV pulses, produces light in the highly desirable visible range, it is difficult to optimize and to use with very short pulses owing to the large group velocity difference between the pump at 400 nm and

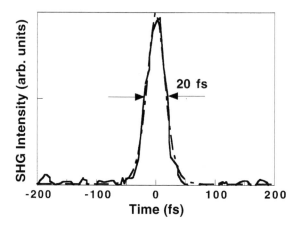

Fig. 6.14. Ultrashort 20 fs pulse obtained at 1.3 μm from a parametric amplifier pumped at 800 nm

the signal and idler waves. The pulses must be retimed after each crystal and the pulse duration is limited to 100 fs or so.

There is an alternative to the use of parametric fluorescence as the source of a broadband seed for the OPG, which is to use a single-filament white-light continuum [6.41]. The spectral part of the continuum fulfilling the phase-matching condition will be amplified, and gives rise to a high-power coherent radiation. The main advantage is that the diffraction-limited nature of the single-filament continuum can be transferred to the OPG output and lead to a high-power, diffraction-limited, tunable beam.

The same principle can be applied to an 800 nm pump wavelength [6.42]. In such a configuration, the OPG output is tunable from 1 μm to 3.2 μm for BBO crystals, and can be extended to 4 μm with KTP and up to 5 μm with $KNbO_3$. When 800 nm pulses are used as the pump, much shorter pulses can be obtained because of the very low group velocity difference between the pump and signal waves. Longer crystals can be used without sacrificing the performance. An example of 20 fs pulses obtained around 1.3 μm in a type I BBO OPG is shown in Fig. 6.14. The smaller group velocity difference also leads to higher efficiencies and energies over 100 μJ per pulse at 1 kHz are obtained [6.42]. These pulses can always be converted to the visible by mixing the OPG output with the residual 800 nm radiation in an extra BBO crystal. Working at 800 nm has the advantage of strongly reducing the effects of dispersion issue, and produces shorter pulses with overall efficiencies equivalent to a 400 nm pumped OPG.

Longer wavelengths can be produced by difference-frequency mixing in a crystal transparent in the infrared [6.43]. For instance, mixing a signal at 1.5 μm with an idler at 1.7 μm, produced simultaneously in a first OPG, would produce a radiation around 13 μm. Typical crystals used to produce radiation around 10 μm are $AgGaS_2$, $AgGaSe_2$ and $ZnGeP_2$. The efficiency is of course

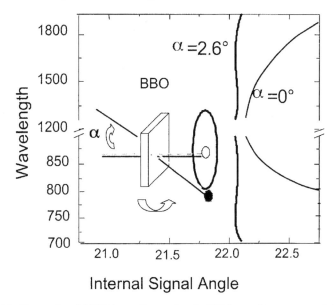

Fig. 6.15. Example of NOPA configuration in BBO nonlinear crystal (see text)

quite low since only one photon at $10 \, \mu m$ is produced per pair of photons around $1.6 \, \mu m$. With the much lower energy of a $10 \, \mu m$ photon, this translates into lower energies for the same number of photons. Furthermore, the large group velocity difference between $1.6 \, \mu m$ and $10 \, \mu m$ waves makes it very difficult to produce short pulses at $10 \, \mu m$.

As seen in Fig. 6.11, the angle dependence of phase matching can be used to tune the OPA output wavelength over a large band. Because of the group velocity dispersion between the pump, signal and idler waves, only a relatively narrow spectrum can be produced for a given angle. Using noncollinear phase matching can change this [6.50]. In a noncollinear optical parametric amplifier (NOPA) the pump and the signal are incident on a nonlinear crystal, with an angle between the two beams. This angle provides an extra parameter, which can be optimized to provide an almost vertical phase matching curve. A typical example obtained for BBO is given in Fig. 6.15. The pump wavelength is 532 nm and the angle between the pump and the signal beam is 2.6°. The OPA is then equivalent to an ultra-broadband amplifier. When seeded by a single-filament spectral continuum, amplified spectrum width over 300 nm can be obtained. The amplifier keeps the phase properties of the seeding pulse and, if the phase dispersion is controlled, such a broadband pulse can be recompressed to duration as short as 4.5 fs [6.51]. With spectra that broad, tunability is, of course, meaningless. Note that NOPAs are not restricted to 400-nm pumps [6.52] nor to BBO, but it requires the right couple of phase matching and incident angles, which cannot be found in all crystals.

6.5 Conclusion

In summary, we have seen that it is possible to reduce the durations of pulses, to increase their energy or to change their wavelength. These techniques are being widely used to push the limits of ultrashort pulse generation. New wavelengths in the vacuum ultraviolet can be obtained using high-harmonic generation in a gas jet. Theoretical studies already predict that these pulses can be compressed down to attosecond durations. On the other hand, peak powers in the petawatt range are now a reality and average powers in the 100 W range should be obtained soon. Although these manipulations open the door for many applications, one should always remember that any change in the pulse characteristics is accompanied by a decrease of the pulse quality. For instance, compression leads to wings in the pulse, and parametric generation increases pulse-amplitude instabilities.

6.6 Problems

1. Setup a fast photodiode near a Ti:sapphire crystal used as an amplifier (regenerative or multipass). Block the input to the amplifier (by blocking the input beam in a multipass amplifier or switching off the Pockels cell in a regenerative amplifier). Connect the photodiode to a fast ($> 100\,\mathrm{MHz}$) oscilloscope using a $50\,\Omega$ load. Record the fluorescence signal from the Ti:sapphire crystal by looking at the oscilloscope screen. Then unblock the input to the amplifier. Compare the signal with the previous one. Can you deduce from this measurement the extraction efficiency of the amplifier?
   Try reducing the number of passes (by blocking the beam after a variable number of passes in a multipass amplifier or by adjusting the timing in a regenerative amplifier) and check the effect on the extraction efficiency.
2. Setup an autocorrelator at the output of an amplification chain that includes a regenerative amplifier. Adjust the compressor to get the shortest pulses and record the position of the grating. Then reduce the number of round trips in the regenerative amplifier by one. Check the pulse duration. Why did the pulse duration increase? Try to adjust the compressor to get the shortest pulses. Measure the displacement between the previous position and this new one. From the grating-pair dispersion formula given in [6.1] or [6.4], try to compute the dispersion per round trip in the regenerative amplifier.
3. We assume that a Ti:sapphire regenerative amplifier is pumped by a 100-mJ 532-nm pulse focused on an area 3 mm in diameter. The crystal absorption is 90% at the pump wavelength. Starting from the fact that a regenerative amplifier will almost never reach the saturation level and keeps working in the small gain, calculate the number of round trips needed to amplify a 1-nJ pulse to 10-mJ. This amplifier is followed by a multipass

amplifier pumped by another 100-mJ pump pulse over the same 3-mm diameter. Compute the small signal gain of the amplifier and the saturated gain for a 10-mJ input at 800 nm. What is the extracted energy after 1 pass, 2 passes, and 3 passes? Do not forget that each pass of the pulse to be amplified decreases the excited population and reduces the gain.

References

[6.1]   G.P. Agrawal: *Nonlinear Fiber Optics* (Academic Press, London 1989)
[6.2]   A.M. Johnson, R.H. Stolen, W.M. Simpson: Appl. Phys. Lett. **44**, 729 (1984)
[6.3]   W.J. Tomlinson, R.H. Stolen, C.V. Shank: J. Opt. Soc. Am. B **1**, 139 (1984)
[6.4]   E.B. Treacy: Phys. Lett. **29A**, 539 (1968)
[6.5]   W.J. Tomlinson, W.H. Knox: J. Opt. Soc. Am. B **4**, 1404 (1987)
[6.6]   A.M. Weiner, J.P. Heritage, A.M. Weiner, W.J. Tomlinson: IEEE J. Quant. Electr. **QE 22**, 682 (1986)
[6.7]   J.D. Kafka, B.H. Kolner, T. Baer, D.M. Bloom: Opt. Lett. **10**, 176 (1985)
[6.8]   E.M. Dianov, A.Ya. Karasik, P.V. Mamyshev, A.M. Prokorov, D.G. Fursa: Sov. J. Quant. Electr. **17**, 415 (1987)
[6.9]   B. Zysset, W. Hodel, P. Beaud, P. Weber: Opt. Lett. **11**, 156 (1986)
[6.10] A.S.L. Gomes, W. Sibett, J.R. Taylor: Opt. Lett. **10**, 338 (1985)
[6.11] M. Nisoli, S. De Silvestri, O. Svelto: Appl. Phys. Lett. **68**, 20 (1996)
[6.12] A.E. Siegman: *Lasers*, (University Science Books, Mill Valley, CA 1986)
[6.13] L.M. Frantz, J.S. Nodvick: J. Appl. Phys. **34**, 2346 (1963)
[6.14] R.L. Fork, C.V. Shank, R.T. Yen: Appl. Phys Lett. **41**, 223 (1982)
[6.15] A. Migus, C.V. Shank, E.P. Ippen, R.L. Fork: IEEE J. Quant. Electr. **QE 18**, 101 (1982)
[6.16] W.H. Knox, M.C. Downer, R.L. Fork, C.V. Shank: Opt. Lett. **9**, 552 (1984)
[6.17] I. Duling III, T. Norris, T. Sizer II, P. Bado, G. Mourou: J. Opt. Soc. Am. B **2**, 616 (1985)
[6.18] V.J. Newell, F.W. Deeg, S.R. Greefield, M.D. Fayer: J. Opt. Soc. Am. B **6**, 257 (1989)
[6.19] S. Szatmari, F.P. Schafer, E. Muller-Horsche, W. Muckenheim: Opt. Commun. **63**, 305 (1987)
[6.20] P. Georges, F. Estable, F. Salin, J.P. Poizat, P. Grangier, A. Brun: Opt. Lett. **16**, 144 (1991)
[6.21] J.E. Murray, W.H. Lowdermilk: J. Appl. Phys **51**, 3548 (1980)
[6.22] J.P. Chambaret, C. Le Blanc, A. Antonetti, G. Cheriaux, P.F. Curley, G. Darpentigny, F. Salin: Opt. Lett. **21**, 1921 (1996)
[6.23] C. Le Blanc, P. Curley, F. Salin: Opt. Commun. **131**, 391 (1996)
[6.24] D. Strickland, G. Mourou: Opt. Commun. **62**, 419 (1985)
[6.25] O.E. Martinez: IEEE J. Quant. Electr. **QE 23**, 59 (1987)
[6.26] J. Squier, F. Salin, G. Mourou, D. Harter: Opt. Lett. **16**, 324 (1991)
[6.27] F. Salin, J. Squier, G. Mourou, G. Vaillancourt: Opt. Lett. **16**, 1964 (1991)
[6.28] A. Sullivan, W.E. White: Opt. Lett. **20**, 192 (1995)
[6.29] B.E. Lemoff, C.P.J. Barty: Opt. Lett. **18**, 1651 (1993)
[6.30] G. Cheriaux, P. Rousseau, F. Salin, J.P. Chambaret, B. Walker, L.F. Dimauro: Opt. Lett. **21**, 414 (1996)

194   F. Salin

[6.31]   G. Mourou, C.P. Barty, M.D. Perry: Phys. Today 51, 22 (1998)
[6.32]   K.Yamakawa, M. Aoyama, S. Matsuoka, T. Kase, Y, Akahane, H. Takuma: Opt. Lett. **23**, 1468 (1998) and personal communication for a 500 TW record
[6.33]   V. Bagnoud, F. Salin: Appl. Phys. B, **70**, S165 (2000)
[6.34]   A. Courjaud, R. Maleck-Rassoul, N. Deguil, C. Honniger, F. Salin: ASSL2002
[6.35]   A Liem, D. Nickel, J. Limpert, H. Zellmer, U. Griebner, S. Unger, A. Tünnermann, G. Korn: Appl Phys B **71** 889, (2000)
[6.36]   R.R. Alfano, S.L. Shapiro: Phys. Rev. Lett. **24**, 584 (1970)
[6.37]   Y.R. Shen: The Principles of Nonlinear Optics Wiley, New York (1984)
[6.38]   A. Yariv, P. Yeh: Optical Waves in Crystals Wiley, New York (1984)
[6.39]   R.W. Boyd: Nonlinear Optics (Academic Press, Boston 1992)
[6.40]   V.G. Dmitriev, G.G. Gurzadyan, D.N. Nikogosyan: Handbook of Nonlinear Optical Crystals, 2nd edn., Series in Optical Sciences, 64 (1991)
[6.41]   H. Liu, J. Yao, A. Puri: Opt. Commun. **109**, 139 (1994)
[6.42]   F. Salin, F. Estable, F. Saviot: Ultrafast Phenomena IX, Springer-Verlag, 194(1994)
[6.43]   J. Ringling, O. Kittelman, F. Noack, G. Korn, J. Squier: Opt. Lett **18**, 2035 (1994)
[6.44]   F. Seifert, V. Petrov, F. Noack: Opt. Lett. **19**, 837 (1994)
[6.45]   V.V. Yakovlev, B. Kohler, K.R. Wilson: Opt. Lett. **19**, 2000 (1994)
[6.46]   M. Reed, M.K. Steiner-Shepard, D.K. Negus: Opt. Lett. **19**, 1855 (1994)
[6.47]   M. Nisoli, S. De Silvestri, V. Magni, O. Svelto, R. Danielus, A. Piskarskas, G. Valiulis, V. Varanavicius: Opt. Lett. **19**, 1973 (1994)
[6.48]   V. Petrov, F. Noack: J. Opt. Soc. Am. B **12**, 2214 (1995)
[6.49]   S. Fournier, F. López-Martens, C. Le Blanc, E. Baubeau, F. Salin: Opt. Lett. **23**, 627 (1998)
[6.50]   G. Gale, M. Cavallari, T. Driscoll, F. Hache: Opt. Lett. **20**, 1562 (1995)
[6.51]   T. Kobayachi, A. Shirakawa: Appl. Phys. B **70**, S339 (2000)
[6.52]   R.Maleck-Rasoul, A. Courjaud, F. Salin: OSA TOPS, **68** (2002)

7

How to Measure the Characteristics of Laser Pulses

L. Sarger and J. Oberlé

With 22 Figures

7.1 Introduction

In the previous chapter we showed how to produce and how to manipulate ultrashort laser pulses. But before such pulses are used for experiments, it is necessary to characterize these pulses properly to determine the experimental conditions. This chapter deals with such characterization.

Indeed to extract any kind of data from experimental results, one has to analyze carefully the tools used. This aphorism is of great importance when dealing with (ultra) short light pulses as one does not usually control the required parameters, at least with conventional equipment. In this chapter, we will emphasize some specific parameters, such as

- energy in each pulse or average power;
- shape, duration and spectrum of the laser pulse.

Of course, the wide diversity of lasers in the picosecond and femtosecond range, with different peak powers, repetition rates and stabilities, and the precision required for a given experiment allows a large variety of characterization techniques. All of these techniques exist because of the absence of (sufficiently) fast photodetectors with a large dynamic range, such as one can use for rather longer laser pulses. The speed and dynamic range fall short by orders of magnitude. This is why this field is always a changing domain; many laboratories are using and, at the same time, designing their tools to continue their research. We will present here some simple ideas, as well as commercially available data, to introduce the world of short-pulse characterization. Of course, a laser experiment requires measurement of other parameters such as beam sizes, beam shapes, wavelength and beam overlap in pump–probe experiments. These will not be addressed here, as they do not basically differ from what is needed for CW laser experiments. We will first focus on that energy and power of the laser beam before addressing the very special aspect of temporal characterization.

7.2 Energy Measurements

Measurement of the energy of a laser pulse (in joules, or more often, in microjoules) from either a single-shot or a high-repetition-rate system is always done using a pyroelectric device. In such a material, an increase in the temperature will give a proportional electrical response. This is quite similar to the piezoelectric effect, where such an electrical signal is related to a mechanical constraint. These behaviors are obviously linked through the same material coefficient – susceptibility – and thus are present at the same time. Here, as the thermal loading is a problem, there is a strong need to wait until the material temperature returns to its original value before each new measurement. This type of detector must be used either in the single-shot mode or, at maximum, for a very low-repetition-rate system at moderate energy, of the order of few millijoules. The state of the art for such pyroelectric sensors is rather fast detectors, but working only in the AC mode at low repetition rate.

The electrical response, generally a voltage, is strictly proportional to the temperature rise due to the energy deposited by the laser pulse in the material. As this is a thermal effect, it is essentially due to the overall absorbing power of the material, with or without an adequate coating. The spectral domain is thus exceptionally wide, as shown in Fig. 7.1, and extends essentially with a flat response function from the deep UV to the far infrared, covering the region in which all the measurements have to be made. The linearity of these detectors is excellent and their calibration is a fairly easy task.

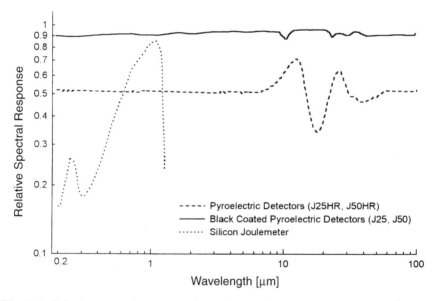

Fig. 7.1. Relative spectral response of pyroelectric detectors and joulemeter (Molectron)

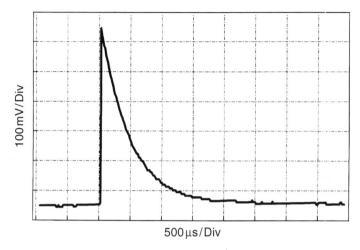

Fig. 7.2. Typical voltage output from a joulemeter

The impulse response of the detector-head–electronics system corresponds to the integral over time of the laser pulse intensity, and the peak value of the signal is proportional to the pulse energy. After that, thermal exchanges take place and the signal decreases, with a long time constant reflecting the return to equilibrium. Typical electronic responses (voltage) are similar to the one shown in Fig. 7.2 and can be seen on a remanent or storage oscilloscope during laser optimization, or, after optimization, on a peak detector used to perform a calibrated measurement.

Typical sensitivities of pyroelectric detectors are in the range of milliamps per watt, with a thermal time constant of the order of few milliseconds. Thus they are well suited for ultrashort laser pulses from amplified systems at repetition rates up to kilohertz, giving an electronic response of the order of a few volts per millijoule.

7.3 Power Measurements

For CW mode-locked laser systems at high repetition rate, the measurement of the average power is the only way to deduce the pulse amplitude, and requires a good knowledge of the temporal aspect of the laser pulse train, i.e. the duration, repetition rate and stability of the laser output. Thus one prefers to integrate the signal fully over time using a thermoelectric photodetector, where the electrical signal reflects the equilibrium temperature reached by a fully absorbing material under quasi-permanent laser illumination. As it also relies on absorption, it has to be virtually insensitive to the laser wavelength. The rather long time constant of this detector gives, after calibration, an average value of the laser power, as long as the power dissipated inside the detector head is mainly due to the laser flux. Special care has to be taken when the power to be measured is very low, but fortunately this is not often the case even for the weakest laser oscillator.

As in the energy measurement case, the detector surface has to be covered with an appropriate coating to absorb all the impinging power. On its back, a set of thermodetectors measures the voltage associated with the steady-state temperature. The average sensitivity is of the order of $100\,\mu$V per watt, with a time constant of a fraction of a second. Examples of specifications for the power meter are given in Table 7.1.

This kind of detector can be easily replaced during the laser optimization process by simple photoelectric diode. Using photodiodes with a large sensitive area and a response time of the order of few nanoseconds greatly improves the comfort during laser tweaking. Indeed, the sensitivity of the order of amperes per watt really speeds up the preliminary alignment, but as these detectors are wavelength-sensitive (as with any photoelectric process) and have a rather low damage threshold, they only can be used on a small part of the laser beam and, in any case, have to be backed up with a real power meter for a valid measurement.

Table 7.1. Example specifications for the power meters

Wavelength range	190 nm–11 μm
Power range	0.1 mW–200 W
Max power density	25 kW/cm^2
Damage resistance	Up to 5 J/cm^2
Sensitivity variation	$\pm 2\%$,
Response time	0.5 s–10 s

7.4 Measurement of the Pulse Temporal Profile

The characterization of the temporal profile of the laser pulse is the basis of any laser optimization and, moreover, of any measurement in ultrashort-pulse laser techniques. Unfortunately, photoelectric response times are seriously limited when one is dealing with femtosecond or even picosecond laser pulses. Although novel photodetectors with picosecond response have been developed in the last decade, we will describe in more detail the optical methods based on autocorrelation techniques to measure directly not only the amplitude but also the phase of the laser pulse. These techniques are the cornerstone of any lab involved with ultrashort pulses and have to be fully understood, in all their variety in terms of precision and limitations.

7.4.1 Pure Electronic Methods

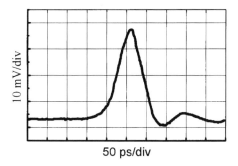
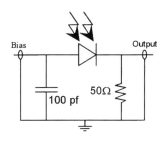

Fig. 7.3. Typical electrical signal given by a photodiode (Antel AR-S2) in response to a femtosecond laser pulse, and the corresponding electric circuit

7.4.1.1 Photodiodes. The direct photoelectric effect, with the photon energy larger than the band gap of a semiconductor, is currently used at least to visualize the laser pulse train. The finite time required to completely clean up the depletion zone of all the free carriers induced by the laser field usually prevents the true measurement of the pulse envelope. Even using an ultrasmall active area (a few microns square) and a high reverse voltage, a limit close to tens of picoseconds for both the leading and the trailing edge of the electric pulse can be found. Using either heterostructures such as PIN (a sandwich of intrinsic semiconductor between n- and p-doped semiconductors) or avalanche photodiodes can only be useful for long laser pulses and high repetition rates. In fact, the visualization itself requires an oscilloscope with a bandwidth so large that only a sampling technique can give a correct answer. This technique is currently used for CW mode-locked argon lasers and even Nd-YAG oscillators at a repetition rate of a few tens of megahertz. These systems deliver relatively long laser pulses ($> 100\,\text{ps}$), with a stable triggering signal from the active mode-locker.

Figure 7.3 displays a typical signal obtained using an ultrashort laser pulse on a photoelectric detector. Clearly, the appearance of the pulse envelope is greatly affected on this timescale by the overall electronic circuit surrounding the detector itself. The impedance imbalance is a real difficulty and will greatly modify the trailing edge, even hiding satellite pulses for example.

Recently, new high-speed MSM (metal-semiconductor-metal) photodetectors have been developed and commercialized. These MSM detectors consist of two contact pads and interdigitated Schottky contact fingers, which form the active area of the device, on a semiconductor layer. MSMs operate on the same principle as PIN photodetectors, but the interdigitated design minimizes parasitic resistances and the planar structure reduces the capacitance, making them suitable for high-speed operation. Commercially available MSM detectors have impulse responses below 10 ps.

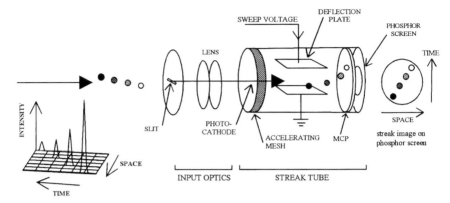

Fig. 7.4. Operating principle of streak tube (Hamamatsu)

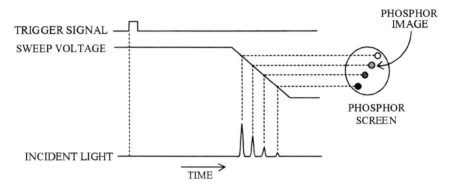

Fig. 7.5. Timing of operations of streak tube (at time of sweep) (Hamamatsu)

7.4.1.2 Streak Camera. A more complex photoelectric instrument can be also used to characterize the pulse envelope. Although its use (and its price) can only be justified in time-resolved spectroscopic experiments, where it is the only option, its characteristics allow a good analysis of laser pulses in the range of a picosecond and above. This device will measure ultrafast light phenomena, and its name goes back to early developments when high-speed rotating drums where used as shutters in standard cameras. These cameras will streak reflected light on a film.

The basics of the streak camera is depicted in Figs. 7.4 and 7.5. The incident light pulse, or eventually a sequence of pulses, is converted into photoelectrons inside a photocathode very similar to the one found in a photomultiplier tube. The resulting electron flux is collimated, strongly accelerated and deflected by an ultrahigh-speed sweep. The impact of the electrons on a phosphor screen is spread out along an axis calibrated in time by the sweep itself. The light intensity of this trace reflects the electron density and thus is a true image of the laser pulse.

The time resolution is related to the spatial electron (photon) profile, and usually a slit at the entrance window, orthogonal to the ultimate "time axis", defines the lateral extent. The laser light pulse impinges on this slit and a lens images it onto a photocathode. The temporal profile of the electrons in the beam reflects the temporal evolution of the light intensity on the slit. An electric field due to an accelerating voltage between 2 and 5 kV accelerates the electrons into the sweeping field, whose sweep rate is a function of the required time resolution. The relative timing between the electron flux and the sweep has to be precisely set, with more precision as one goes to higher time resolution and higher-speed sweeps. Thus a really fast electronic system is required in order to provide sufficiently good timing, and generally this is the most important difficulty. Part of the beam is detected prior to the streak tube for synchronization purposes. Either the synchronization signal can be provided by an independent PIN photodiode which triggers a built-in sweep or through a photo-switch in a simple high-voltage circuit that connects directly to the streak tube's deflection plates to produce a jitter-free timing.

Usually one has to reduce the photocathode emissivity to keep the space charge as low as possible. Then the overall luminosity of the phosphor screen is not sufficient and it is found more practical to insert a microchannel plate, acting as an electron multiplier, before the phosphor plate. A net gain of 10^3 for this image intensifier is usually sufficient and the final image can easily be read by standard CCD techniques using a frame grabber and a computer.

The analysis relies on two basic facts:

- the electron creation times (at the photocathode) is linked to their deflection angle and thus to their position on the time axis;
- the number of photoelectrons, i.e. the image of the laser envelope, is linked to the "streak image" density.

Then a simple densitometric reading of the "streak image" gives a correct representation of the incoming laser pulse.

In order to calibrate the device easily, one generally shines, not a single laser pulse onto the input slit, but rather a sequence of four or five pulses coming out of a simple reflective scheme (a high-loss Fabry–Pérot interferometer with a spacing greater than the coherence length of the light) excited by the pulse itself. The light pulse bounces back and forth and leaks out of the device at a known rate and amplitude. As the geometrical length and the reflectivity can be precisely measured, one has a cheap calibration tool not only in time but also in amplitude.

Two different ways of recording the streak images can be implemented, depending on the type of laser and/or the type of measurement. Either a single shot "single sweep" can give the required data or an averaged set of data can be recorded in the synchroscan mode.

Synchroscan Mode. The high-speed sweep is periodically repeated and carefully set to be synchronous with the incoming laser pulses. This is very similar to a sampling approach, with a very powerful accumulating effect; each pulse of the train reaches the very same position on the time axis, as long as the jitter problem is solved. Figure 7.6 shows the advantages of such a synchroscan mode compared to a single shot of the same laser sequence.

The current state of the art of this technique for pulse characterization corresponds to a time resolution, at its best, of the order of hundreds of femtoseconds. But the fact that it can then be associated with spectroscopes and have sweep capabilities ranging from tens of picoseconds to a few milliseconds, with a dynamic range of the order of 1000, are unique features for time-resolved spectroscopic measurements.

7.4.2 All-Optical Methods

7.4.2.1 Introduction. All the preceding devices we have described are obviously limited to light pulses longer than a few hundreds of femtoseconds (streak camera) or picoseconds, at their best. Other methods are thus very much needed to characterize the ultrashort pulses now easily available, which are often much shorter than 100 fs. The techniques we are to describe here are indirect approaches, and one has not only to take into account the technique itself but also to use a model to retrieve pulse shapes from the experimental data.

The general principle uses two different ingredients. First of all, these techniques have their basis in the time–space transformation, as it takes only a picosecond for a light pulse to travel 300 microns in air, and this length is easy to measure and calibrate. The other key point is the correlation function described in mathematics textbooks. Given two time-dependent functions $F(t)$ and $F'(t)$, then if one is already known, say $F'(t)$, the measurement of $G(\tau)$, defined as

$$G(\tau) = \int_{-\infty}^{+\infty} F'(t)F(t-\tau)\,\mathrm{d}t, \qquad (7.1)$$

directly gives the other one. Here $G(\tau)$ is the first-order correlation function. Higher orders can be also defined; all can give the same results as long as one knows the test function $F'(t)$.

In the approach to laser pulses, unfortunately, this test function cannot be synthesized on this short timescale and the pulse is used as its own test function. In this case we use the so-called "autocorrelation functions". By various means [7.1–2], one splits the laser pulse under measurement into two replicas $F(t)$ and $F(t-\tau)$, to be compared within an interferometric setup able to construct the autocorrelation function. The arm imbalance between the two paths inside the interferometer in Fig. 7.7 acts as the delay τ in the mathematical expressions.

7.4.2.2 Optical Correlation. In interferometric techniques, one easily accesses the instantaneous electromagnetic field $\boldsymbol{E}(t)$, which plays the role of

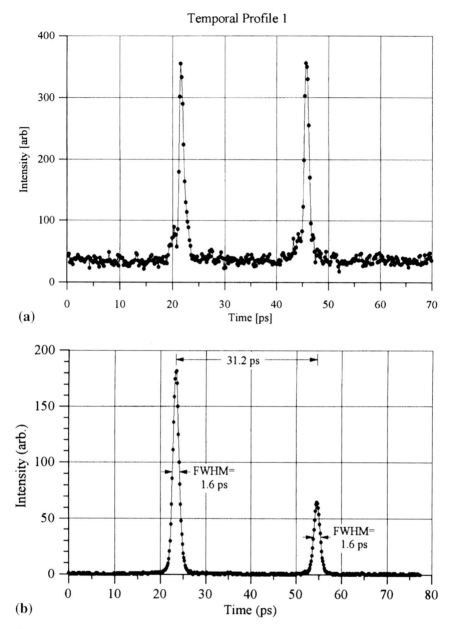

Fig. 7.6. Typical temporal resolution of a streak camera measured in (**a**) single-shot and (**b**) synchroscan mode (courtesy of ARP Photonetics)

$F(t)$. Usually only the laser field intensity $I(t) = \langle E(t) \cdot E^*(t) \rangle$, the temporal evolution of the pulse, is to be characterized, and it can be computed using the Maxwell equations. The interferometric output of the setup shown in Fig. 7.7

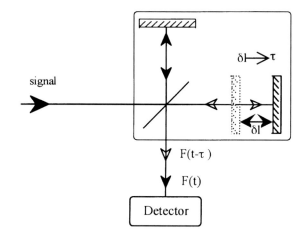

Fig. 7.7. Basic principle of interferometric cross-correlator

is given by $I_1(t)$ as a function of the delay τ:

$$I_1(\tau) = \int_{-\infty}^{+\infty} |\boldsymbol{E}(t) + \boldsymbol{E}(t - \tau)|^2 \, \mathrm{d}t. \tag{7.2}$$

This can be expressed using the first-order autocorrelation function $G(\tau)$:

$$I_1(\tau) \propto 2 \int I(t) \, \mathrm{d}t + 2G(\tau). \tag{7.3}$$

Using only autocorrelation functions, it can be shown that the full knowledge of $\boldsymbol{E}(t)$ requires the measurement of all the successive $G_n(\tau)$ (or equivalent $I_n(\tau)$). For example, the second-order correlation function can be written as

$$I_2(\tau) = \int_{-\infty}^{+\infty} \left| [\boldsymbol{E}(t) + \boldsymbol{E}(t - \tau)]^2 \right|^2 \, \mathrm{d}t \tag{7.4}$$

and so on.

Experimentally, all these higher-order correlation functions can be obtained using multiphoton processes. For example, the second-order function can be mimicked using either two-photon absorption or second-harmonic generation, and the third-order one through a Kerr experiment or third-harmonic generation.

As already said, the first-order autocorelation function is included in the output of the Michelson interferometer and the width of the function $I_1(t)$ is related to the coherence length of the laser pulse. It is by no means a unique measurement of the duration of the pulse. As has been well known since Fourier, this signal is the strict analog of the light spectrum. Nevertheless, a white-light spectrum gives a very short coherence length due to the lack of coherence but is not a signature of an ultrashort pulse. Yet, on the other

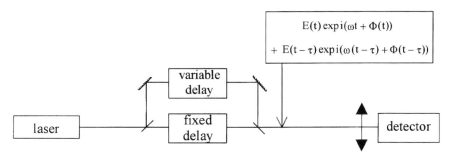

Fig. 7.8. Basic scheme of second-order autocorrelator

hand, if one knows that the incoming light is really a laser pulse, the relationship between the coherence length and the pulse duration can be computed assuming the actual phase of the field $\boldsymbol{E}(t)$.

There is usually a need to go on through all the orders of the G_n functions, but practically, and hopefully, one has to stop at the second order and assume a given realistic pulse envelope to analyze the experimental results. The results obtained in this restricted approach will appear fully symmetric but do not reflect for sure the exact reality. These assumptions have to be cross-checked with more sophisticated approaches, to be described later.

We will first focus on measurements of high-repetition-rate systems, as they are often the start of any laser system using an oscillator. In this case, one only has access to an average of a very large number of light pulses. Then, we will describe the single-shot approach, which is mainly applied to high-peak-power amplified sources.

7.4.2.3 High-Repetition-Rate Second-Order Autocorrelator.
The classical way to get the required second-order function is to start with and adapt the first-order technique already introduced (Fig. 7.8).

Practically, one can use the two-photon process which takes place in a semiconductor detector excited at a wavelength below the band-gap cutoff. Only a high peak power such as in laser pulses can gives a two-photon signal which can be directly recorded on an oscilloscope as a function of the delay τ. Alternatively, a phase-matched doubling crystal, properly filtered to provide only the second-harmonic field, can extract from the interferometer output the correct answer. This harmonic field, recorded by a photodetector – a photomultiplier, or a simple photodiode if the average signal is sufficient – as a function of the delay will give essentially the same result.

Either of these two processes can be written as a function of I_2 (7.4), where $\boldsymbol{E}(t) + \boldsymbol{E}(t-\tau)$ is either the actual driving force for two-photon absorption or the output field from the doubling crystal at the harmonic frequency. If one records the whole signal, one gets a response proportional to

206   L. Sarger and J. Oberlé

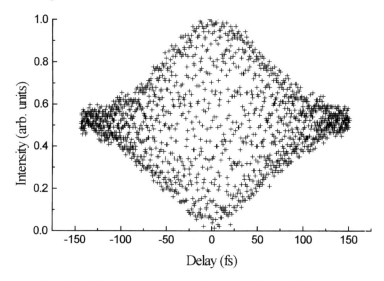

Fig. 7.9. First-order autocorrelation of 800 nm 100 fs pulse

$$I_2(\tau) = \int_{-\infty}^{+\infty} \left| \{E(t)\exp i[\omega t + \Phi(t)] \right.$$
$$\left. +E(t-\tau)\exp i[\omega(t-\tau)+\Phi(t-\tau)]\}^2 \right|^2 dt \tag{7.5}$$

so therefore

$$I_2(\tau) = \int_{-\infty}^{\infty} \left| 2E^4 + 4E^2(t)E^2(t-\tau) \right.$$
$$+4E(t)E(t-\tau)[E^2(t)+E^2(t-\tau)]\cos[\omega\tau+\Phi(t)-\Phi(t-\tau)]$$
$$\left. +2E^2(t)E^2(t-\tau)\cos[2(\omega\tau+\Phi(t)-\Phi(t-\tau))] \right| dt. \tag{7.6}$$

Interferometric Autocorrelations.

Figures 7.9 and 7.10 are the corresponding recordings of the first-order and the second-order functions for the same laser pulse of 100 fs at 800 nm (the output of a Kerr lens mode-locked Ti:sapphire laser); details of these two figures are shown in Fig. 7.11. As one can see, the normalized first-order signal goes classically from 0.5, far from the coherence length, where the pulse and its replica successively give their own signal, to 1 when they fully interfere at zero delay. For the second order, correspondingly, the signal increases to

$$I_2(\tau=0) = 2^4 \int \boldsymbol{E}^4(t)\,dt \tag{7.7}$$

at the coherent superposition, while far from the superposition it only reaches

$$I_2(\tau\to\infty) = 2 \int \boldsymbol{E}^4(t)\,dt, \tag{7.8}$$

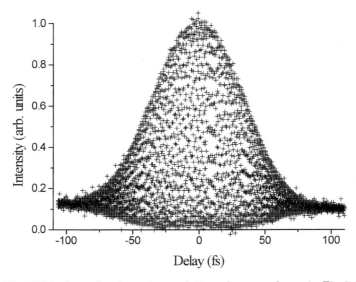

Fig. 7.10. Second-order autocorrelation of some pulse as in Fig. 7.9

giving a contrast of 8:1 (Fig. 7.10).

The shapes of both fringe envelopes are very sensitive to the actual pulse amplitude but, furthermore, are a particularly complicated function of the instantaneous frequency of the laser pulse. For example, as fringes only appear when the optical frequencies of interfering beams are equal, any possible chirp along the pulse will produce a very complicated fringe set, keeping nevertheless the same contrast ratio at $\tau = 0$ and $\tau \to \infty$ for both interferograms. For example, in a pulse affected by self-phase-modulation, the optical frequency changes mainly at the pulse peak, giving a narrower interferometric trace, while at the end and beginning of the pulse it is mostly unchanged and will still produce fringes in the wings. This effect is more evident in the correlation function than in the spectrum as classically observed. The overall appearance can then be interpreted only in terms of pulse quality, and any further analysis is far too complicated. One can use a simpler signal obtained by averaging the second-order interferometric autocorrelation.

Intensity Autocorrelations.

As the full signal is not so easy to analyze, one used to average over the (interferometric) delay either by using a numerical filter to wash out the fringes or, experimentally, by sweeping the delay fast enough that the detection time constant performed the same task. This much simpler signal appears as

$$S_I = 2 \int I^2 \, \mathrm{d}t + 4 \int I(t)I(t - \tau) \, \mathrm{d}t, \tag{7.9}$$

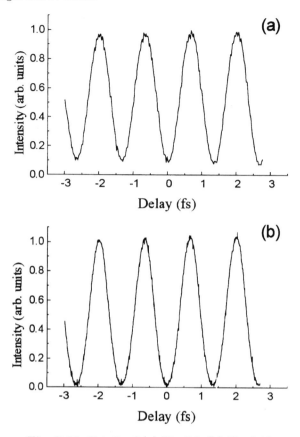

Fig. 7.11. Details of (**a**) Fig. 7.9, (**b**) Fig. 7.10

directly proportional to the second-order autocorrelation function $G_2(\tau)$ given by

$$G_2(\tau) = \frac{\int I(t)I(t-\tau)\,dt}{\left|\int I^2\,dt\right|},\qquad(7.10)$$

where I is the pulse intensity. This corresponding curve – often called improperly the "intensity autocorrelation with background" due to the first term in (7.9) – is shown in Fig. 7.12, and has a contrast of only 3:1. It can be exploited in order to get the actual pulse width.

If the integration constant is not too high, the signal far from the central peak will be greatly affected by any amplitude noise in the laser pulse. As it is well known with modern laser oscillators that one has first to worry about the laser stability before trying to reduce and measure the pulse width, this special behavior gives the interferometric setup a superiority over any other scheme "without background" such as is often seen in the literature.

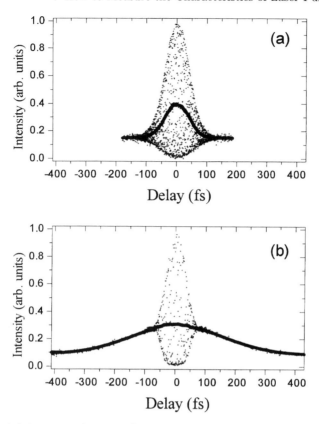

Fig. 7.12. (a) Intensity (*solid line*) and interferometric (*points*) autocorrelation traces for a Kerr lens mode-locked Ti:sapphire laser pulse. (b) Intensity (*solid line*) and interferometric (*points*) autocorrelation traces of the pulse after passing through a 5 cm thick glass plate

Basic Experimental Setups.

At present, only few practical designs for second-order autocorrelators are currently used. They differ from each other in the beam splitters, the delay line and, more often, the detection system. An overview of two approaches is presented in Fig. 7.13. They are both based on a modified Michelson interferometer with a 50/50 beam-splitter, and two cube corners are used to avoid coupling back to the laser source. One of the mirrors moves back and forth in a fast scan on a miniature translation stage, while the other can be moved using a translation stage to precisely calibrate the delay τ. In the arrangement of Fig. 7.13b the fully colinear beams will interfere all along their paths (within the coherence length), while the arrangement of Fig. 7.13a uses a noncolinear arrangement, where the interference field is strongly restricted. The corresponding results are shown in Figs. 7.14a and 7.14b. The second-order process

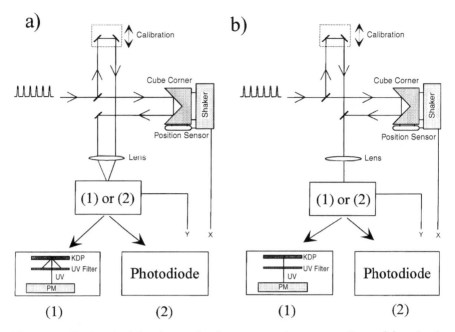

Fig. 7.13. Basic principle of second-order autocorrelator: noncolinear (**a**) and colinear (**b**) arrangements

needed requires a reasonable peak power, and a short-focal-length lens is used differently in both arrangements.

Both systems can be equipped with one of the following:

(1) a nonlinear crystal, a high-pass filter and a photodetector to detect the second harmonic;
(2) a photodiode with a band gap greater than the photon energy.

For detection (i.e. (2) above), one has to choose a suitable photodetector. For example, an AsGaP photodiode will give excellent results for wavelengths above 600 nm [7.3], while a standard silicon photodiode will do the same for the near infrared ($> 1.2\,\mu$m). The focusing lens in both setups focuses inside the depletion zone and only the setup of Fig. 7.13b can provide the interference fringes already described. The other setup directly gives the averaged intensity autocorrelation. The sensitivity may be so high that one has to reduce the pulse power to stay out of the saturation in the case of a photovoltaic detector. One can instead use an inexpensive LED whose band gap is adequate, but in this case one has then to worry about the extremely small size of the emitting zone, which is used here as a photodetector [7.4].

For a doubling-crystal approach, one has to choose, cut and tune the crystal axis in order to produce frequency doubling effectively. In the fully colinear case (Fig. 7.13b), of course, the interference fringes are present in the funda-

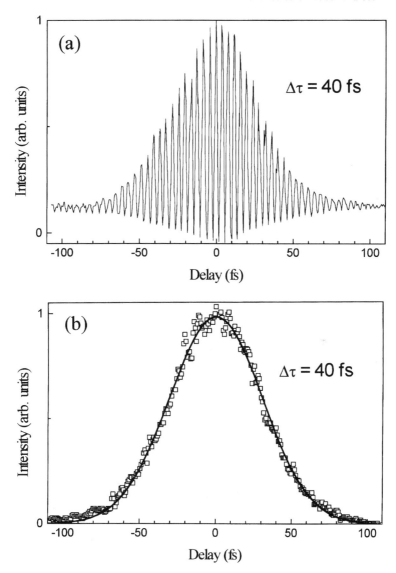

Fig. 7.14. Intensity autocorrelation traces for a self-mode-locked Cr:forsterite laser: (**a**) interferometric autocorrelation trace obtained using the setup shown in Fig. 7.13b(1), (**b**) intensity autocorrelation without background obtained using the setup shown in Fig. 7.13a(1)

mental as well as in the harmonic wavelength and a high-pass filter has to be used to select only the second-order process. In the noncollinear arrangement, the cut is obviously different and only a harmonic signal can be generated along the bisector of the incoming beams. Here, only a spatial filter can extract the required signal, so that it appears without background. The corresponding signal is then given by

$$S_{II} = 4 \int I(t)I(t - \tau)\,\mathrm{d}t. \qquad (7.11)$$

The choice of doubling crystals is generally an easy task. BBO or KDP crystals can be phase-matched over a wide range of laser wavelengths, but the actual thickness is of major importance. In fact, as the interaction length increases, the group-velocity mismatch, particularly with femtosecond pulses, can greatly affect the harmonic generation and thus the recorded signal. To overcome this difficulty, one can either use a very short crystal length, of the order of $100\,\mu$m, or focus strongly into the entrance face of the crystal with a microscope lens to get a short confocal parameter.

Different scanning systems for the moving cube corner can be used, depending on the laser repetition rate and/or the level of averaging required. The most frequently used systems is either a DC motor and a translation stage or a speakerphone. These systems can be equipped with any kind of motion or position sensor to synchronize the measurement system, although the inherent fringes present in these setups can easily be used as an automatic position control. Nevertheless, the speed of motion and time constants have to be carefully chosen to retrieve a nonintegrated signal in the measurement.

Comments.

- Experimentally, if one is measuring pulses shorter than 50 fs, one has to be careful to avoid any pulse broadening before and inside the correlator. A pellicle beam splitter and a thin lens have to be selected and, in any case, will give a practical limit for the accuracy. Obviously, when a two-photon photodiode is used, the protective resin or the plastic window included in the diode housing have to be removed.

- Interferometric autocorrelations present the advantage of a direct calibration of the time axis by the fringes, but the drawback is a constant background precluding any high-dynamic range measurements. For some experiments, the signal to noise ratio is not sufficient to provide a correct dynamic range for pulse characterization. For example, an eventual prepulse at a level of 10^{-6}, undetectable in second-order autocorrelations, can be strongly detrimental for any experiment in high field plasma physics. Moreover, asymmetric pulses in time cannot be revealed by second-order autocorrelations. Recently, it has been demonstrated that a third-order autocorrelation, between the second harmonic of the investigated pulse and the pulse itself, can solve this issue and lead to dynamic range increase of the measurement [7.5].

Signal Analysis: Pulse Width. The intensity autocorrelation trace has an experimental full width at half maximum (FWHM) $\Delta\tau$ (easily calibrated using the speed of light: $0.3\,\mu\text{m/fs}$). As already mentioned, we have to select some realistic pulse shape and compute the expected signal in the intensity autocorrelation function. For example, if the pulse intensity has a full width at half maximum of Δt, how large will be the autocorrelation function $\Delta\tau$? Table 7.2 lists some mathematical results obtained using various pulse shapes.

Table 7.2. Mathematical results (after J.C. Diels [7.2])

I(t)	Δt	I(ω)	$\Delta\omega$	$\Delta\omega\cdot\Delta t$	$G_2(\tau)$	$\Delta\tau$	$\Delta\tau/\Delta t$
e^{-t^2}	1.665	$e^{-\omega^2}$	1.665	2.772	$e^{-\frac{\tau^2}{2}}$	2.355	1.414
$\mathrm{sech}^2(t)$	1.763	$\mathrm{sech}^2\left(\frac{\pi}{2}\omega\right)$	1.122	1.978	$\frac{3[\tau\cosh(\tau)-\sinh(\tau)]}{\sinh^3(\tau)}$	2.720	1.543
$\dfrac{1}{e^{t/(t-A)}+e^{-t/(t-A)}}$ $A=1/4$	1.715	$\dfrac{1+1/\sqrt{2}}{\cosh\left(\frac{15\pi}{16}\omega\right)+1/\sqrt{2}}$	1.123	1.925	$\dfrac{1}{\cosh^3\left(\frac{8}{15}\tau\right)}$	2.648	1.544
$A=1/2$	1.565	$\mathrm{sech}\left(\frac{3\pi}{4}\omega\right)$	1.118	1.749	$\dfrac{3\sinh(\frac{8}{3}\tau)-8\tau}{4\sinh^3\left(\frac{4}{3}\tau\right)}$	2.424	1.549
$A=3/4$	1.278	$\dfrac{1-1/\sqrt{2}}{\cosh\left(\frac{7\pi}{16}\omega\right)-1/\sqrt{2}}$	1.088	1.391	$\dfrac{2\cosh(\frac{16}{7}\tau)+3}{5\cosh^3\left(\frac{8}{7}\tau\right)}$	2.007	1.570

Obviously, the best (most optimistic) result can be obtained for a sech^2 pulse, with a reduction factor $\Delta\tau/\Delta t = 1.54$. This is one of the main reasons why all the scientific community agrees to choose this standard, regardless of the actual pulse shape; the latter has to be analyzed by more sophisticated methods. The "rule of thumb" will then be the ratio $\Delta\tau = 1.54\Delta t$.

Table 7.3 lists also the product $\Delta t\,\Delta\omega$ for each different function. This also has to be cross-checked using the laser pulse spectrum for that particular function.

7.4.2.4 Single-Shot Second-Order Autocorrelator. For an amplified system, after tuning the oscillator and measuring the laser pulse width with one of the systems described above, one has to tune the output of the system, generally at a very low repetition rate. As many different processes affect the pulse before this final stage, a single-shot measurement is usually needed.

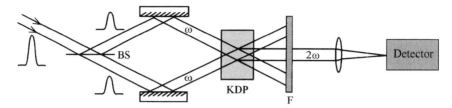

Fig. 7.15. Principle of a single-shot noncolinear interferometric autocorrelator. The incident beam (at frequency ω) is split into two equal parts by the beam-splitter (BS). The intensity distribution across the second-harmonic signal (generated by the KDP crystal) gives the autocorrelation function

As the delay cannot be incremented, a geometrical arrangement should be used to transform the autocorrelation function into a spatial intensity distribution, which can be recorded with a CCD matrix. The use of a noncolinear interferometer [7.6], as represented in Fig. 7.15, provides the required temporal scan.

In the setup shown in Fig. 7.15, the beam is split into two equal parts using a beam-splitter (BS). Beam crossing is performed inside the doubling crystal at a small angle Φ (Fig. 7.16). A high-pass filter and a lens relay this linear crossing zone into a CCD detector.

As the two beams are noncolinear and with a beam diameter in the overlap region much larger than the pulse length ($c\Delta t$), a certain effect ensures that the second harmonic is generated only when both pulses coincide in time and space. As shown in Fig. 7.16, if the two beams are uniform, with $I_1(t)$ and $I_2(t)$ the temporal shapes of the beams at a given distance x_0 from the center, the radiated field at the harmonic frequency is proportional to $I_1(t-\tau)I_2(t+\tau)$, where the delay τ is given geometrically by

$$\tau = \frac{nx_0 \sin(\Phi/2)}{c}. \tag{7.12}$$

As a result, the filtered signal observed at the linear detector is a function of position along the crossing line and is proportional to

$$S(x) = \int I_1(t+\tau)I_2(t-\tau)\,dt, \tag{7.13}$$

and so it is directly proportional to the second-order autocorrelation function $G_2(2\tau)$ of the incident pulse

$$G_2(2\tau) \propto \int I(t)I(t-2\tau)dt. \tag{7.14}$$

The calibration is easily performed by inserting a calibrated delay ΔT (such as a glass plate) in one of the two arms of the autocorrelator. As a consequence, the translated image on the CCD detector corresponds to the translation of the second-order autocorrelation function $G_2(2\tau)$ by $\Delta T/2$.

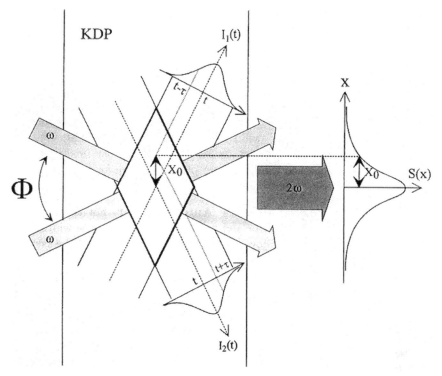

Fig. 7.16. Detail of Fig. 7.15. The second-harmonic signal at frequency 2ω is generated by the KDP crystal, oriented for phase-matched type I SHG for the two beams intersecting at an angle Φ. The crossing of the pulse wavefronts induces second-order process and transforms autocorrelation in time into a spatial intensity distribution along the x axis

7.5 Spectral Measurements

Basically, the spectral measurement of an ultrashort pulse does not require any special tools. The spectrum associated with femtosecond laser pulses only requires a moderate resolution but one has to control carefully the oscillating background of the laser. In fact, most of the conventional oscillators can also run fully CW and, as a result, will introduce a sharp line into the spectrum. The correct optimization of the laser source requires one, then, to monitor this behavior. In some commercial products, the response of a Fabry–Pérot interferometer to either the CW or the pulsed regime is used as an automatic control of the laser pulse.

As already said, the FWHM of the laser spectrum, $\Delta\nu$, can be related to the measured pulse width Δt through autocorrelation techniques. If one assumes a transfom-limited laser pulse, the product $\Delta t \, \Delta\nu$ can be accurately computed as a function of the actual pulse shape (see Table 7.3).

For the "standard" sech^2 pulse shape, this "magic" product is (in appropriate units)

$$\Delta t \, \Delta \nu = 0.315. \qquad (7.15)$$

Any deviation from this value will reflect a phase problem completely missed in all the preceding measurements. It has been shown that completely different laser behaviors can be obtained starting from the same setup. In an antiresonant colliding-pulse dye laser, in either passive or hybrid mode-locked operation, the same autocorrelation trace can be obtained [7.6], generated by completely different spectra. In this case, gain and saturation dynamics strongly modify the pulse phase and there is a need for a more complete characterization.

We will now describe new approaches to retrieve the pulse in terms of amplitude and phase and then to compute the expected spectrum.

7.6 Amplitude–Phase Measurements

In the simple approach of the temporal characterization presented in section 7.4.2, the real pulse shape is hidden within an autocorrelation function. Although these autocorrelation techniques have been widely used in all the diagnostic approaches, they are far to qualify unambiguously the laser pulse electric field (amplitude and phase). The complete characterization of ultrashort pulses is of great importance in many femtosecond experiments, such as those involving coherent and optimal control, where complicated pulses must be generated using elaborate pulse-shaping techniques. Although some improvement can be made and few assumptions can be made to overcome this limitation, the inherent symmetry of the interferometric setup remains. For example, one can infer the actual pulse shape by analytically comparing the recorded interferometric autocorrelation trace and the theoretical one with a given standard shape (either Gaussian or sech^2 with a given phase), and thus have a better control. This will find some limitations, first experimentally, due to signal over noise ratio, and then theoretically, as the analytical shape is not clearly established [7.8]. An alternative combining interferometic autocorrelation and spectrum has also been proposed and demonstrated convincing results [7.9].

A more rigorous approach can be found in the simultaneous measurement of the amplitude and phase of the laser pulse. Two different approaches have been demonstrated either in the frequency domain, where the time of arrival of a given frequency is recorded, or in the time domain, where frequencies are recorded at a given time. The first approach, referred to as FDPM (Frequency Domain Phase Measurement) [7.10–13], is equivalent to a sonogram usually used in acoustic wave analysis, while the second is widely nicknamed as FROG (frequency resolved optical gating) [7.14–17].

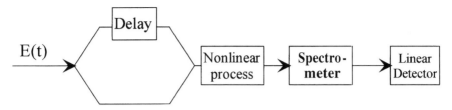

Fig. 7.17. Principle of FROG technique. The pulse to be measured is split into two replicas, which are crossed in a nonlinear optical medium to create a signal field, as in autocorrelation. The signal pulse is then spectrally resolved

7.6.1 FROG Technique

In the FROG approach shown in Fig. 7.17 an nth-order correlation function is spectrally resolved. The nonlinear device can be either a Kerr gate (3rd-order process) or more commonly a doubling crystal, in which case the technique is called "SHG FROG".

A FROG trace is obtained in the (ω, τ) plane, where the projection along the frequency axis gives the classical autocorrelation function. The amplitude and phase of the incident laser field $E(t)$ are then retrieved from the FROG trace by using an iterative algorithm (usually convergent in realistic cases).

For example, in SHG FROG, the frequency-doubled light of the autocorrelation forms the signal beam:

$$E_{\text{sig}}^{\text{SHG}}(t, \tau) = E(t)E(t - \tau). \tag{7.16}$$

This signal is imaged onto the entrance slit of an imaging spectrometer, where the relative time delay (τ) between the two replicas is parametrized along the slit. The spectrometer disperses the light in a direction perpendicular to the slit, resulting in a two-dimensional image with delay time and frequency as axes. The linear detector (a CCD array) detects the intensity (the FROG trace):

$$I_{\text{FROG}}(\omega, \tau) = \Big| \int_{-\infty}^{+\infty} E_{\text{sig}}(t, \tau) \exp(\mathrm{i}\omega t) \, \mathrm{d}t \Big|^2. \tag{7.17}$$

This approach has been extensively studied and documented [7.16]. Recently, a very useful experimental setup was proposed, which greatly simplifies the procedure. The so-called single-shot GRENOUILLE device uses a Fresnel biprism, which replaces the beam splitter and the delay line, followed by a thick doubling crystal, acting as both the nonlinear-optical time-gating element and the spectrometer. Spectral resolution proceeds from the output light of the crystal, which is angularly resolved. By use of cylindrical lenses a SHG FROG trace is directly formed on the CCD camera. The amplitude and phase of the laser field are retrieved using the same FROG algorithm.

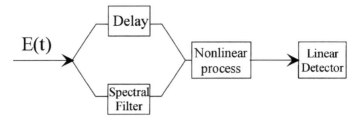

Fig. 7.18. A portion of the pulse spectrum is selected by a spectral filter in one arm of the interferometer. The cross-correlation of the selected portion with the input pulse is generated by crossing the two in a nonlinear medium (e.g. an SHG crystal)

7.6.2 Frequency Gating

One can also spectrally filter the laser pulse before gating it with the original pulse, in an arrangement of the kind shown in Fig. 7.18 [7.10].

At each frequency ω_i, one scans the delay τ to find the position of the maximum of the cross-correlation τ_i, while measuring the amplitude $A(\omega_i)$. In the (ω, τ) plane, one obtains directly the derivative of the spectral phase as a function of frequency:

$$\tau(\omega_i) = \frac{\partial \Phi}{\partial \omega}(\omega_i). \tag{7.18}$$

This function is easily integrated and gives the characteristic of the pulse in the frequency domain. There is here no need of any algorithm, and most of the time this is sufficient to fully chararacterize the pulse.

Here, experimentally, the spectral filtering is not trivial, as one has to filter out a given frequency without altering its chronology. This is done using a zero dispersion delay line, highlighted (ZDL) in the sonogram scheme presented in Fig. 7.19.

This device, discussed in Chap. 6, usually stretches or compresses the impinging laser pulse depending on the respective spacing of the lenses and gratings. When precisely tuned, i.e., the first grating image exactly corresponds to the second grating position, the output pulse exactly reproduces the input one, whereas in the symmetry plane (at the Fourier plane), all the pulse frequencies are uncoupled and can be safely selected using a simple slit. By moving the slit across this plane, a given frequency is selected and nonlinear process, such as an up-conversion with a reference pulse (the unfiltered one), directly gives the time of arrival. A sequential scanning of the slit position and the delay for the up-conversion map the pulse parameters, as presented in the Fig. 7.20. Any deformation of the contour lines can be related to deviation from a prefect Fourier-transformed laser pulse in a very similar way as the FROG trace does. Obviously, a thinner slit gives a better spectral resolution but a worse timing precision. This apparent limitation can be easily overcome as many numerical techniques of (a known) signal analysis can be used [7.18].

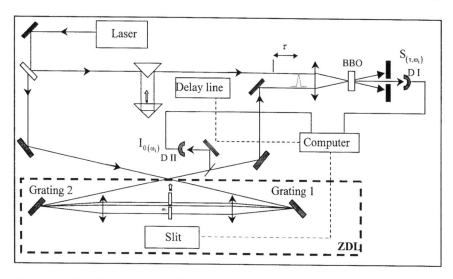

Fig. 7.19. FDMP technique. The enclosed part highlights the zero dispersion line (ZDL) used to frequency-filter the laser pulse. D I, D II: detectors.

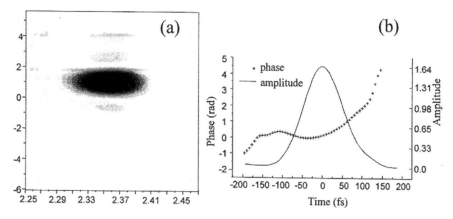

Fig. 7.20. (a): Sonogram of a Ti:Sa oscillator as recorded with the FDMP technique. The satellite along the delay axis is due to the cross correlation between a Dirac (the reference laser pulse) and a filtered pulse with a square window slit; (b): Amplitude and phase of a Ti:Sa oscillator corresponding to the sonogram

Basically, the sonogram technique is not easily seen as a single-shot technique as the FROG one.

7.6.3 Spectal Interferometry and SPIDER

Recently, an old point of view, referred to as shearing interferometry, was proposed as a powerful alternative to pulse measurements. Here, the spectral

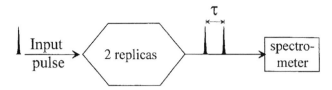

Fig. 7.21. Schematic of the spectral interferometry technique

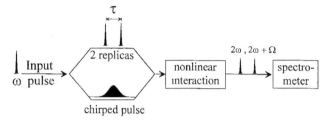

chirped pulse

Fig. 7.22. Schematic of the SPIDER technique: The two up-converted pulses have slightly different frequencies.

interferogram of two delayed pulses (delay by a time τ) can reveal the phase and amplitude characteristic:

$$S_{(\omega)} = |\tilde{E}_{0(\omega)} + \tilde{E}_{(\omega)} e^{i\omega\tau}|^2 = \tilde{I}_{0(\omega)} + \tilde{I}_{(\omega)} + 2\sqrt{\tilde{I}_{0(\omega)} \tilde{I}_{(\omega)}} \cos(\varphi_{(\omega)} - \varphi_{0(\omega)} + \omega\tau). \tag{7.19}$$

In the basic principle of this linear technique, presented in Fig. 7.21, the pulse to be characterized is combined with a reference-delayed pulse (delay τ) at the input slit of a grating spectrometer. A Fourier analysis of the spectral interferogram gives the differences is spectral amplitude (by the contrast) and spectral phase (by the beating frequency along the spectrum) between the two pulses. As always, the need for a reference pulse greatly reduces the convenience of this approach, as one needs, to prepare the reference pulse form the pulse under measurement [7.19–20].

Some improvements have been implemented, and the power of this technique is now fully demonstrated in commercially available instruments. The acronym SPIDER (Spectral Phase Interferometry for Direct Electric field Reconstruction) describes the following setup, where the two spectrally interfering pulses are also spectrally sheared of a known amount. Thus, all the ambiguities coming form the replica approach vanish and quasi-single-shot measurements are routinely performed and enable the precise tuning of the low repetition rate of amplified femtosecond pulses. [7.21–24].

The scheme of the instrument is presented in Fig. 7.22. First, the measured pulse is split in three parts by a thin plate. The two reflections, delayed by a time τ (due to the etalon thickness), are up converted in a doubling crystal with the remaining pulse on which a controlled chirp has been applied through spectral dispersion (stetcher or glass sample). Thus, the two up-converted pulses, with different center frequencies (2ω and $2\omega + \Omega$), exhibit spectral shear interference fringes that are analyzed by a spectrometer. The spectral interferogram will then be written as:

$$S(2\omega) \propto \left| \tilde{E}_{0(2\omega+\Omega)} + \tilde{E}_{(2\omega)} e^{i2\omega\tau} \right|^2$$
$$= \tilde{I}_{0(\omega+\Omega)} + \tilde{I}_{(2\omega)} + 2\sqrt{\tilde{I}_{0(\omega+\Omega)}\tilde{I}_{(2\omega)}} \cos(\varphi_{(2\omega)} - \varphi_{0(2\omega+\Omega)} + 2\omega\tau),$$
$$(7.20)$$

and thus the amplitude and phase of the laser pulse are provided by a standard Fourier analysis.

References

[7.1]   J.C. Diels, J. Fontaine, W. Rudolph: Rev. Phys. Appl. **22**, 1605 (1987)
[7.2]   J.-C. Diels, W. Rudolph: Ultrashort Laser Pulse Phenomenon: Fundamentals, Techniques and Applications on a Femtosecond Time Scale, Academic Press, Boston (1996)
[7.3]   J.K. Ranka, A.L. Gaeta, A. Baltsuka, M.S. Pshenichnikov, D.A. Wiersma: Opt. Lett, **22**, 1344 (1997)
[7.4]   D.T. Reid, M. Padgett, C. McGowan, W. Sleat, W. Sibbett: Opt. Lett **22** 233 (1997)
[7.5]   J.P. Chambaret, C. Dorrer, M. Franco, A. Mysyrowicz, B. Prade, S. Ranc, J.F. Ripoche, J.P. Rousseau: "High-Dynamic Range Measurements of Unltrshort Pulses," UltraFast Optics Conference (1999)
[7.6]   A. Brun, P. Georges, G. Le Saux, F. Salin: J. Phys. D **24**, 1225 (1991)
[7.7]   N. Jamasbi, J.C. Diels, L. Sarger: J. Modern. Opt. **35**, 1891 (1988)
[7.8]   J.-H. Chung, A.M. Weiner: IEEE J. Sel. Top. Quantum Elect. 7, 656 (2001)
[7.9]   J.W. Nicholson, J. Jaspara, W. Ruldolph, F.G. Omenetto, A.J. Taylor, Opt. Lett **24** 1774 (1999)
[7.10] J. Chilla, O. Martinez: IEEE J. Quantum Elect. **27**, 1228 (1991)
[7.11] J. Chilla, O. Martinez: Opt. Lett. **16**, 39 (1991)
[7.12] D.T. Reid, I.G. Cormack: Opt. Lett. **27**, 658 (2002)
[7.13] V. Wong, I.A. Walmsley: J. Opt. Soc. Am. B **14**, 944 (1997)
[7.14] R. Trebino, D.J. Kane: J. Opt. Soc. Am. A **11**, 2429 (1993)
[7.15] G. Taft, A. Rundquist, M.M. Murname, H.C. Kapteyn, K.W. Delong, R. Trabino, I.P. Christov: Opt. Lett. **7**, 743 (1995)
[7.16] K.W. DeLong, D.N. Fittinghoff, R. Trebino, B.K. Kohler, K. Wilson: Opt. Lett. **19**, 2152 (1994)
[7.17] R. Trebino, K.W. DeLong, D.N. Fittingholf, J.N. Sweetser, M.A. Krumbugel, B.A. Richman, D.J. Kane: Rev. Sci. Inst. **68**, 3277 (1997)
[7.18] S. Rivet; "Carachérisation complète d'un faisceau laser impulsionnel femtoseconde: mise en évidence et analyse du couplage spatio-temporel dans la propagation linéaire et non linéaire", PhD thesis, Univesité Bordeaux 1 (2001) http://147.210.235.3/proprietes.html?numero_ordre=2448
[7.19] L. Lepetit, G. Cheriaux, M. Joffre: JOSA B **12** (12), 2467 (1995)
[7.20] C. Froehly, A. Lacourt, J.C. Vienot, J. Opt. (Paris) **4**, 183 (1973)

[7.21] C. Iaconis and I.A. Walmsley: Opt. Lett. **23**, 792 (1998)

[7.22] C. Dorrer: JOSA B **16** (7), 1160 (1999)

[7.23] L. Gallmann, D.H. Sutter, N. Matuschek, G. Steinmeyer, U. Keller, C. Iaconis, I.A. Walmsley: Opt. Lett. **24**, 1314 (1999)

[7.24] C. Dorrer, B. deBeauvoir, C. Le Blanc, J.-P. Rousseau, S. Ranc, P. Rousseau, J.-P. Chambaret, F. Salin: Appl. Phys. B **70** (7), S77 (2000)

Spectroscopic Methods for Analysis of Sample Dynamics

C. Rullière, T. Amand and X. Marie

With 42 Figures

8.1 Introduction

In the previous chapters of this book, we dealt with ultrashort laser pulses by themselves, without paying attention to their possible uses. Since you are now familiar with the methods of generation of ultrashort laser pulses, with their manipulation and with the associated difficulties, it is now time to consider an important topic: what can we do with these ultrashort laser pulses?.

One important application domain of ultrashort laser pulses is the behavior analysis of a sample perturbed by such a light pulse. Let us explain what exactly we mean by "sample" and "behavior analysis" by giving some typical examples. The sample can be a semiconductor, in which the light pulse will create charge carriers by excitation into the conduction band. In this case, after being created, the charge carriers may migrate or annihilate, and behavior analysis will provide information on the migration or annihilation dynamics of the charge carriers. The sample can also consist of a molecule or of an ensemble of molecules, which after absorption of photons from the light pulse will be changed photochemically into other kinds of molecules: the photoproducts. Behavior analysis will lie in observation of the transformation of the molecules in order to identify the photoproducts or the intermediate chemical species in the transformation pathway, or in order to study the associated reaction rates. But the sample can also be a simple piece of glass, which under the action of the light pulse will see its refractive index temporarily changed and, as a consequence, its optical properties too. In this case, behavior analysis will consist of measuring these changes and observing how long they last.

This chapter will deal with experimental methods which use ultrashort laser pulses to make such behavior analysis possible.

However, in this chapter we will assume generally that the laser-pulse durations are long compared to the dephasing time and that the analysis is made after the perturbed systems have lost their phase. Oscillating dipoles, for example, will oscillate in phase until dephasing processes induce phase loss. If analysis is made during the in-phase conditions, coherent processes may be superposed on incoherent processes. This very interesting aspect will be described seperately in the next chapter because it deserves specific comments and explanations. The techniques described in the present chapter are associated with mainly with incoherent processes.

First of all, why are ultrashort laser pulses interesting to use for behavior analysis of light-induced perturbations?

During the perturbation of a sample, it is generally difficult to make measurements. It is usually necessary to wait till the end of the perturbation before analyzing the consequences of the perturbation. However, if some processes are very fast compared to the duration of the perturbation, these processes will be hidden during the perturbation so that only those processes which are slower than the pulse duration will be observed. Some slower processes can be the consequences of faster processes. In this case it will be difficult to understand the origin of the slower processes in the absence of information on the faster processes. Therefore, the shorter the perturbation, the easier it will be to understand the faster processes. Clearly, the unique property of ultrashort laser pulses, namely their very shortness, makes them ideally suited for the initial perturbation of a sample.

But sample evolution should also be analyzed as a function of time. Fine-scale analysis requires detection methods with response times which are as short as possible, so as to lose as little information as possible. Using ultrashort laser pulses for the analysis will therefore be very suitable for overcoming this response-time problem. Since the response time is limited by the pulse duration, very short pulses will improve the time resolution.

Last but not least, ultrashort laser pulses can have a very large peak power, even for weak pulse energies. Thus, intensity effects may appear at moderate energy, so that data can be taken on the sample while remaining below its damage threshold. This specific aspect of ultrashort laser pulses must be kept in mind.

Short pulse duration and high peak power are the two main characteristics of the ultrashort laser pulses that have been decisive in the behavior analysis of samples under light perturbation. Let us now describe the various methods for behavior analysis, and let us give a few pedagogic examples of applications showing the potential of these methods.

8.2 "Pump–Probe" Methods

8.2.1 General Principles

These methods need at least two ultrashort light pulses. The first one, the so-called "pump" pulse, perturbs the sample at time $t = 0$. The second one, delayed with respect to the "pump" pulse, crosses the perturbed sample and acts as a probe (this is the so-called "probe" pulse).

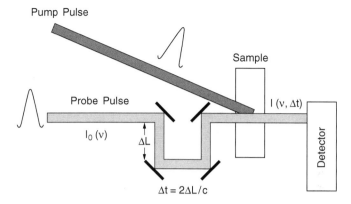

Fig. 8.1. Typical scheme of a "pump–probe" experiment

The action of the pump pulse on the sample may be analyzed in two different ways: (1) by comparing the modifications of the probe pulse characteristics (intensity, phase and wave vector) after crossing the sample, before and after the action of the pump pulse: we are speaking of the time-resolved absorption technique; (2) by observing new effects created by the probe itself before and after the action of the pump pulse: we are speaking of Raman scattering spectroscopy, laser-induced fluorescence and coherent anti-Stokes Raman spectroscopy (CARS), for example. The following sections of this chapter will describe these different methods.

8.2.2 Time-Resolved Absorption in the UV–Visible Spectral Domain

With this method, changes in the absorption spectrum of a sample after perturbation by an ultrashort pulse can be observed and measured.

Changes in the absorption spectrum may involve the spectral characteristics (new transitions appearing under perturbation) and/or the absorbance characteristics (increase or decrease of the optical density). The electronic absorption spectrum is the signature of a sample. Spectral analysis of changes such as new electronic transitions will give the signature of the new electronic states created by the pump pulse (for example photochemical products), while temporal analysis will yield the dynamics of the transformations (for example chemical reaction rates). The principle of this method is shown in Fig. 8.1.

At time $t = 0$, the pump pulse perturbs the sample. At time $t + \Delta t$ the probe pulse crosses the sample, Δt being tunable by means of an optical delay line. The detector then measures the probe pulse intensity before ($I_0(\nu)$) and after ($I(\nu, \Delta t)$) the perturbed sample. Applying the Beer–Lambert law, we can write

$$I(\nu, \Delta t) = I_0(\nu) \times 10^{-\varepsilon_\nu N(\Delta t) l}, \qquad (8.1)$$

where ε_ν is the absorption coefficient of the sample at frequency ν, $N(\Delta t)$ is the population absorbing at time Δt at frequency ν and l is the length of sample excited.

In fact, the quantity of interest is the measured optical density (OD), defined as

$$\text{OD}\ (\nu, \Delta t) = \log \frac{I_0(\nu)}{I(\nu, \Delta t)} = \varepsilon_\nu N(\Delta t)l. \tag{8.2}$$

For a fixed delay Δt, at each frequency ν a quantity will be measured which is proportional to ε_ν: we thus construct the absorption spectrum of the excited sample at time Δt.

For a given frequency ν, varying Δt will yield a quantity proportional to $N(\Delta t)$: we can now construct the population dynamics. If the population N decays exponentially, the decay time τ can be measured by varying Δt according to

$$\ln \text{OD}\ (\Delta t) = \ln N(0)\varepsilon_\nu l - \Delta t/\tau. \tag{8.3}$$

The above considerations form the basis of the method. Let us now present its practical aspects.

8.2.2.1 Probe Characteristics. The probe can have a broad spectrum, or it can be quasi-monochromatic. Let us recall that in any case, as was thoroughly explained in the previous chapters, the spectral width will be limited by the relation $\Delta\nu\,\Delta\tau = \text{constant}$. Probe pulses with narrow spectral ranges can be generated from dye lasers (CPM mode), tunable solid-state lasers (Ti^{3+}:sapphire, etc.), OPOs, OPAs or OPGs, as explained in previous chapters. The probe wavelengths should be in a spectral domain in which it is known (or expected) that some species created by excitation will be present in the sample at a certain time Δt after excitation and will have resonant electronic transitions in the probe spectral domain. When such a probe pulse is used, one assumes that the behavior of the sample has already been characterized by other preliminary studies which have indicated the frequency domain of interest. When this spectral domain is not known, it is best to use a probe which is spectrally as broad as possible.

In this case two possibilities exist: (1) continuum generation, and (2) broadband laser emission.

(1) *Continuum Generation.* Focusing a high-peak-power light pulse in a medium (liquid, glass, etc.) generates a continuum of light (Sect. 4.3.5) with a very broad spectral domain, which may even appear as white light [8.1]. The origin of this process is well described elsewhere [8.2], and is mainly governed by self-phase-modulation and stimulated Raman emission. The rather good directivity of this white-light pulse makes it possible to use it as a spectrally broad probe and to record transient absorption at different wavelengths simultaneously.

As to the medium used to generate this light continuum, each scientist generally has his or her own method, as in cooking. But the most popular

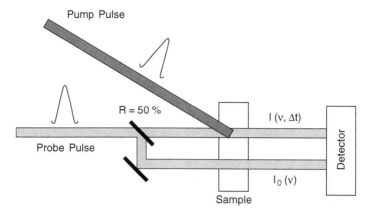

Fig. 8.2. Dual-beam pump–probe experiment

media are D_2O, H_2O, ethylene glycol, phosphoric acid and quartz glass. The main problem is to produce a spectral distribution which is as flat as possible. Depending on the medium, the wavelength and the peak power available to generate the continuum, it may prove necessary to use filters (liquids or solids) to flatten this spectral profile [8.3].

(2) *Broadband Dye Laser.* Another method which is sometimes employed is to generate amplified spontaneous emission (ASE) in a laser medium, generally a dye (see Chaps. 3 and 4). Used as a probe, ASE is spectrally less broad than a continuum, but it usually has a stronger intensity, which can be an advantage in some cases. However, the shape and duration of these probe pulses, which depend strongly on the pumping conditions, need to be controlled very precisely. Nevertheless, it can be a useful alternative to continuum generation in those cases which do not need a very broad spectral range.

8.2.2.2 Detection Systems. To improve the sensitivity of the measurement, it is better to record the spectral distribution of the probe simultaneously in the presence and in the absence of the perturbation of the sample at each laser shot. This is done, as shown in Fig. 8.2, by splitting the probe beam into two equal beams: the first partial beam crosses the perturbed part of the sample while the other beam crosses an unperturbed part of the sample. These two beams are then directed onto the slit of a spectrograph at two different heights, and are read by a double photodiode array or by a CCD camera coupled to a microcomputer. The optical density $OD(\nu, \Delta t)$ can thus be measured directly as a function of the frequency.

When simultaneous measurements at different wavelengths are not necessary, the detectors may consist of two photodiodes (or photomultipliers) coupled to a lock-in amplifier.

8.2.2.3 Experimental Tricks to Avoid Artifacts. Even though these experiments are based on a very simple idea, experimental artifacts may still arise. They are related to the probe and pump pulse polarizations, to the GVD along the optical pathways, to noise of various origins and to the detection system.

(1) *Polarizations of the Pump and Probe.* If, for example, the exciting beam is linearly polarized, the sample will be perturbed preferentially along the polarization direction. If the beam is resonant with an electronic transition, the absorbing entities whose transition moment is oriented along this polarization direction will be excited preferentially, and the spatial distribution of perturbed entities will be anisotropic. If these entities can reorient themselves, a spatial randomization will occur which will be controlled by the mean reorientation time. If the probe is polarized parallel to the pump beam, it will probe preferentially the excited entities whose transition moment is oriented in the polarization direction, and the measurements will be perturbed by the reorientation process. In the same way, if the probe beam is polarized perpendicularly to the pump beam, the measurement will be affected by the molecules which have reoriented to the polarization direction of the probe. The dynamics measured for perpendicular and parallel polarizations of the probe beam will not be the same. The dynamics will depend not only on the lifetime of the population of the perturbed entities, but also on their reorientation time. Depending on the process of interest, the respective polarizations will have to be adjusted. For example, to measure the dynamics free from reorientation effects, the angle between the pump and probe polarizations should be set at $54.6°$ (the so-called "magic" angle in the literature [8.4]). On the other hand, in order to measure exclusively the effects of reorientation, one must measure the dynamics for parallel ($OD_\parallel$) and perpendicular ($OD_\perp$) polarizations independently. The depolarization ratio, defined as $R(t) = [OD_\parallel(t) - OD_\perp(t)]/[OD_\parallel(t) + 2OD_\perp(t)]$, is proportional to $\exp(-t/\tau_{or})$ and gives the reorientation time τ_{or} [8.5].

(2) *Probe and GVD.* Not all the probe frequencies travel through the optical components at the same speed, including the medium used for continuum generation, because of GVD. As a consequence, the different spectral components do not all reach the sample at the same time: the blue wavelengths arrive at the sample later than the red wavelengths. Therefore, if the delay between the pump beam and the probe beam is changed, the shape of the recorded spectrum may also change. Suppose that the pump beam arrives at the sample later than the probe beam. In this case the probe beam will not be perturbed. If this delay is changed a little, the red wavelengths could still arrive before the pump beam and not be affected. But now the blue wavelengths could be in phase with the pump beam, and would thus be perturbed. Only the blue part of the spectrum would then reflect the perturbation. Changing this delay more and more would shift the interaction into the red part of the spectrum.

It is possible to remove this effect, by using gratings for example, as in a stretcher/compressor (see Chap. 6), and to rephase all the spectral components of the probe on the sample. But this procedure is somewhat delicate and it is better to correct the spectra numerically by using appropriate software. This can only be done if the dispersion for the continuum is well known. To obtain an exact measurement of the GVD, it should be measured by several different methods.

The first method consists of measuring two-photon interaction. The second one uses frequency mixing in a nonlinear crystal. The third one uses the electronic optical Kerr effect (OKE).

The first method is described in [8.3]. The absorbance of a sample of benzene diluted in methanol is probed at different probe wavelengths. The depletion process is a coherent two-color, two-photon absorption involving absorption from the pump *and* from the probe. This instantaneous process requires a perfect temporal coincidence between the two pulses (pump and probe) to reach maximum efficiency. Changing the delay time between the pump and probe makes it possible to determine at what time the maximum signal is reached, i.e. the time at which the pump and probe pulses coincide perfectly for a given wavelength. In this way, one obtains the dispersion curve, which can then be used for the numerical corrections of the spectra.

The second method consists of obtaining the sum frequency of the probe and the pump using a nonlinear crystal. The crystal should be chosen to be as thin as possible so as to avoid dispersion effects. In this method the maximum signal at the sum frequency is obtained when the two pulses coincide perfectly, thus yielding the dispersion curve.

The third method uses the optical Kerr effect in a liquid such as hexane. A cell containing hexane is inserted between two crossed polarizers. The pump pulse, polarized at 45° relative to the polarizers, is directed into the cell and induces instantaneous birefringence (more details about this method will be given later in this chapter). The probe pulse is passed through the polarizers. When the probe wavelengths are exactly synchronized with the pump pulse in the cell, the induced birefringence allows these components to be transmitted and detected. By changing the time delay between the pump and the probe, the dispersion curve can be measured. The cell and the entrance polarizers should be as thin as possible to avoid GVD in these components and to increase the sensitivity of the method.

In our opinion the third method is the simplest, because n-hexane is strongly nonlinear, resulting in very efficient transmission and detection mechanisms.

(3) *Noise and Reference Beam.* To measure transient spectra, the direct measurement of the ratio

$$I_0(\nu)/I(\nu, \Delta t),$$

as described in Sect. 8.2.2, is not sufficient. Indeed, some parts of the recorded signal are not directly related to the transient absorption and should be ac-

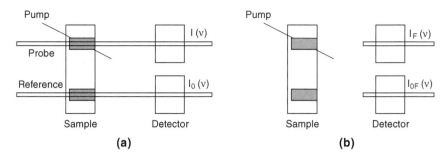

Fig. 8.3. Illustration of the different contributions to the recorded signals in pump–probe experiments. (**a**) The total signal recorded. (**b**) The contribution of the sample emission to the recorded signals in (**a**)

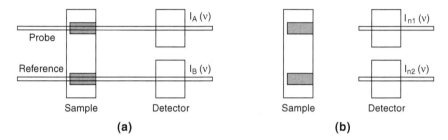

Fig. 8.4. Illustration of the different contributions to the recorded signals in pump–probe experiments. (**a**) Recorded signals allowing one to correct for spectral differences between the probe and test beams in the absence of excitation. (**b**) The contribution of noise (ambient light and electronic noise) to the recorded signal

counted for (e.g. ambient light, electronic noise and fluorescence or emission by the excited sample). In any case, the spectral profile of the two beams cannot be exactly the same, owing to their different optical paths. These effects should be taken into account as shown in Figs. 8.3 and 8.4.

Let us define the following quantities.

- $I(\nu)$ and $I_0(\nu)$ are the spectral distribution of the recorded signals in the presence of the pump, probe and reference beams in the sample (Fig. 8.3a).
- $I_F(\nu)$ and $I_{0F}(\nu)$ are the spectral distributions of the recorded signals when only the pump is present in the sample, and the probe and reference beams are absent (Fig. 8.3b). In this case the detectors record only sample emission and noise.
- $I_A(\nu)$ and $I_B(\nu)$ are the spectral distributions of the recorded signals in the absence of the pump beam but in the presence of the probe and reference beams (Fig. 8.4a). In this case the detectors record the spectral distributions of the probe and reference beams superposed to noise.

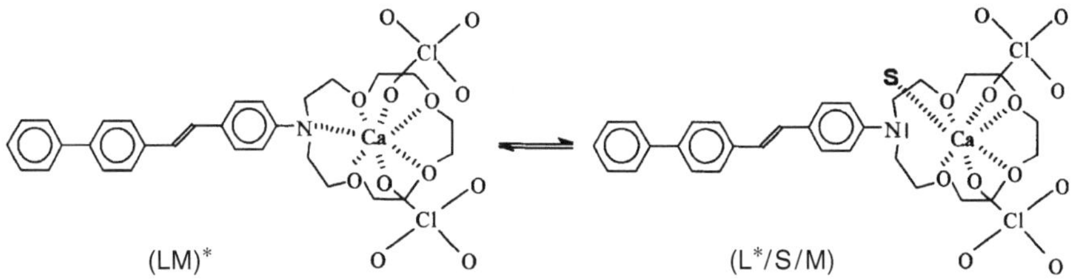

Fig. 8.5. Structure of a molecule involved in a cation photoejection mechanism: The PDS-crown (4-N-(1-monaza-4,7,10,13-tetraoxacyclopentadecane)-4'-phenyl-stilbene) molecule [8.6, 8]

- $I_{n1}(\nu)$ and $I_{n2}(\nu)$ are the recorded signals in the absence of all of the beams (Fig. 8.4b). In this case the detectors record only electronic and ambient noise.

With these definitions and for a given delay Δt, the real optical density, corrected for all parasitic light and all possible changes of the spectral distribution along the different optical paths, can be calculated according to

$$\text{OD}(\nu, \Delta t) = \log\left(\frac{I_0(\nu) - I_{0F}(\nu)}{I(\nu) - I_F(\nu)}\right)\left(\frac{I_B(\nu) - I_{n2}(\nu)}{I_A(\nu) - I_{n1}(\nu)}\right). \qquad (8.4)$$

Note that probe and reference beams also induce emission from the sample.

(4) *Detector Linearity.* Some problems may appear which are due to short pulses, especially in detectors using a light amplifier. Indeed, a great number of photons arriving on the photocathode within a short time interval may cause a saturation effect due to the limited current available, and this may affect the detection linearity. To overcome this problem, the pulse to be processed should be injected into a long optical fiber before entering the spectrograph. Propagation through the fiber will increase the pulse length by GVD by a sufficient amount to be in the linear regime of the detector. Some care must be taken to inject the right part of the pulse into the fiber.

8.2.2.4 Example of Application: Cation Ejection. To illustrate the possible applications of this technique, we shall describe here an example related to transient absorption changes during a photochemical reaction involving the breaking of a dative bond. Such events may occur in the molecules shown in Fig. 8.5.

In this kind of molecule, the emission properties are different depending on the presence or absence of Ca^{2+} cations in the solution [8.6]. In the absence of calcium ions in the solution, this molecule has an emission transition at about 450 nm. In the presence of calcium ions in the solution, the molecule in its ground state traps a Ca^{2+} cation inside the ether-crown, where it forms a dative bond with the nitrogen atom. In this state, the emission of the molecule is shifted to the blue, to 375 nm.

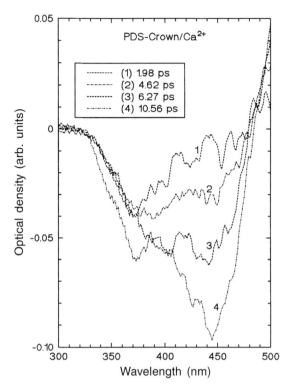

Fig. 8.6. Gain bands for different time intervals after excitation of a solution of 10^{-4} M PDS-crown and 5 mM $Ca(ClO_4)_2$ in BuCN at $20°$ C (see [8.7])

By absorbing a photon, the dative bond can break, releasing the Ca^{2+} ion and putting the molecule into a state in which it ignores the calcium ion. The emission at 450 nm, corresponding to the breaking of the dative bond, should show up on the experimental emission curve, recorded as a function of time. Thanks to pump–probe techniques, this chemical process can be observed. As a matter of fact, if the gain is sufficient, amplification of light may occur at the emission transition. This behavior is shown in Fig. 8.6. Just after excitation, the band at 450 nm is absent and the gain spectrum corresponds to the spectrum of a molecule with a calcium ion bonded to the nitrogen atom. But as time passes, this bond is broken at a certain rate, and the band corresponding to the molecule without a calcium ion bonded to the nitrogen atom begins to appear and grows until the process is terminated. As long as this band at 450 nm is growing, the process of chemical-bond breaking is going on. In the case of this specific molecule, the breaking process has a rate of 8 ps^{-1} [8.7].

This example shows how powerful pump–probe experiments are for observing the kinetics of excited states or the evolution of their spectral properties. Let us now present some techniques used in the IR spectral range.

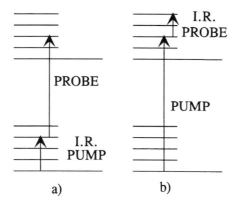

Fig. 8.7. Possible methods using IR pulses for recording behavior of modes of (**a**) ground or (**b**) excited states

8.2.3 Time-Resolved Absorption in the IR Spectral Domain

8.2.3.1 Principle of the Method. IR methods are based on the same principles as UV–visible methods, the only difference lying in the laser sources and the detection systems.

The IR sources may be OPOs, OPAs or diode lasers, as described in earlier chapters. Detectors must have a sensitivity extending into the IR spectral range (photodiode, etc.) and are usually coupled to IR spectrographs or to IR monochromators. As shown in Fig. 8.7, the excitation may lie in the IR, while the probing of the perturbed system takes place in the UV–visible range, or else the perturbation of the sample may occur in the UV–visible range, while the induced perturbations are probed in the IR. The choice of pump and probe frequencies will depend on the information we are seeking from the sample.

The probing of vibration modes in the ground state of a molecule follows Fig. 8.7a, while the probing of vibration modes of excited states follows Fig. 8.7b.

However, IR sources and detectors usually require a technology which is completely different from than used in the UV–visible domain. An interesting alternative method was presented ten years ago [8.8], and is described below.

8.2.3.2 Time-Resolved IR Spectroscopy Without IR Detection. This method, illustrated in Fig. 8.8, may appear paradoxical since it performs time-resolved spectroscopy in the IR domain using neither IR pulses nor IR-sensitive spectrographs or detection systems.

A continuous diode laser, emitting at an active IR vibration frequency ω_{IR} of the sample, is directed through the sample and then passes through a nonlinear crystal. A mode-locked Nd^{3+}/YAG laser is used for the synchronous pumping of two dye lasers. The first one (pump dye laser), which delivers the pump pulse, perturbs the sample and its frequency ω_{pump} is adjusted

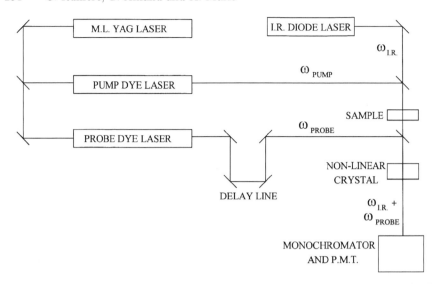

Fig. 8.8. Experimental setup for time-resolved IR spectroscopy (adapted from [8.8])

to an electronic transition of the sample, for example. The pulse emitted by the second dye laser, at frequency ω_{probe}, in the visible domain, is sent through a delay line and is superposed on the diode laser beam in a nonlinear crystal. The phase-matching angle of the crystal is adjusted to generate the frequency $\omega_{signal} = \omega_{probe} + \omega_{IR}$ by sum-frequency mixing. In the crystal, a pulse is generated in the visible domain. This pulse can be detected easily by conventional photodetectors (photomultiplier, etc.), and its intensity is proportional to the probe dye laser intensity and to the diode laser intensity after it has crossed the sample ($I_{signal} \propto I_{probe} \times I_{diode}$). The crystal acts as a booster for the diode laser intensity: if a small change occurs in the diode laser beam intensity after crossing the sample, it is multiplied by the probe laser intensity. Furthermore, the detected signal intensity will depend on the time delay between the pump pulse and the probe pulse. The photons of the diode laser which mix with the probe photons inside the crystal will be the ones that crossed the sample at a particular time after the sample was excited. Then, if after excitation some changes occur in the sample which affect the IR absorption, the intensity of the diode laser beam will be changed after crossing the sample and it will modulate the signal generated inside the nonlinear crystal. The intensity of the generated signal will therefore be controlled by the dynamics of the sample. In other words, by changing the optical delay, the IR activity of the probed vibration mode can be measured directly as a function of time.

As an example, an interesting application is described below.

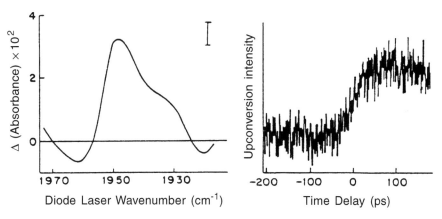

Fig. 8.9. *Left*: time-resolved infrared spectrum of MbCO (myoglobin–CO) taken 150 ps after excitation ($\Delta t = 150$ ps). Positive absorbance (Δ(absorbance) > 0) indicates an increase of IR transmission. *Right*: time-resolved transmission related to up-conversion intensity (see text) at the peak of the transient IR absorption spectrum (at 1949 cm^{-1}) [8.8]

8.2.3.3 Photodissociation of Hemin–CO and Myoglobin–CO. Hemin and myoglobin carbonyls have an active IR vibration near 1960 cm^{-1}. After excitation of a hemin–CO molecule, the CO group is dissociated, resulting in a decrease of the IR absorbance corresponding to the active CO mode. Using the setup described above, the photodissociation kinetics in these compounds can be recorded directly, as explained in [8.8] and illustrated in Fig. 8.9. In this figure, the recorded kinetics directly illustrate the dissociation on a picosecond timescale.

8.2.4 Pump–Probe Induced Fluorescence

Here we describe another application of the pump–probe method. In the previous section, the technique consisted of measuring the changes induced by the excited sample in the probe beam, thus providing information about the perturbed sample. In the present technique, the probe itself induces fluorescence emission from the excited sample and thereby yields information about the perturbed sample.

8.2.4.1 Application to "Femtochemistry": Photodissociation of ICN. Ever since 1949, when for the first time G. Porter observed the formation of a photoproduct on a millisecond timescale (thanks to the first flash photolysis experiment), chemists have been dreaming of observing directly the chemical reactions leading to the formation of photoproducts after photon absorption. In other words, of observing transition states in the chemical path, revealing the intermediate steps between a formed and a broken chemical bond. Thanks

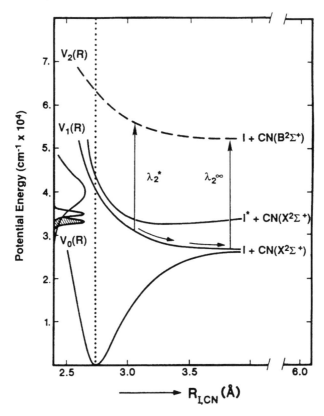

Fig. 8.10. Schematic illustration of femtosecond transition-state spectroscopy (FTS) and potential-energy surfaces of ICN [8.9]

to ultrashort laser pulses it is now possible to observe these transition states of the chemical transformation directly. One book would not be enough to describe all the results obtained in this field, and it is therefore beyond the scope of the present chapter to do so. The reader who is particularly interested in this topic can refer to [8.10], which gives an up-to-date overview of this rapidly expanding field. Nevertheless, it is interesting to present a pedagogic example illustrating the method: the photodissociation of ICN [8.9]. The chemical reaction is as follows:

$$ICN + h\nu \rightarrow (ICN)^* \rightarrow I + CN.$$

The potential wells for different ICN states as a function of the interatomic distance $R_{I,CN}$ are shown in Fig. 8.10. The experiment can be summarized as follows. The pump pulse brings the ICN molecule onto the $V_1(R_{I,CN})$ potential, which is dissociative. The interatomic distance $R_{I,CN}$ then increases until dissociation is reached and the I and CN fragments separate.

With the pump–probe method, it is possible to probe the different transition states corresponding to different interatomic distances $R_{I,CN}$ along this reaction coordinate. Indeed, if for a given delay Δt after excitation the probe beam is in resonance with the $V_1(R_{I,CN}) \rightarrow V_2(R_{I,CN})$ transition, $CN(B^2\Sigma^+)$ excited fragments will be created by the probe, as depicted in Fig. 8.10, and will emit fluorescence (transition from $CN(B^2\Sigma^+)$ to $CN(X^2\Sigma^+)$ states) The $CN(B^2\Sigma^+)$ fluorescence intensity will at first be proportional to the transition state population on the $V_1(R_{I,CN})$ potential curve, and then, depending on the probe wavelength, it will be proportional to the population corresponding to a well-defined interatomic distance $R_{I,CN}$. Choosing a particular probe wavelength will determine a specific observation window on the $V_1(R_{I,CN})$ curve, as well as a class of ICN molecules with a specific value of $R_{I,CN}$. Varying Δt will probe the dynamics of this population through the selected spectral window. By recording the fluorescence intensity of the $CN(B^2\Sigma^+)$ population, one can follow the movement of the ICN molecules on the photodissociation potential curve $V_1(R_{I,CN})$.

Figure 8.11 illustrates this behavior for different wavelengths of the probe beam. At high wavelengths (top left) the corresponding value of $R_{I,CN}$ is far from the equilibrium value of $R_{I,CN}$ in the ground state, and the fragments are well separated. A rise time of the $CN(B^2\Sigma^+)$ fluorescence of the order of 250 fs is observed, which corresponds to the photodissociation rate. But at lower probe wavelengths (bottom right), the corresponding value of $R_{I,CN}$ is close to the equilibrium distance in the ground state: the first observed step of the photodissociation consists in a rapid sliding of the ICN molecule along the potential curve. We arrive at the remarkable result that one single, specific transition state with a very short lifetime, of the order of a few ten femtosecond, is observed directly through the selected spectral window.

This example shows how the pump–probe method can reveal directly such processes as a photochemical transformation, along with the different transition states in the chemical pathway, thereby demonstrating how powerful this method is.

8.2.5 Probe-Induced Raman Scattering

Just as in UV, visible and IR absorption spectra, Raman spectra also characterize a sample in a given state. Observing the evolution of the Raman spectrum after perturbation will then give information about the transformation (and its associated structure) of a sample. In this section, we shall describe a technique in which the probe induces Raman scattering in the sample.

8.2.5.1 Time-Resolved Raman Technique. The principle of the method is displayed in Fig. 8.12. It is basically very simple. The pump pulse perturbs the sample at time $t = 0$. The probe pulse, at frequency ω_p, is incident on the sample at time Δt and creates Raman scattering at frequencies $\omega_{Ri} =$

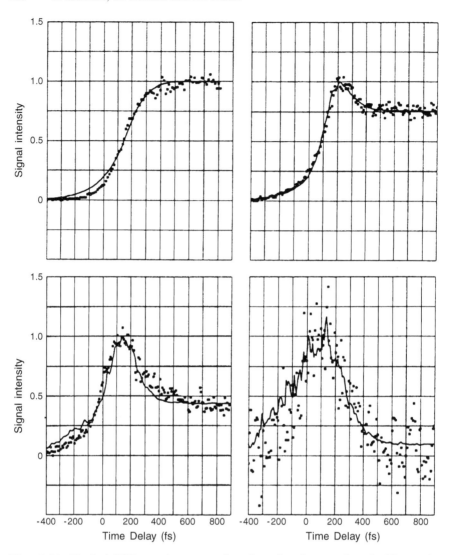

Fig. 8.11. Typical ICN transients as a function of probe wavelength. The probe wavelengths were (*top left*) 388.9 nm, (*top right*) 389.8 nm, (*bottom left*) 390.4 nm and (*bottom right*) 391.4 nm [8.9]

$\omega_{\mathrm{p}} \pm \sum_i \omega_{vi}$ (where ω_{vi} are the frequencies of the Raman modes of the sample), corresponding to the Raman spectrum of the sample at time Δt. The Raman spectrum is recorded by a detection system and is studied as a function of Δt.

8.2.5.2 Experimental Tricks. The quantum yield of Raman scattering is very weak and as much care should be used in this case as in "classical" Ra-

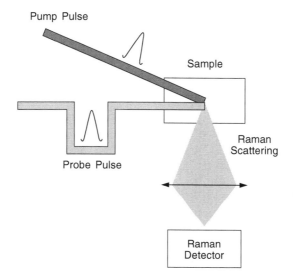

Fig. 8.12. Experimental setup for time-resolved Raman spectroscopy

man spectroscopy (removing probe beam and pump beam scattering by appropriate filters, using double or triple monochromators, sensitive detection, etc.). But in time-resolved Raman spectroscopy these problems are emphasized since the contribution to the Raman signal from the excited part of the sample is hidden under an important contribution from the solvent and from the unexcited part, even if the probe is in resonance with excited states. For example, it is generally difficult to obtain a concentration of excited molecules larger than 10^{-3} M. Extracting the contribution of the excited sample from the total signal requires a high reproducibility of the laser from shot to shot, since the signal will have to be recorded once without excitation and once with excitation. To overcome this problem, one can use a probe with a frequency ω_p close to one of the electronic transitions of the excited sample; in resonance conditions, the Raman scattering due to the excited molecules is greatly enhanced. However, to tune ω_p to the resonance frequency requires a previous knowledge of this frequency; this can be obtained by a preliminary recording of the absorption spectrum of the excited states using the time-resolved absorption technique described in Sect. 8.2.2.

Furthermore, to determine the spectral resolution, one must take into account the pulse duration. Spectral details may be hidden by the intrinsic bandwidth of the laser pulse. Time resolution can only be improved at the expense of spectral resolution, so that a compromise must be found for each specific experiment. Let us now present an application of this technique.

8.2.5.3 The Photochemical Sigmatropic Shift of 1,3,5-Cycloheptatriene. Among many other examples, a very nice application of the probe-

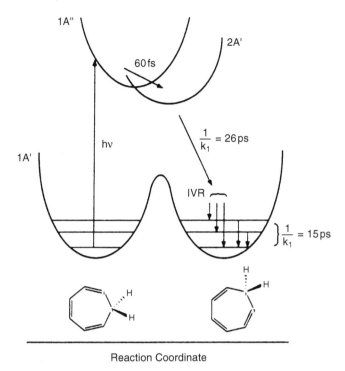

Fig. 8.13. Schematic reaction coordinate for the photochemical (1,7)-sigmatropic shift of 1,3,5-cycloheptatriene [8.11]

induced Raman technique is the photochemical sigmatropic shift of 1,3,5-cycloheptatriene [8.11]. In this molecule, as shown in Fig. 8.13, the double bonds change places along the heptatriene ring after photoexcitation, and during the process the molecule passes through a transition state (state $2A'$) in which the vibration pattern is completely changed compared to that of the ground state. In the ground state $1A'$, a Raman-active mode appears at the frequency $1536 \, cm^{-1}$, corresponding to the double-bond stretch. When the molecule is excited into the $2A'$ state, this mode disappears during the lifetime of the $2A'$ state, and then reappears when the $1A'$ state is populated again from the $2A'$ state. The dynamics of this photochemical process can be obtained by studying the dynamics of this particular Raman-active mode.

The time-resolved Raman spectra shown in Fig. 8.14 illustrate the recovery dynamics of the ground state population after photoexcitation, via the intensity of the Raman-active mode at $1536 \, cm^{-1}$. Some experimental details are given in the figure caption. This example is a good illustration of the convenience of this technique for elucidating photochemical processes.

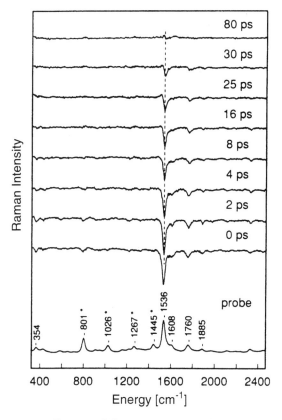

Fig. 8.14. Resonance Raman difference spectra of the photochemical (1,7)-sigmatropic rearrangement of 1,3,5-cycloheptatriene excited at 284 nm. A ground state spectrum of cycloheptatriene is given at the bottom of the figure for comparison. Lines due to the cyclohexane solvent are marked with asterisks. The ground-state (probe) spectrum has been divided by a factor of 3 [8.11]

8.2.6 Coherent Anti-Stokes Raman Scattering (CARS)

As mentioned above, Raman scattering has a very low quantum yield. When the excited sample is fluorescent, or if there is a high level of parasitic light (as in flames), Raman scattering is difficult to detect with a good precision. In comparison with spontaneous Raman scattering, coherent anti-Stokes Raman scattering (CARS) offers various advantages, the main ones being that the CARS signal is much more intense than the spontaneous Raman signal, and, furthermore, it is collimated in a well-known and precise direction. Taking advantage of this, a CARS signal can be observed with an excellent sensitivity even in the presence of a strong luminescent or ambient-light background [8.12].

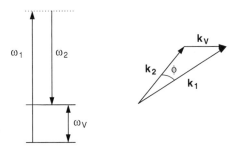

Fig. 8.15. Illustration of energy and momentum conservation laws for CARS, for two beams at frequencies ω_1 and ω_2

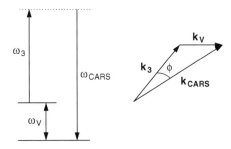

Fig. 8.16. Illustration of energy and momentum conservation laws for CARS, for the two beams at frequencies ω_3 and ω_{CARS}

Before presenting some experimental details concerning this technique, let us recall the principle on which it is based.

8.2.6.1 Basic Principle and Technique. If a sample has a vibration mode at frequency ω_V, this mode can be excited coherently by using two light pulses of central frequencies ω_1, ω_2. The frequencies (ω_1 and ω_2) and the wave vectors ($k_1 = n(\omega_1)\omega_1/c$, $k_2 = n(\omega_2)\omega_2/c$), where c is the speed of light in vacuum and $n(\omega)$ the refractive index of the material at frequency ω, must obey the energy and momentum conservation laws expressed by the following equations and depicted in Fig. 8.15:

$$\omega_1 - \omega_2 = \omega_V \quad \text{and} \quad k_1 - k_2 = k_V. \tag{8.5}$$

If, during the coherence time of the excited level which has been populated, a third pulse of frequency ω_3 is sent into the sample, a coherent pulse is generated at frequency $\omega_{\text{CARS}} = \omega_3 + \omega_V$ which travels in the direction shown in Fig. 8.16.

This process is nonlinear, and the CARS signal intensity depends on the third-order nonlinear optical susceptibility $\chi^{(3)}(\omega_{\text{CARS}}; \omega_1, -\omega_2, \omega_3)$ and on the intensity of the three pulses as

$$I_{\text{CARS}} \propto \left|\chi^{(3)}(\omega_{\text{CARS}}; \omega_1, -\omega_2, \omega_3)\right|^2 I(\omega_1)I(\omega_2)I(\omega_3). \tag{8.6}$$

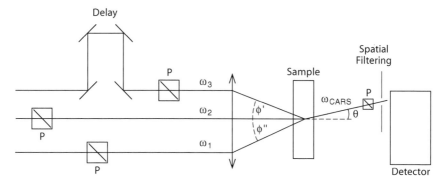

Fig. 8.17. Typical experimental setup for CARS measurements. P: polarizers

The signal at the frequency $\omega_{CARS} = \omega_3 + \omega_V$ may be very intense if the peak powers of the pulses are high. Since, owing to the conservation of momentum, this signal is generated in a well-defined direction (Fig. 8.16), it is possible to use spatial filtering to select only the signal of interest, among the parasitic light or fluorescence emitted in all directions. Thanks to this technique, the Raman spectrum of a sample can be recorded by tuning the pulse frequency ω_2, even under experimental conditions in which a "normal" Raman scattering signal would not be detectable.

The CARS signal also depends on the fraction of the population of the studied vibration which is "in phase", since the signal is related to the macroscopic quantity $\chi^{(3)} \propto \gamma^{(3)} \langle n \rangle$ ($\langle n \rangle$ is the average density of oscillators and $\gamma^{(3)}$ the molecular hyperpolarizability). Recording the signal as a function of the delay between the ω_1, ω_2 and ω_3 pulses will yield the dephasing time T_2 of the vibration. The experimental setup is shown in Fig. 8.17.

The CARS technique associated with ultrashort pulses is also an interesting tool to measure the dephasing time of the vibration population, as will be described in the next section.

8.2.6.2 Dephasing-Time Measurements in Benzonitrile. Before presenting some typical results, let us recall that the CARS signal is sensitive to polarization. Indeed, depending on the polarization directions of the ω_1, ω_2 and ω_3 pulses, different components of the susceptibility tensor $\chi^{(3)}(\omega_{CARS}; \omega_1, -\omega_2, \omega_3)$ may be observed using the proper polarization analysis [8.13]. It is beyond the scope of this book to present this technique in detail, but it is explained very well in [8.14]. In short, the idea is that using different combinations of pulse polarizations, one can, for instance, observe selectively the isotropic or the anisotropic components of the dephasing relaxation. A typical example is shown in Fig. 8.18 for the vibration mode of the C≡N triple bond of benzonitrile [8.15]. The dynamics of the CARS signal are plotted for several different polarization directions of the incoming beams and for differ-

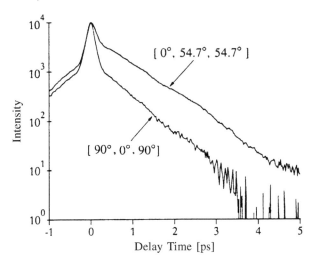

Fig. 8.18. Typical polarized time-resolved CARS, observed using neat benzonitrile. The angles in parentheses are the polarization directions with respect to the polarization of the ω_1 beam of the ω_2, ω_3 and ω_{CARS} beams respectively. The isotropic decay is measured for the $[0°, 54.7°, 54.7°]$ configuration and the anisotropic decay for the $[90°, 0°, 90°]$ configuration [8.15]

ent analyzer angles. This way, either the pure vibrational component or the vibrational plus rotational component of the dephasing time of the studied mode can be measured [8.15].

8.2.6.3 Time-Resolved CARS of Excited States. Another interesting application of the CARS technique is to record the CARS spectrum of an excited sample and to observe the changes of the vibrational mode in the excited state and the evolution of the population of the excited state. It can be an interesting alternative to the pump–probe absorption method described in Sect. 8.2.2, in cases in which different species are present in the sample, but in which it is impossible to get information about the different species because of accidental overlapping of electronic transitions, or even because of absorption which is too weak to be detected. Indeed, especially for large molecules, the widths of Raman bands are generally narrower than those in electronic absorption spectra, and overlap happens only occasionally. In this case CARS spectra, rather than the electronic absorption spectra, should be used as a signature of the different species whose population dynamics are to be recorded. Figure 8.19 shows an experimental setup in which the frequencies of the ω_1 and ω_3 pulses are identical. This particular configuration, called "degenerate" CARS [8.12], is simpler. Although, the dephasing time of the vibration mode cannot be obtained using this setup, it will yield the CARS spectrum of the vibration and its population dynamics.

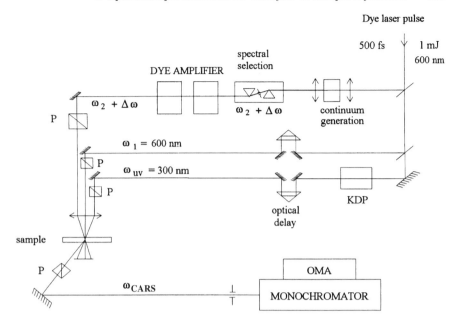

Fig. 8.19. Experimental setup for studying time-resoved CARS of excited states

In the degenerate (two-color) polarization-sensitive CARS technique, two pulses are required to generate the CARS signal: one pulse (the so-called "pump beam") with a fixed frequency ω_1, and a second pulse (the so-called "Stokes beam") with an adjustable frequency ω_2, such that $\omega_1 - \omega_2 = \omega_V$, the molecular vibration frequency to be observed. CARS experiments were performed using the apparatus shown in Fig. 8.19 in the following way [8.16].

The pulse of frequency ω_1 corresponding to 600 nm, generated by the laser system, was split into three parts by means of several partially reflecting mirrors. One part of the beam (10 % of the total energy at 600 nm) was used as the ω_1 pump beam to generate the CARS signal. Another part (about 40 % of the total energy at 600 nm) was focused into a 2 cm water cell by a 15 cm focal-length lens to produce a 2 ps light continuum extending from ~ 340 to ~ 900 nm.

This continuum was used as the Stokes beam with the variable frequency ω_2. A spectral selector, composed mainly of 60° prisms (TF5 heavy flint) and an adjustable slit, was used to select the desired spectral region without changing the time delay. The group velocity dispersion was minimized by adjusting the dispersive elements. A spectral bandwidth range of 2700 cm^{-1} was selected so as to study the CARS spectrum of the sample over a wide spectral range (frequency $\omega_1 - \omega_2 = \omega_V$ from 800 to 2500 cm^{-1}). The spectrally selected part of the continuum was then amplified by a two-stage dye amplifier.

The last part of the beam (50 % of the total energy at 600 nm) was frequency-doubled in a KDP crystal ($\omega_{UV} = 300$ nm) and used to excite the sample and to populate the excited state under study.

As described above, polarizers and analyzers were used to control the polarization directions of the various incoming pulses, and were carefully controlled and adjusted by means of wave plates and high-quality Thomson–Glan prisms. The ω_{UV}, ω_1 and ω_2 pulses were focused onto the sample by a 15 cm focal-length lens. The energies of these pulses could be adjusted by rotating the wave plates in front of the Thomson–Glan prisms. The relative delays between the ω_{UV}, ω_1 and ω_2 pulses could be adjusted with optical delay lines to record the dynamics of the CARS signal. The CARS pulse, generated in a different direction from the incoming pulses, was spatially filtered by an aperture, sent through a polarizer and focused on the slit of the detector spectrograph.

The CARS spectrum was then recorded according to the following procedure. Without UV excitation of the sample, the CARS signal was optimized and recorded by adjusting the propagation directions and the delays between ω_1 and ω_2. The UV pulse was then sent onto the sample and the CARS signal was recorded again for a given delay Δt_{UV} between the ω_{UV} pulse and the ω_1 and ω_2 pulses. The same procedure was repeated for different delays Δt_{UV} to study the time evolution of the CARS spectrum of the excited sample.

The calculation of the CARS spectral parameters can be performed by using a CARS fitting program which optimizes the vibrational wavenumber, bandwidth and amplitude for each Raman band [8.17]. A typical application of this experimental setup is described below.

8.2.6.4 Photoinduced Formation of Cations Studied by Time-Resolved CARS: 1,4 Diphenylbutadiene (DPB).

Among the possible processes which can occur after excitation of DPB, direct photoionization and production of ion radicals have been observed at nanosecond time resolution in polar solvents [8.18]. However, owing to the nanosecond time resolution, the formation rate of the radical ions could not be measured directly, since ultrashort laser pulses are needed to observe this process. Transient absorption experiments with subpicosecond time resolution were made, revealing the formation of the radical ion. Unfortunately, owing to the strong overlap between the radical-cation absorption bands and the main, broad $S_1 \to S_n$ absorption band, it was very difficult to deduce a reasonably accurate formation rate (rise time) for the radical cations in polar solvents [8.16]. In this situation, the time-resolved CARS technique appears well suited to overcome the overlap problem, since it should be able to resolve the corresponding Raman bands.

Typical transient resonance CARS spectra of excited and unexcited DPB dissolved in ethanol (measured in the 1100–1800 cm^{-1} anti-Stokes region) are shown in Fig. 8.20. After UV excitation, new Raman bands, which are due to this UV excitation, develop on top of the Raman solvent bands. The main new band near 1580 cm^{-1} is due to the radical cation formed after excitation.

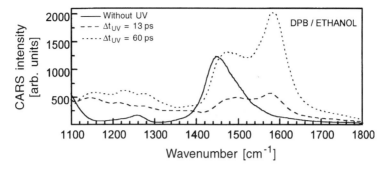

Fig. 8.20. Transient polarization-sensitive CARS spectra of DPB at different delays (Δt_{uv}) after UV excitation. The nonresonant CARS signal has been suppressed using polarization selection. The solvent and ground-state contribution is shown by the spectrum obtained without UV excitation. The DPB was dissolved in ethanol at concentration $c = 2.5 \times 10^{-3}$ M [8.16]

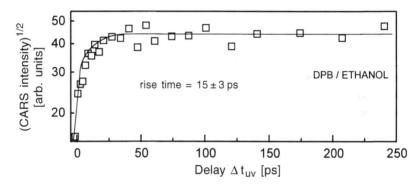

Fig. 8.21. Kinetics of the square root of the intensity of the DPB CARS signal in the $1580\,cm^{-1}$ band, after correction for the reabsorption process by scaling to the solvent bands [8.16]. The DPB was dissolved in ethanol at concentration $c = 2.5 \times 10^{-3}$ M

Its kinetics can be recorded, as shown in Fig. 8.21, and these yield a value for the formation time of the radical cation of DPB in the range 15 ± 3 ps, as well as an extremely slow decay time (> 1 ns), which is correlated with the time evolution of the DPB$^{\bullet+}$ radical cation [8.16].

This example shows that the time-resolved CARS technique is a powerful tool to study photophysical behavior when techniques such as transient absorption or spontaneous Raman spectroscopy are unable to do so. However, even if this technique appears very interesting, from the practical point of view there are some experimental tricks and conditions which should be known, and should be used routinely in the lab. Let us now introduce this aspect.

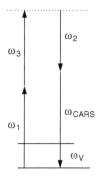

Fig. 8.22. Illustration of the nonresonant contribution in CARS experiments

8.2.6.5 Experimental Tricks. There are three fundamental issues which must be dealt with in order to obtain time-resolved CARS spectra: these are the phase-matching conditions, the non-resonant contribution to the CARS signal and the electronic resonance conditions.

(1) *Phase-Matching Conditions.* As explained above, a CARS signal is generated only if the law of momentum conservation is respected, meaning that inside the sample phase-matching must exist between the different incoming pulses. When a spectrally broad incoming pulse of frequency $\omega_2 \pm \Delta\omega_2$ is used to generate CARS spectra over a wide spectral range, phase-matching conditions cannot be strictly respected for all the spectral components, so that the available spectral range is reduced. The conditions under which phase-matching applies must then be minimized. This problem can be solved by making the samples as thin as possible. When the sample is a liquid solution, a jet is recommended. For example, a jet with a thickness of $300\,\mu\text{m}$ can yield simultaneous recordings of CARS spectra over a $700\,\text{cm}^{-1}$ spectral range, as shown in Fig. 8.20.

(2) *Nonresonant Contribution.* When the difference frequency $\omega_1 - \omega_2 = \omega_V$ corresponds to a specific vibration mode of the sample, the CARS signal is called "vibration-resonant" and is enhanced because of an increase of the third-order nonlinear susceptibility $\chi^{(3)}(\omega_{\text{CARS}}; \omega_1, -\omega_2, \omega_3)$ in the neighborhood of resonance frequencies. However, in all cases, a nonresonant signal which is superposed on this resonant signal is generated, as shown in Fig. 8.22, because the laws of phase-matching and energy conservation are respected for this pattern of interaction between four photons.

The CARS signal will be of the form:

$$I_{\text{CARS}} \propto \left|\chi^{(3)}\right|^2 = \left|\chi^{\text{NR}(3)} + \chi^{\text{R}(3)}\right|^2, \tag{8.7}$$

where $\chi^{\text{NR}(3)}$ and $\chi^{\text{R}(3)}$ are the nonresonant and the resonant contributions respectively.

It is well known that the nonresonant contribution affects the spectral shape of CARS spectra, and this effect makes CARS spectra difficult to analyze [8.19]. However, it is possible to suppress this nonresonant part of the CARS signal by adjusting the polarization of the pump (ω_1) and Stokes (ω_2) pulses, as well as the orientation angle of the analyzer with respect to these two pulses. This is called "polarization-sensitive CARS spectroscopy", as mentioned above, and has been widely discussed in the literature [8.13]. The basic principles of this technique are as follows.

Taking into account the angles φ (angle between the polarizations of the ω_1 and ω_2 pulses) and ψ (angle between the polarization of the ω_1 pulse and the analyzer), the nonresonant susceptibility can be expressed as $\chi^{NR} = (\cos\varphi\cos\psi - \rho^{NR}\sin\varphi\sin\psi)\chi^{NR}_{1111}$, where ρ^{NR} is the depolarization ratio [8.19].

For an isotropic medium such as a liquid, and provided that the electronic transitions of the solvent are far away from the ω_1, ω_2 and ω_{CARS} frequencies, the Kleinman relation requires that

$$\rho^{NR} = \frac{\chi^{(3)NR}_{1221}}{\chi^{(3)NR}_{1111}} = \frac{1}{3}. \tag{8.8}$$

Under these conditions the nonresonant contribution can be suppressed if $\tan\varphi\tan\psi = 3$. This means that to suppress the nonresonant contributions to the CARS signal, the angles between the polarizations must be adjusted to $\varphi = 60°$ and $\psi = -60°$. Under these conditions, a CARS band appears as a Lorentzian, free from nonresonant contributions.

(3) *Electronic Resonance Conditions.* It is well known that Raman scattering is enhanced in the neighborhood of electronic transitions (resonant Raman). The same applies to CARS. Since the ratio between the concentrations of excited and unexcited of molecules is usually small, it can be useful to take advantage of this property to enhance the contribution to the CARS signal from the excited part of the sample. Pump–probe absorption experiments, as described earlier, can help to determine the electronic resonance conditions by giving the absorption spectra of the excited states. These spectra being known, it is then easy to adjust the ω_1, ω_2 and then ω_{CARS} frequencies in the desired spectral absorption regions. Under these conditions, the contribution to the CARS signal from the excited part becomes very intense compared to the contribution from the non-excited part. A typical example in which advantage is taken of this effect is given in [8.16].

8.3 Time-Resolved Emission Spectroscopy: Electronic Methods

Just like IR absorption spectra or Raman spectra, emission spectra provide a signature of a sample. The changes which occur in these spectra after the sample has been perturbed by a light pulse can give information about the evolution of the sample. It is therefore very important to observe the evolution of the emission spectrum and the associated dynamics to analyze sample behavior; the emission lifetime and spectral changes should be measured. Two kinds of methods exist to obtain the time-resolved emission of a sample: electronic methods using only electronic devices, described in this section, and optical methods exploiting the correlation between the exciting pulse and the observed emission, which are described in Sect. 8.4.

8.3.1 Broad-Bandwidth Photodetectors

The principles of the methods using broad-bandwidth photodetectors for characterization of short pulses have already been described in Chap. 7. They consist of collecting the emission of the sample, after dispersion in a spectrograph, in a photodetector (photodiode or photomultiplier) coupled to an oscilloscope or any other recording system. The time resolution is limited by the bandwidth of the electronic system and can reach a few picoseconds. The sensitivity of these systems is limited, however, especially with photodiodes, and they can only be applied to strongly emissive samples.

8.3.2 The Streak Camera

The streak camera, the principle of which was explained in Chap. 7, is an useful tool to use for emission and dynamics measurements. The time resolution routinely reaches a picosecond, and may reach a few hundreds of femtoseconds with up-to-date equipment. The main limitation is the dynamic range of single-shot measurements. Accumulation of many shots greatly improves this dynamic range, but the jitter from one shot to the next in the camera triggering remains a limitation. However, using optoelectronic devices to do the triggering can greatly reduce this jitter, sometimes even to less than 1 ps [8.20]. The main advantage of the streak camera is that it obtains the whole emission spectrum simultaneously with the associated dynamics at each emission wavelength, after dispersion in the focal plane of a spectrograph.

8.3.3 "Single"-Photon Counting

This technique is especially interesting when a high-repetition-rate laser source is available. It is based on the following principle.

Suppose that a molecule absorbs a photon at time $t = 0$. Depending on the emission probability per unit time, this photon will be re-emitted at time Δt. Let us assume that Δt can be measured. If it is measured many times, different values of Δt will be obtained, but some values will be obtained more often than others. The number of times a specific value is observed is related to the emission probability per unit time. A histogram of the number of times

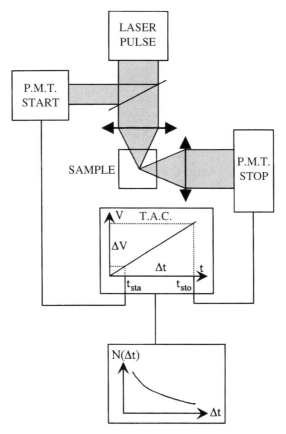

Fig. 8.23. Single-photon counting apparatus (P.M.T. = photomultiplier tube)

$n(\Delta t_i)$ a photon is re-emitted at time Δt_i will reconstruct the decay curve of the emission sample. By using the single-photon counting technique, this histogram can be recorded directly, as shown in Fig. 8.23.

A high-repetition-rate pulsed-laser source sends photons onto the sample. One part of the laser pulse is sent to a fast photodetector coupled to a constant-fraction discriminator. The generated electronic signal, called the "start" signal, is sent to a time–amplitude converter (TAC) to trigger the latter device. The level of the signal is adjusted in order to minimize the jitter between the laser pulse and the electronic devices. The photons emitted by the sample are detected by a photomultiplier tube (PMT). The excitation level is adjusted in such a way that a single event is seen by this PMT, generating a "stop" signal which is sent to the TAC. With this procedure, the TAC can easily record the time interval between the absorption of a photon by the sample and the re-emission of a photon. By making this measurement a great number of times and storing each result in a multichannel analyzer, the decay profile of the excited sample can be constructed. The decay profile will in turn yield the associated lifetime. With this method, the dynamics of the measurement (signal/noise ratio) may be very high if the repetition rate of the laser source is high. A rate of a few MHz is quite common, and gives access to dynamics ranging over as much as 10 orders of magnitude.

The time resolution is limited by the laser pulse duration and by the response functions of the electronic devices, in particular of the photomultiplier "stop". The response function can be measured directly with a purely scattering sample. An accurate measurement of the response function allows sophisticated deconvolution procedures to reach time resolutions of the order of ten picoseconds. Making this measurement at different emission wavelengths yields the emission spectra for different time intervals after excitation. This method is the so-called "global analysis", and it will give the spectral evolution as well as the dynamics of the sample.

8.4 Time-Resolved Emission Spectroscopy: Optical Methods

In Chap. 7, it was shown that measurements of the autocorrelation function are possible by optical methods, and from these measurements the pulse duration can be deduced provided the pulse shape is known. The same principle can be used to measure the emission dynamics of a sample, by measuring cross-correlations between the excitation pulse and the emission decay. Let us now present some typical optical methods.

8.4.1 The Kerr Shutter

8.4.1.1 Principle of the Method.
This method was used for the first time in 1969 by *Duguay* [8.21]. This method uses the ability of isotropic samples (liquids or glasses) to become anisotropic under the action of an applied electric field (Kerr effect). The electric field associated with an intense light pulse induces the Kerr effect (optical Kerr effect). The experimental setup is shown in Fig. 8.24a.

A Kerr sample (a cell containing a liquid such as CS_2, or a piece of a material such as glass) is inserted between two crossed polarizers (P1 and P2). A probe pulse cannot cross this arrangement, but if an intense linearly polarized light pulse (the opening pulse) is sent into this Kerr cell, it induces an anisotropy. The linear polarization (after the P1 polarizer) of the probe will change after crossing the Kerr cell so that transmisson through polarizer P2 will be possible. In other words, the optical shutter is open. It will be open as long as the induced anisotropy holds. The time aperture depends on the "pump" (opening) pulse duration and on the relaxation time of the anisotropy. The setup can be used to study the Kerr characteristics of a sample, as will be

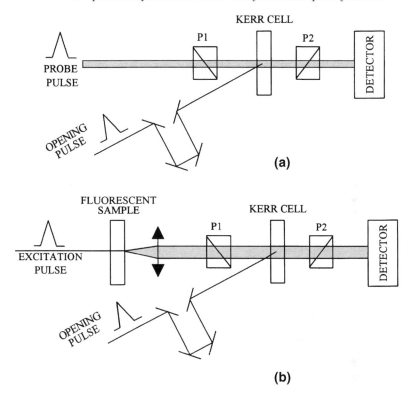

Fig. 8.24. (a) Schematic illustration of Kerr shutter. (b) Time-resolved emission technique using an optical Kerr gate. P1 and P2 are polarizers

described at the end of this chapter, but it can also be used for time-resolved emission, as described below.

Suppose now, as shown in Fig. 8.24b, that the "probe" light is the fluorescence emission of a sample placed in front of the P1 polarizer and excited at time $t = 0$. If the optical shutter is open at time Δt, the intensity of the transmitted light will be proportional to the number of photons emitted at time Δt after excitation. Recording the transmitted intensity as a function of Δt will then yield the decay time of the sample. Moreover, if the fluorescence is dispersed by a spectrograph, the fluorescence spectrum at time Δt will be recorded, so that the spectral evolution of the fluorescence after excitation can also be observed.

The main advantage of this experimental setup is that the whole emission spectrum is obtained directly at each time interval after excitation. Reconstruction of the decays at different emission wavelengths is then easy. However, this very simple experiment has some limitations, and a few specific requirements should be taken into consideration, as described below.

8.4.1.2 Experimental Tricks

(1) *Respective Polarizations of "Probe" and "Pump" Pulses.* The transmission efficiency of the optical shutter is greatest when the polarization of the incoming light in the Kerr cell is at 45° to the polarization of the "pump" (opening) pulse. This means that the linear polarization of the opening pulse should be oriented at 45° to the entrance polarizer [8.21].

(2) *Leakage Through Polarizers P1 and P2.* Even with good-quality polarizers (Glan are recommended) and precise adjustment, it is impossible to obtain complete extinction between crossed polarizers. The leakage through the polarizers may even be very significant compared to the true signal due to the opening of the shutter. Let us assume, for example, a transmission efficiency of 10 % (which is a very good value), an extinction ratio between polarizers of 10^{-5}, a fluorescence lifetime of $\tau_f = 5$ ns and a time aperture of $\tau_{op} = 1$ ps. Then the ratio between the signal due to leakage and the signal due to the opening of the gate will be equal to

$$R = \frac{10^{-5} \times \int_0^\infty e^{-t/\tau_f}\, dt}{10^{-1} \times \int_0^\infty e^{-t/\tau_{op}}\, dt} = 0.5. \tag{8.9}$$

This simple (but realistic) estimate shows that leakages may not be negligible. It is therefore necessary to measure the fluorescence leakage (without the opening pulse) and the total signal (with the opening pulse) independently and then to subtract the two quantities, in order to measure the true signal recorded during the opening time and corresponding to the fluorescence emission at time Δt.

(3) *Spectral Dispersion of the Transmission Function.* The transmission efficiency is not the same for all wavelengths, particularly near a resonance of the Kerr material; this is mainly due to the spectrally dependent Kerr constant. This fact must be taken into account, for instance by recording the emission spectrum of a standard solution and introducing a spectral correction function.

(4) *Parasitic Light from the Opening Beam.* The intensity of the opening pulse is generally very strong compared to the transmitted signal. Scattering of this pulse through the system is difficult to avoid and can perturb the experiment. One way to overcome this problem is to choose, if possible, an opening pulse at a wavelength which lies well outside the fluorescence domain of interest.

(5) *Spectral Limitation.* The main spectral limitation is due to the absorption bands of the Kerr cell (for example, CS_2 begins to absorb near 400 nm). The intrinsic cutoff frequency of the Kerr cell has to be taken into account.

(6) *Time Resolution.* This is limited by the optical response of the Kerr medium. Liquids generally have a short electronic response (quasi-instantaneous), but they also have a "slow" response due to reorientation dynamics, usually in the range of a few picoseconds, which affects the time resolution.

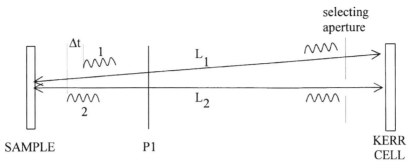

Fig. 8.25. Illustration of the effect of spatial dispersion on the response time. Photons 1 and 2 emitted at times separated by Δt may arrive at the same time at the Kerr cell if $L_1 - L_2 = c\Delta t$

Glasses, and solid materials in general, do not have this reorientation problem and their response time is exclusively of electronic origin. Depending on the requirements, this limitation may have to be taken into account.

The response time also depends on the GVD, owing to the influence of the various optical elements. These effects have to be corrected for. The geometry of the excitation and of the collimation also affect the response time. The excited volume should be as small as possible to avoid spatial dispersion, but it is also recommended to collimate the collected light with a small aperture, so as to avoid photons emitted at different times but following different optical paths (as shown in Fig. 8.25) arriving fortuitously at the same time through the Kerr cell. The sample and the Kerr cell should therefore be as thin as possible and one should reduce the aperture of optical collection.

8.4.2 Up-conversion Method

The up-conversion technique, which we shall describe below requires a nonlinear crystal, a spectrometer and a photon counting system. This optical sampling technique is very well suited to low-energy laser pulses with a high repetition rate. Time resolutions of less than 100 fs can be achieved, and the spectral range of this detection technique is wider in the infrared region than that of photocathodes [8.22, 23].

8.4.2.1 Principle of the Technique. The up-conversion method is an optical sampling technique which is based on the nonlinear properties of a dielectric crystal ($LiIO_3$, LBO or BBO for instance). Mode-locked picosecond or sub-picosecond dye or Ti:sapphire lasers are normally used as excitation sources. Figure 8.26 shows a diagram of the experimental setup.

The pulsed laser beam is divided into two parts. The first part excites the sample. The induced sample emission is collected by an optical system of lenses (or parabolic mirrors) and focused on the nonlinear crystal. The

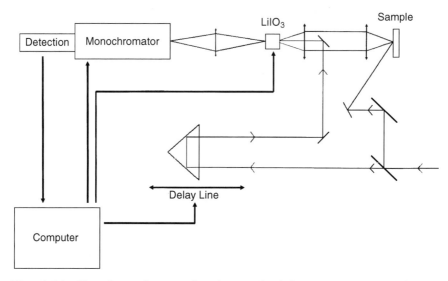

Fig. 8.26. Experimental setup for time-resolved luminescence using the up-conversion technique

second part of the laser beam passes through an optical delay line and is also focused on the nonlinear crystal. When the laser beam and the emission signal are present simultaneously in the nonlinear crystal, frequency mixing occurs, which results in the generation of an up-converted signal. For a given detection axis, the up-converted frequency is determined both by the angle between the optical axis of the crystal and the incoming beams, and by the optical frequencies of these incoming beams (Fig. 8.27a). The nonlinear crystal thus acts as an optical gate which is opened only when the laser pulse is present in the crystal. By varying the optical path of the laser pulse which opens this optical gate, the photoluminescence signal is sampled at different time delays (Fig. 8.27b). The up-converted signal is then dispersed by a monochromator and detected by a photomultiplier linked to a photon-counting system.

8.4.2.2 Frequency Mixing in the Nonlinear Crystal. The generation of the sum frequency results from the nonlinear properties of the dielectric susceptibility of the crystal. The lithium iodate ($LiIO_3$) crystal, which is one of the most widely used nonlinear crystals for the up-conversion technique, is a negative uniaxial crystal (i.e. $n_e < n_o$, where n_e and n_o are the extraordinary and the ordinary refractive indices). When spatial and temporal overlap of the beams is achieved inside the crystal, phase-matching conditions must be fulfilled for sum-frequency generation to occur. These conditions can be summarized as follows:

$$K_\Sigma = K_S + K_L \quad \text{and} \quad \omega_\Sigma = \omega_S + \omega_L. \tag{8.10}$$

(a)

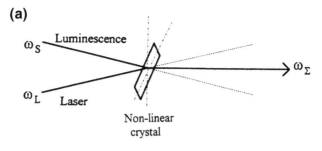

(b)

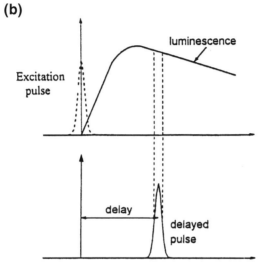

Fig. 8.27. Principle of the temporal and spectral sampling of the luminescence signal by the up-conversion technique see text

These equations just express the photon energy and wave vector conservation rules. The subscripts Σ, S and L correspond to the up-converted, the photoluminescence and the laser waves respectively. For the sake of simplicity, we assume in the following that the incoming beams are colinear. Equations (8.10) can be satisfied if the up-converted beam polarization is extraordinary. In the usual configuration of type I frequency mixing, the incoming beams are ordinary polarized. In this case, (8.10) can be rewritten in the following way:

$$\cos^2 \theta = \frac{1/n_e^2(\lambda_\Sigma) - 1/n_{\text{eff}}^2(\theta, \lambda_\Sigma)}{1/n_e^2(\lambda_\Sigma) - 1/n_o^2(\lambda_\Sigma)}, \tag{8.11}$$

with n_{eff} defined by

$$\frac{n_{\text{eff}}(\theta, \lambda_\Sigma)}{\lambda_\Sigma} = \frac{n_o(\lambda_S)}{\lambda_S} + \frac{n_o(\lambda_L)}{\lambda_L}, \tag{8.12a}$$

and λ_Σ by

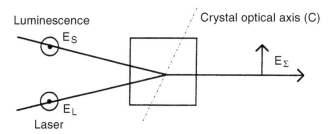

Fig. 8.28. Beam polarization in type I frequency mixing

$$\frac{1}{\lambda_\Sigma} = \frac{1}{\lambda_S} + \frac{1}{\lambda_L}. \tag{8.12b}$$

Here, θ is the angle between the incoming beams and the optical axis of the crystal. Note that the crystal acts as a polarizer: only the part of the luminescence which is ordinary polarized is converted (Fig. 8.28).

8.4.2.3 Quantum Efficiency and Spectral Selectivity of the Nonlinear Crystal.

A good approximation of the quantum efficiency is given [8.24] by

$$\eta_{qu}(\Delta K = 0) = \frac{2\pi^2 d_{eff}(\theta)\beta L^2 (P_P/A)}{c\varepsilon_0^2 \lambda_S \lambda_\Sigma n_o(\lambda_L) n_{eff}(\theta, \lambda_\Sigma)}, \tag{8.13}$$

with d_{eff} the effective nonlinear susceptibility for a given angle θ, P_P the laser beam power, A the surface area of the laser beam on the nonlinear crystal (the signal beam (S) cross-sectional area in the crystal is assumed to be smaller than A), L the interaction length within the crystal, ε_0 the vacuum permittivity and c the speed of light.

$\Delta K = 0$ corresponds to the strict phase-matching conditions: $\Delta K = K_\Sigma - K_S + K_L = 0$. Typical values for $\eta_{qu}(\Delta K = 0)$ are of the order of 1×10^{-3}. If $\Delta K \neq 0$, the conversion efficiency decreases according to the law.

$$\eta_{qu}(\Delta K) = \eta_{qu}(\Delta K = 0)\frac{\sin^2(\Delta K\, L/2)}{(\Delta K\, L/2)^2}. \tag{8.14}$$

For a given angle between the incoming beams and the crystal axis and for a given laser wavelength, frequency up-conversion occurs within a limited spectral width ΔE of the photoluminescence spectrum. Assuming a monochromatic laser source, the spectral acceptance ΔE for the LiIO$_3$ crystal is given by

$$\Delta E(\text{meV}) = \frac{3.5 \times 10^{-2}}{L(\text{cm})(\partial K_L/\partial \omega_L - \partial K_S/\partial \omega_S)(\text{s cm}^{-1})}. \tag{8.15}$$

In an experiment in which the output direction $\boldsymbol{K}_\Sigma$ is fixed, the crystal acts as a monochromator, since for a given angle it converts only the small spectral domain defined by (8.15). However, another severe limitation in the spectral resolution occurs when ultrafast laser sources are used. Equations (8.10) and (8.11) can be satisfied, for a given up-converted wavelength, by different pairs of signal and laser wavelengths; as a consequence, the converted spectral domain in the up-conversion technique is primarily determined by the laser spectral width.

8.4.2.4 Time Resolution of the Up-conversion Technique. The time resolution of the setup is basically limited by the temporal width of the laser pulse which opens the optical gate. However, if the laser source delivers subpicosecond pulses, the group velocity dispersion (GVD) induced by the nonlinear crystal and by the various optical elements may alter this time resolution. For case of the type I mixing (O + O → E) discussed here, the broadening of the temporal response of the system induced by a group-velocity mismatch in the nonlinear crystal is given by [8.22]

$$\Delta t = L\,(\mathrm{cm}) \left(\frac{\partial K_\mathrm{L}}{\partial \omega_\mathrm{L}} - \frac{\partial K_\mathrm{S}}{\partial \omega_\mathrm{S}} \right) (\mathrm{s\,cm}^{-1}). \qquad (8.16)$$

In this case, a thin nonlinear crystal (typically 1 or 2 mm thick) should be used, along with parabolic mirrors (instead of lenses) for the collection of the luminescence, in order to optimize the time resolution.

8.4.2.5 Calibrations

(1) *Angular Calibration.* An angular calibration is required in order to know the crystal angle θ which corresponds to the generation and detection of the up-converted signal wavelength λ_Σ for a given laser wavelength λ_L and a given detection direction. These angular calibration curves are obtained experimentally by frequency mixing, in the setup, between the laser and a quasi-continuous spectrum delivered by an incandescent lamp. The data for $\lambda_\Sigma = f(\theta, \lambda_\mathrm{L})$ are then recorded in a file used by the computer program which controls the whole experiment. During the experiment, the nonlinear-crystal angle and the monochromator grating angle will thus move together according to this calibration file. The monochromator acts as an optical filter which rejects stray light coming from the frequency-doubling of the laser (which occurs at $\lambda_\mathrm{L}/2$).

(2) *Temporal Calibration.* Temporal calibration consists of defining the time origin of the setup. This is achieved when the optical path of the excitation–collection setup is equal to the optical path of the delay line. By performing an autocorrelation experiment with the light scattered from the sample surface, this temporal origin can be determined. The laser pulse width can be measured at the same time (in the subpicosecond regime, the GVD in the setup must be taken into account).

8.4.2.6 Example of Application. Figure 8.29 shows a typical example of time-resolved photoluminescence spectroscopy using the up-conversion technique, in this case for the study of the electronic properties of low-dimensional semiconductor heterostructures [8.25]. The sample is a 3 nm GaAs multiple quantum well. The laser source is a mode-locked Ti:sapphire laser which delivers 1.2 ps (FWHM) pulses at a repetition rate of 80 MHz. The excitation laser beam is circularly polarized (σ^+). The exciton luminescence components of opposite helicities I_+ and I_- are detected by the up-conversion technique. Figures 8.29a,b correspond to different excitation conditions (non-resonant and resonant respectively with the heavy-hole exciton absorption peak). When a magnetic field (**B**) is applied perpendicular to the growth direction, quantum beats are observed in the luminescence components. These quantum beats correspond to the Larmor precession frequency ω of free electrons (Fig. 8.29a) and that of the heavy-hole exciton Ω (Fig. 8.29b). Figure 8.29c shows the oscillations of the luminescence intensity component I_+ under resonant excitation (dashed line) and of the luminescence polarization $P_L = (I_+ - I_-)/(I_+ + I_-)$ under nonresonant excitation (full line), under the same magnetic field $B = 3$ T. These time-resolved data yield the effective electron Landé g factor and the exciton exchange energy with very high accuracy (inset of Fig. 8.29c).

8.5 Time-Resolved Spectroscopy by Excitation Correlation

In 1981, *Von der Linde* et al. [8.26] proposed a new time-resolved spectroscopy technique based on a correlation principle. We briefly describe this technique as applied to the particular case of time-resolved photoluminescence of semiconductor materials.

In this technique, called population mixing or excitation correlation (EC), the laser pulses are divided into two separated pulse trains, delayed with respect to each other and modulated separately at two different frequencies. The photoluminescence signal at the sum of or at the difference between the two modulation frequencies is detected by a conventional photodetector. The observed signal gives the cross-correlation of the population levels involved in the photoluminescence process; the temporal resolution of the system is limited by the laser pulse width.

The main advantage of this technique lies in its simplicity (no ultrafast electronics are required). It allows a direct detection of the photoluminescence signal and is therefore a very sensitive technique. The main drawback comes from its principle: the correlation signal does not reflect directly the time evolution of the individual carrier populations. A model is required in order to obtain this information.

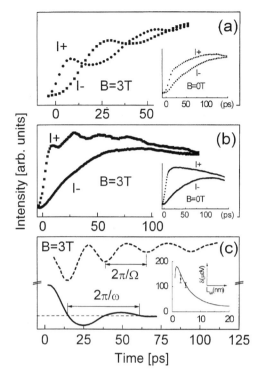

Fig. 8.29. Luminescence intensity dynamics of multiple quantum well after $\sigma^{\pm}$ polarized excitation. (a) The excitation energy is nonresonant and $B = 3\,T$ (inset, $B = 0T$); (b) the excitation energy is resonant with the heavy-hole exciton energy and $B = 3\,T$ (inset, $B = 0\,T$); (c) oscillations of the luminescence intensity component I_{+} under resonant excitation (*dashed line*) and of the luminescence polarization under nonresonant excitation (*full line*) for $B = 3\,T$. Inset: the well-width dependence of the measured exciton exchange energy δ [8.25]

8.5.1 Experimental Setup

Figure 8.30 shows the experimental setup. The excitation source is usually a mode-locked picosecond or subpicosecond dye or Ti:sapphire laser. The laser beam is divided into two beams by a polarizing beam-splitter (PBS1). The first beam, which is horizontally polarized, passes through a delay line. The second one, vertically polarized, follows a fixed optical path. Each beam is modulated at a different frequency, ω_1 for beam 1 and ω_2 for beam 2, by two optical choppers (CH1 and CH2 respectively). The two beams are then mixed again in polarizing the beam-splitter 2 (PBS2) and focused on the sample. The induced luminescence is collected by the lens L2, dispersed by a monochromator and detected by a conventional slow-response photodetector (a silicon photodiode for instance).

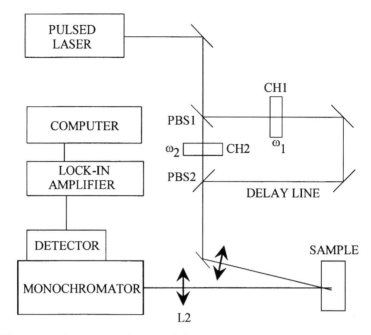

Fig. 8.30. Experimental setup of the excitation correlation experiment

8.5.2 Interpretation of the Correlation Signal

In the correlation excitation technique, the two pulse trains, delayed with respect to each other by a time interval Δt, excite the sample (a semiconductor, for example). The electron and hole densities created by the first pulse at time t are labeled $N_A(t)$ and $N_B(t)$; $N_A(t + \Delta t)$ and $N_B(t + \Delta t)$ are the photo-generated densities created by the second pulse. The photoluminescence I_F due to the radiative recombination of these carriers is usually given by the following bimolecular law:

$$I_F(t, \Delta t) \propto (N_A(t) + N_A(t + \Delta t))$$
$$\times (N_B(t) + N_B(t + \Delta t)), \tag{8.17}$$

neglecting the carrier density at equilibrium (without any laser excitation).

The average photoluminescence $\langle I_F(t, \Delta t) \rangle$ detected by the photodetector is then given by

$$\langle I_F(t, \Delta t) \rangle = I_1(\Delta t) + I_2, \tag{8.18}$$

with

$$I_1(\Delta t) \propto \int_{-\infty}^{+\infty} N_A(t + \Delta t) N_B(t) \, dt + \int_{-\infty}^{+\infty} N_A(t) N_B(t + \Delta t) \, dt \tag{8.19}$$

and

$$I_2 \propto \int_{-\infty}^{+\infty} N_A(t)N_B(t)\,dt + \int_{-\infty}^{+\infty} N_A(t+\Delta t)N_B(t+\Delta t)\,dt, \qquad (8.20)$$

where only $I_1(\Delta t)$ depends on Δt.

The signal $I_1(\Delta t)$ is thus proportional to the cross-correlation of the electron (or hole) density generated by the first pulse and the hole (or electron) density generated by the second one. The EC technique is also applicable when the carrier lifetime depends not only on the bimolecular recombination process but also on another mechanism of a different order. For instance, in the simplest case, if we assume exponential relaxation times τ_A and τ_B for the N_A and N_B populations, the correlation signal can be written as

$$I_1(\Delta t) \propto N_A(0)N_B(0)[\exp(-\Delta t/\tau_A) + \exp(-\Delta t/\tau_B)]. \qquad (8.21)$$

In this example, the ratio of the correlation signal dependence versus the time delay Δt between the excitation pulses directly reflects the combined temporal variations of the electron and hole densities.

The difficulty of the EC technique is due to the fact that the signal I_2 is usually much greater than the signal $I_1(\Delta t)$ (see (8.19) and (8.20)). I_2 must therefore be subtracted from $\langle I_F(t,\Delta t)\rangle$ by a heterodyne signal-processing technique. The suppression of the I_2 component is achieved through the independent frequency modulation of the two excitation beams. The photodetected signal is subsequently modulated at the frequencies given by the Fourier spectrum ω_1, ω_2, $\omega_1 + \omega_2$, $\omega_1 - \omega_2$, $\omega_1 + 2\omega_2$, ..., $n\omega_1 + m\omega_2$ ($n, m \in \mathbb{B}$). The photoluminescence signal modulated at ω_1, $I(\omega_1)$, corresponds to the radiative recombination of carriers generated by the first pulse train, modulated at ω_1; symmetrically, the photoluminescence signal modulated at ω_2, $I(\omega_2)$, corresponds to the radiative recombination of carriers generated by the second pulse train, modulated at ω_2. All the other Fourier components $I(n\omega_1 + m\omega_2)$ correspond to the radiative recombination of the carriers generated by a mixing of the two pulses, i.e. to $I_1(\Delta t)$.

In practice, the photodetector signal is processed by a lock-in amplifier with a reference signal at $\omega_1 + \omega_2$ or $\omega_1 - \omega_2$.

8.5.3 Example of Application

Relaxation processes which compete with the radiative recombination mechanism can be measured with the EC method. It has been used both in the picosecond and in the subpicosecond regime. Initially, *Von der Linde* et al. [8.26] used the technique to study the cooling process of electron–hole pairs in bulk GaAs (Fig. 8.31). The decrease of the correlation signal as a function of the delay time Δt is a measure of the temporal dependence of the hot-carrier populations. The same kind of study has been performed in GaAs/AlGaAs quantum well structures [8.27]. The EC technique has also been applied on the subpicosecond domain with time resolutions of down to 40 fs [8.28]. We

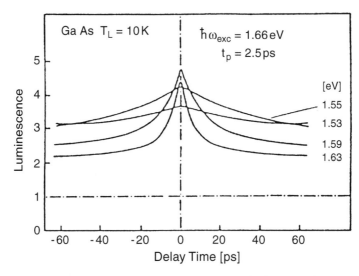

Fig. 8.31. Luminescence correlation traces in bulk GaAs for different detection energies, for an investigation of the cooling of hot carriers by phonon emission. T_L is the lattice temperature ([8.26] © Springer-Verlag 1982)

have presented the excitation correlation technique in the framework of the photoluminescence of semiconductor materials, but it has also been used in a transmission geometry for the study of organic molecules [8.29].

8.6 Transient-Grating Techniques

Transient-grating techniques are powerful tools, with which the dynamic response of a perturbed sample can be tested and which can give information about changes in the refractive index, changes of the absorption and the associated population dynamics, as well as about thermal relaxations, electronic responses, and nonlinearities in the optical responses. The aim of this section is not to give a complete review of these techniques, since they are well documented in [8.30]. We would just like to emphasize the general principle and to give a few typical examples.

8.6.1 Principle of the Method: Degenerate Four-Wave Mixing (DFWM)

The basic principle of the method is shown in Fig. 8.32 and can be described as follows. Two synchronized pulses (called the k_1 and the k_2 pulse) propagating in two slightly different directions $(\boldsymbol{k}_1, \boldsymbol{k}_2)$ interfere in the sample to form a grating by spatially modulating its optical properties. Depending

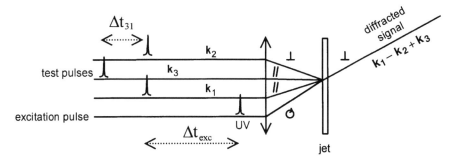

Fig. 8.32. Schematic diagram of a DFWM experiment. To study excited samples, the excitation pulse is sent into the sample before the arrival of the test pulses

on the experimental conditions, the absorption (amplitude grating) or the refractive index (phase grating) can be modulated. After laser pulse interaction, the amplitude of the modulation decreases: the dynamics of the modulation (i.e. the grating) will depend on the population lifetime (for an amplitude grating) and/or on the dynamics of the changes in the refractive index (for a phase grating). A third beam, k_3, propagating in direction $\boldsymbol{k_3}$, is sent into the sample and is diffracted by the grating in directions $\boldsymbol{k_0}$, which satisfies the conservation of momentum: for example, $\boldsymbol{k_0} = \boldsymbol{k_1} - \boldsymbol{k_2} + \boldsymbol{k_3}$ as shown in Fig. 8.32. By changing the optical delay of the k_3 pulse with respect to the k_1, k_2 pulses, one can measure the diffracted intensity I_D, which depends on the modulation amplitude, as a function of time.

The diffracted intensity is equal to

$$I_\mathrm{D}(\omega) = \frac{\omega^2 d^2}{\varepsilon_0^2 n_0^4 c^4} |\chi^{(3)}|^2 \left(\frac{1 - e^{-\alpha d}}{\alpha d}\right) e^{-\alpha d} \, I_1 I_2 I_3, \qquad (8.22)$$

where $I_\mathrm{D}, I_1, I_2, I_3$ are respectively the diffracted intensity and the intensities of the k_1, k_2 and k_3 pulses, ω is the pulse frequency, α and d are the absorbance and the length of the sample, ε_0 is the permittivity and n_0 the refractive index of the sample, c is the speed of light and $\chi^{(3)}$ is the third-order nonlinear optical susceptibility. As indicated in (8.22), the diffracted intensity depends on the third-order nonlinear susceptibility $\chi^{(3)}$ of the sample. If $\chi^{(3)}$ changes, the dynamics of $\chi^{(3)}$ can be studied by observing the changes in the diffracted intensity. Moreover, by changing the polarizations of the various incoming beams and selecting a specific diffracted beam using an analyzer, different components χ_{ijkl} of the third-order nonlinear optical susceptibility $\chi^{(3)}$ can be observed [8.30].

As with other techniques, one can also study the behavior of an excited sample. In this case, a fourth pulse (the excitation pulse) is sent into the sample before the arrival of the k_1, k_2 and k_3 pulses. The relaxation of the excited sample can then be studied by changing the delay time Δt_EXC between the excitation pulse and the other pulses.

8.6.2 Example of Application: *t*-Stilbene Molecule

The experimental setup used in these experiments was as shown in Fig. 8.32: the wavelenghts of the k_1, k_2 and k_3 pulses are 600 nm, and the excitation pulse has a wavelength of 300 nm to populate the excited singlet states of the *t*-stilbene molecules. The polarization of the k_3 beam is chosen to be parallel to that of the k_1 beam in order to measure χ_{1212} components of the nonlinear susceptibility [8.31]. To study the properties of the excited states of the sample, the 300 nm pulse is sent into the sample after passing through a delay line. By changing the delay between the 300 nm excitation pulse and the three other pulses (k_1, k_2, k_3), the optical nonlinear properties of the excited molecule can be studied for different time intervals after excitation, up to the lifetime of the excited states. The UV beam was circularly polarized to ensure an isotropic distribution of the excited molecules.

Using this setup, two kinds of kinetics can be measured.

(1) For a fixed delay between the k_1, k_2 pulses and the excitation pulse and for a variable time delay of the k_3 pulse with respect to the other pulses, one can measure the kinetics of $\chi^{(3)}(-\omega; \omega, -\omega, \omega)$ at a given time after excitation. Under these conditions one obtains the dynamic response of $\chi^{(3)}(-\omega; \omega, -\omega, \omega)$.

(2) By fixing the delays between the k_1, k_2 and k_3 pulses and by varying the delay of the UV pulse relative to the other pulses, one can record the evolution of the amplitude of $\chi^{(3)}(-\omega; \omega, -\omega, \omega)$ during the formation and the lifetime of the excited states.

The corresponding physical characteristics for the ground state can be observed without UV excitation, and can then be compared with those of the excited sample.

8.6.2.1 Observation of $\chi^{(3)}(-\omega; \omega, -\omega, \omega)$ in the Excited States. To observe $\chi^{(3)}(-\omega; \omega, -\omega, \omega)$ in the excited states, one must ensure a zero time delay between the k_1, the k_2 and the k_3 pulses [8.32]. The diffracted signal from the k_3 pulse, which is proportional to $|\chi^{(3)}(-\omega; \omega, -\omega, \omega)|^2$, is then recorded with and without excitation of the sample and for different delays Δt_{EXC} of the UV pulse with respect to the other three pulses. Typical results are shown in Fig. 8.33 (curve 1), where the diffracted signal intensity is plotted as a function of the excitation pulse delay time.

For comparison, the inset in Fig. 8.33 shows a typical transient absorption spectrum of excited *t*-stilbene obtained by the pump–probe method (excitation at 300 nm), as well as the typical kinetics of the optical density (curve 2, Fig. 8.33). The inset shows a strong absorption band centered near 580 nm and extending from 450 to 650 nm. An excited-state lifetime of 100 ± 5 ps is deduced from the kinetic intensity curve (curve 2 of Fig. 8.33). This decay time is comparable to the decay time deduced from DFWM experiments.

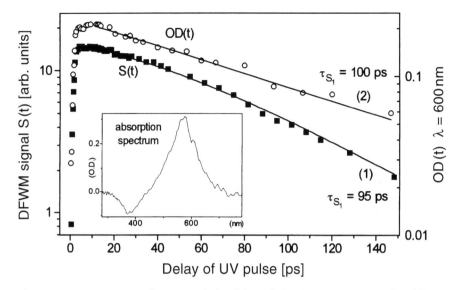

Fig. 8.33. Kinetics as a function of the delay of the UV excitation pulse. (Curve 1) DFWM signal $S(t)$ at wavelength 600 nm. (Curve 2) Optical density $OD(t)$ at wavelength 600 nm. The inset shows the transient absorption spectrum measured 20 ps after excitation. A negative optical density corresponds to a measured gain in the fluorescence spectral region. Solution of t-stilbene in heptane at $c = 10^{-3}$ M [8.32]

These results demonstrate that a diffracted signal from the excited-state population is really observed and that the nonlinear optical properties of the excited t-stilbene molecules can indeed be measured correctly. We have to note, however, that the decay of the DFWM signal $S(t)$ due to the excited population must be corrected for reabsorption effects according to the formula

$$\sqrt{S(\Delta t_{\mathrm{EXC}})} \propto N_{\mathrm{S}_1^*}(0) \exp\left(\frac{-\Delta t_{\mathrm{EXC}}}{\tau_{\mathrm{S}_1}}\right) A(\Delta \tau_{\mathrm{EXC}}), \tag{8.23}$$

with

$$A(\Delta t_{\mathrm{EXC}}) = \frac{1 - \exp[-\alpha(\Delta t_{\mathrm{EXC}})d]}{\alpha(\Delta t_{\mathrm{EXC}})d} \exp\left(\frac{-\alpha(\Delta t_{\mathrm{EXC}} + \Delta t_{31})d}{2}\right) \tag{8.24}$$

and

$$\alpha(\Delta t_{\mathrm{EXC}}) \propto N_{\mathrm{S}_1^*}(\Delta t_{\mathrm{EXC}}) = N_{\mathrm{S}_1^*}(0) \exp\left(\frac{-\Delta t_{\mathrm{EXC}}}{\tau_{\mathrm{S}_1}}\right), \tag{8.25}$$

where $N_{\mathrm{S}_1^*}(\Delta t_{\mathrm{EXC}})$ is the excited-state population for a delay Δt_{EXC} between the UV pulse and the k_1, k_2 pulses; α is the absorption coefficient and d is the thickness of the excited sample. Δt_{31} is the delay time between the k_3 and k_1, k_2 pulses and τ_{S_1} is the lifetime of the excited state S_1^*.

In the measurements described above [8.31], the kinetics were corrected by deducing the decay times using these formulae.

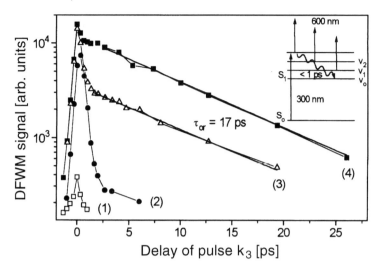

Fig. 8.34. Intensity of the DFWM signal as a function of the delay of the k_3 pulse and for different delays Δt_{EXC} of the excitation pulse. *Curve 1*, without UV excitation. *Curve 2*, $\Delta t_{\mathrm{EXC}} = -500\,\mathrm{fs}$; *Curve 3*, $\Delta t_{\mathrm{EXC}} = 500\,\mathrm{fs}$; *Curve 4*, $\Delta t_{\mathrm{EXC}} = 5.5\,\mathrm{ps}$. The signal was recorded at 600 nm for a *t*-stilbene solution in heptane ($c = 10^{-3}$ M). These results can be exploited to yield information about the enhancement and dynamics of optical non-linearities in the excited states [8.31]

8.6.2.2 Dynamic Response of the DFWM Signal from the Excited States.
The procedure to measure the dynamic response of the DFWM signal is as follows. The time delay between the k_1 and k_2 pulses is adjusted to obtain the greatest possible diffracted signal from the k_3 beam (Fig. 8.34). The delay of the UV pulse with respect to the k_1 and k_2 pulses is fixed. The diffracted signal from the k_3 beam is then recorded for different delays time Δt_{31} between this pulse and the other pulses. So, for a given delay time of the UV pulse, the kinetics of the diffracted k_3 beam, i.e. the response of the excited-state population at this delay, are recorded. The same measurement is repeated for different delays of the UV pulse relative to the k_1 and k_2 pulses. This procedure yields the kinetics of the DFWM signal (which is proportional to $|\chi^{(3)}(-\omega; \omega, -\omega, \omega)|^2$) corresponding to the excited-state populations present at different times after excitation. Under resonant conditions, the signal $S(\Delta t_{31})$ is expected to decay (because of the excited decay of the molecules) according to the following equation [8.30, 33]:

$$\sqrt{S(\Delta t_{31})} \propto N_{\mathrm{S}_1^*}(0) \exp\left(\frac{-\Delta t_{\mathrm{EXC}}}{\tau_{\mathrm{S}_1}}\right)$$

$$\times A(\Delta t_{\mathrm{EXC}}, \Delta t_{31}) \exp\left[-\Delta t_{31}\left(\frac{1}{\tau_{\mathrm{S}_n}} + \frac{1}{\tau_{\mathrm{or}}}\right)\right], \quad (8.26)$$

where τ_{S_n} is the lifetime of the upper excited state S_n^* and τ_{or} is the reorientation time of the excited stilbene in the solvent.

This example shows that DFWM is an alternative tool to pump–probe transient absorption for the characterization of the behavior of excited states. Further examples can be found in [8.34–36].

8.6.3 Experimental Tricks

In practical applications, some care should be taken when performing the kind of experiment described above.

8.6.3.1 Length of Sample. A grating will be generated in each slice of the sample along the propagation direction of the k_1, k_2 pulses. Destructive or constructive interference may occur between the beams diffracted from each of these gratings. In fact, this process is related to the phase-matching conditions inside the sample [8.30]. To avoid this problem, it is best to work with a thin sample. It has been shown that if the length of the sample is of the order of the interfringe distance, one may consider that "thin grating" conditions are satisfied [8.30]. Depending on the angle between the propagation directions of the k_1, k_2 input pulses, these experimental conditions should be easy to calculate.

8.6.3.2 "Thermal Grating" and "Polarization Grating." In the case of resonant conditions (when the frequency ω of the k_1 and k_2 pulses corresponds to an electronic transition of the sample), depopulation of the ground state and population of the excited states occurs. If the polarization directions of the incoming k_1, k_2 beams are parallel, a population grating will be formed by the spatial succession of excited and unexcited zones. During relaxation of the excited states by internal conversion, thermal energy may be deposited in the sample, creating a "thermal" grating and a modulation of the refractive index of thermal origin. This grating will decay because of dynamics related to the thermal diffusion characteristics of the sample; these dynamics are slow compared to the relaxation of the excited molecules. It adds a slow relaxation component, which can obscure other processes and limit the time resolution. We would like to avoid this detrimental effect, by studying the relaxation dynamics free from thermal effects. Fortunately, this can be done very easily by making a "polarization grating" as explained in [8.30]. If the polarizations of the k_1 and k_2 pulses are linear but orthogonal, the amplitude of the resultant field is constant and the thermal effects are the same at all points of space. But the polarization of this field is spatially modulated, creating a so-called "polarization grating". Thanks to the constant amplitude of the resultant field, thermal effects are avoided with this setup.

8.6.3.3 Selecting the Different Components of the Susceptibility Tensor $\chi_{ijkl}^{(3)}$. By varying the polarizations of the k_1, k_2 and k_3 beams, different components of the third-order nonlinear susceptibility can be selected. For example, if the k_1 and k_2 polarizations are orthogonal (polarization grating) and the k_1 and k_3 polarizations are parallel, the diffracted signal will depend only on the component $\chi_{1212}^{(3)}$. If the k_1 and k_3 polarizations are instead orthogonal, the $\chi_{1122}^{(3)}$ component will be measured. By placing an analyzer in the diffracted pulse, other components can be selected for different combinations of the k_1, k_2 and k_3 polarization directions (see for example [8.30]).

8.7 Studies Using the Kerr Effect

In Sect. 8.3.2, we have already described the application of the optical Kerr effect to time-resolved emission, using an optical Kerr shutter. But the Kerr shutter can also be used to study the Kerr effect in a sample: In this case the Kerr cell itself constitutes the sample. The transmission of the probe beam through the system will change as a function of the delay between the probe and the opening pulse, and this change will give information about the dynamics and the amplitude of the Kerr effect in the sample. If the opening pulse is resonant with the sample, one can investigate the behavior of the excited sample.

A typical experimental setup is shown in Fig. 8.35a. It uses two beams, one at 300 nm for the opening pulse, and a continuum for the probe pulse. The 300 nm beam, which is in resonance with electronic transitions of the sample, is linearly polarized and is sent into the sample. It creates an excited-state population, but at the same time, because of its linear polarization, it also creates anisotropy within the sample [8.31, 32, 37]. In this configuration, this beam works at the same time as an "excitation" pulse (excitation of the sample) and as an opening pulse. The induced anisotropy is then tested by the continuum of light which is sent through the excited sample, placed between two crossed polarizers. If the polarization of the 300 nm beam is set at 45° from the entrance polarization (Fig. 8.35b), anisotropy is created by the interaction between the excited-state population and the concomitant hole created in the ground state population. The Kerr effect and its associated third-order optical susceptibility $\chi^{(3)}(-\omega_c; \omega_c, -\omega, \omega)$ can be probed at different frequencies ω_c because of the wide spectral range of the continuum. This method is well adapted to study the dynamics of the Kerr effect in liquids or in molecules in solutions, as can be clearly seen in the striking examples given in [8.38–8.43].

8.7.1 Kerr "Ellipsometry"

In the experimental setup shown in Fig. 8.35a, a dephasing of $\delta\varphi = \omega d(n_{\parallel} - n_{\perp})/c$ occurs between the probe field components polarized parallel and perpendicular to the polarization of the opening pulse, so that transmission

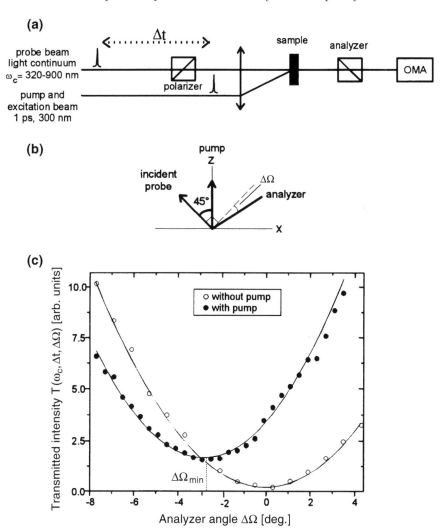

Fig. 8.35. (a) Kerr ellipsometry: experimental setup. OMA: optical multichannel analyser. (b) Polarization scheme of pump and probe beams. (c) Typical transmitted intensity dependence as a function of the analyzer angle $\Delta\Omega$: data corresponding to a 560 nm probe beam with ($\bullet$) and without ($\circ$) the pump beam. The negative value of $\Delta\Omega_{min}$ after excitation indicates a gain band at 560 nm, the sample being DCS (4-dimethylamino 4'-cyanostilbene) in acetonitrile. The experimental points are fitted according to (8.29). (From [8.47])

through the analyzer becomes possible. But the refractive index n may have a real (birefringence) and an imaginary (dichroism) part, depending on the resonance conditions. Taking into account the absorption α of a sample of length d, the transmitted intensity has the following form:

$$I_{\mathrm{T}}(\omega, d) = I_0 \, e^{-\alpha d} \left\{ [\mathrm{Re}(\delta\phi)]^2 + [\mathrm{Im}(\delta\phi)]^2 \right\}. \tag{8.27}$$

Unfortunately, measuring this transmission will not give access independently to the real and the imaginary parts of the nonlinear refractive index, which is an important limitation.

However, using a procedure developed in [8.44], it is possible to obtain the real (birefringence) and the imaginary (dichroism) part of the third-order Kerr-type nonlinear optical susceptibility and their spectral dependence with this same setup. The method, called Kerr "ellipsometry", is as follows.

Let us assume that the opening pulse is linearly polarized along the laboratory z axis and that it propagates along the y axis. Let us also assume that the entrance polarizer polarizes the probe light (the light continuum) at an angle ψ with respect to the pump polarization, and that the analyzer behind the sample is orientated at an angle θ with respect to the pump polarization. According to the Jones matrix procedure, the general expression for the intensity transmitted through the analyzer at frequency ω_c at time Δt after pumping is [8.45]

$$\begin{aligned} I_{\mathrm{T}}(\omega_c, \Delta t) \propto I \, \Big| & \exp\left[-\tfrac{2}{3}\Delta\phi''(\omega_c, \Delta t) + \mathrm{i}\,\Delta\phi'(\omega_c, \Delta t)\right] \cos\psi\cos\theta \\ & + \exp\left[-\tfrac{2}{3}\Delta\phi''(\omega_c, \Delta t) + \mathrm{i}\,\Delta\phi'(\omega_c, \Delta t)\right] \sin\psi\sin\theta \Big|^2, \end{aligned} \tag{8.28}$$

where $\Delta\phi''(\omega_c, \Delta t)$ and $\Delta\phi'(\omega_c, \Delta t)$ are the imaginary and the real parts respectively of the phase delay, and I is the probe intensity. In the classical Kerr gate configuration $\psi \ (= -45°)$ and $\theta \ (= 45°)$ have fixed values. Let the analyzer now be twisted by a small angle $\Delta\Omega$ as shown in Fig. 8.35b. Assuming the weak-signal limit and expanding (8.28) to the second order in $\Delta\phi$ and $\Delta\Omega$, we get a

$$\begin{aligned} I_{\mathrm{T}}(\omega_c, \Delta t) = \; & I \exp\left[-2\phi''(\omega_c, \Delta t)\right] \\ & \times \left\{ [\Delta\phi''(\omega_c, \Delta t) - \Delta\Omega]^2 + [\Delta\phi'(\omega_c, \Delta t)]^2 \right\}. \end{aligned} \tag{8.29}$$

The signal evaluated in (8.29) is a quadratic function of $\Delta\Omega$, whose minimum abscissa $\Delta\Omega_{\min}$ is the pump-induced dichroism $\Delta\phi''(\omega_c, \Delta t)$ and whose minimum ordinate is the magnitude of the pump-induced birefringence $|\Delta\phi'(\omega_c, \Delta t)|$ (Fig. 8.35c). The phase change is the consequence of changes in the refractive indices n_z and n_x, which are characterized by $\chi^{(3)}(-\omega_c; \omega_c, -\omega, \omega)$, the third-order Kerr-type optical nonlinearity of the medium. Measurement of I_{T} for at least three different angles $\Delta\Omega$ provides the angles $\Delta\Omega_{\min} = \Delta\phi''$ and $|\Delta\phi'|$ as solutions of (8.32), without knowledge of the incident probe intensity I, the detector sensitivity or the sample transmission $\exp[-2\phi''(\omega_c, \Delta t)]$ in either the ground or the excited state, in the case of a resonance condition for the opening beam. Recording I_{T} at different frequencies will give the spectral dependence of the dichroism and of the birefringence.

For a resonance condition of the opening beam, the correspondence between the dichroism and the measured excited-state absorption spectra can be found by comparing the theoretical results of [8.44] and [8.45]:

$$\Delta\Omega_{\min}(\omega_{\mathrm{c}}, \Delta t) = \Delta\phi''(\omega_{\mathrm{c}}, \Delta t) = \frac{1}{2}\sum_i \varepsilon_i(\omega_{\mathrm{c}})N_i(\Delta t)r_i(\Delta t)d. \qquad (8.30)$$

The difference between (8.30) and (8.2) is only in the reorientational factor $r_i(\Delta t)$ [8.46], which does not change the shape of the excited-state spectrum. It can only modify the dichroism kinetics, as compared to the transient absorption measured by using the pump–probe method described in Sect. 8.2.

In Kerr ellipsometry experiments, the dichroism angle obtained from the parabolic fit depends only on the nonlinear optical properties of the sample (dichroism angles smaller than 0.2° can be measured easily [8.47]). It is therefore possible to obtain excited-state spectra with a better sensitivity than is possible in pump–probe experiments [8.44]. An illustration is given in Fig. 8.36 which compares the two methods under the same experimental conditions. The similarity between the spectra obtained with the two methods is striking, but the Kerr ellipsometry has a better sensitivity for determining transient excited spectra. The dichroism spectrum in Fig. 8.36b appears as a smoothed form of the optical density spectrum (Fig. 8.36a), so that the gain band appearing at 560 nm and the component at 440 nm inside the main absorption band centered at 490 nm can be properly resolved.

8.8 Laboratory Demonstrations

In this section we would like to describe some pedagogic experiments to perform in the lab as demonstrations. These experiments require laser sources of the kinds which are the most popular in "femtosecond" labs (amplified Ti:sapphire or dye lasers), as well as common nonlinear crystals and ordinary optical items such as lenses, mirrors and optical delay lines. No detectors are required since it is possible *to observe the phenomena directly with the eyes.* This makes group demonstrations possible; the size of the group should not exceed ten persons for reasons of comfort. Naturally, in such a case, it will be necessary before the demonstration to point out the safety rules concerning lasers and to take care of this aspect (verifying the optical pathway and the reflections of the laser beams, checking the safety of electrical power, etc.).

8.8.1 How to Demonstrate Pump–Probe Experiments Directly

The laser system (Fig. 8.37) employed in this experiment is a hybrid mode-locked dye laser (Coherent 702) synchronously pumped by the second harmonic of an actively mode-locked CW-pumped Nd^{3+}:YAG laser (Coherent "Antares-76"). The dye laser operates with Rhodamine 6G as an amplifier medium and pinacyanol chloride as a saturable absorber. The dye laser output, centered at 600 nm (0.5 W CW, pulse duration 500 fs FWHM), is amplified by a dye amplifier (Continuum PTA 60) pumped by the frequency-doubled output (532 nm, 10 Hz, 70 ps, 25 mJ) of the Nd^{3+}:YAG regenerative amplifier

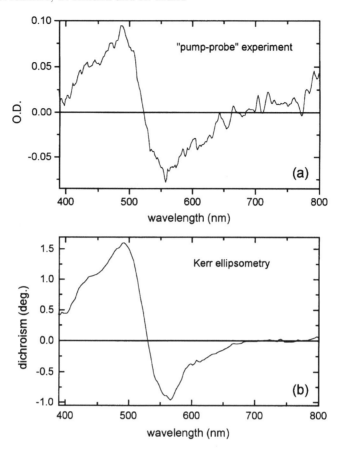

Fig. 8.36. Comparison between pump–probe (**a**) and Kerr ellipsometry (**b**) experiments. The excited spectra were obtained with the same sample (DCS (4-di-methyl-amino 4'-cyanostilbene) in acetonitrile), under the same conditions (low concentration, 10^{-4} M; weak excitation, $3\,\mu$J/pulse)

(Continuum RGA 10). The dye amplifier delivers pulses with an energy of the order of 1 mJ at a 10 Hz repetition rate.

The dye laser pulse at 600 nm is frequency-doubled in a KDP crystal and the energy at 300 nm used in the experiment is $65\,\mu$J. The pulses at 300 nm and 600 nm are separated by a beam-splitter. The UV pulse (300 nm) is used to excite the sample. The 600 nm pulse, after passing through an optical delay line, is focused on a rotating quartz plate (5 mm thickness) to generate the continuum of light used as a probe beam. This continuum is sent through the sample, dispersed by a grating (600 lines/mm) and imaged on a white sheet of paper. As shown in Fig. 8.38, it appears as a rainbow extending from red to blue.

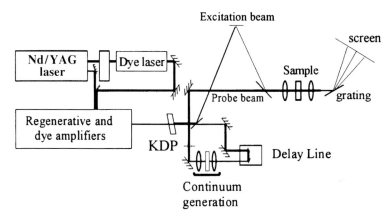

Fig. 8.37. Broadband pump–probe experimental setup

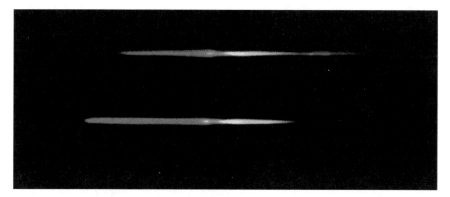

Fig. 8.38. Photograph of probe continuum as it appears on a white screen. *Top*: without excitation beam on sample. *Bottom*: with excitation beam on sample. (For experimental conditions, see text)

The sample is a quartz cuvette (1 cm length) filled with a solution of 1-(p-N,N-dimethylamino)-4-(p-cyanophenyl)-1,3-butadiene (DCB)* dissolved in ethanol (10^{-3} M). This DCB compound is chosen because it shows under UV excitation, strong transient absorption in the blue–green spectral region and a strong gain in the red spectral region. These absorption and gain transitions occur in polar solvents such as ethanol, from the singlet excited state, which has a strong charge transfer character.

The demonstration is then as follows.

(1) While preventing the excitation beam reaching from the sample, show the probe continuum on the paper sheet appearing as a rainbow and observe the limits on the blue and red sides, as shown at the top of Fig. 8.38.

---

* Contact Claude Rullière directly for details of availability of DCB.

(2) Adjust the optical delay line so that the excitation beam reaches the sample 100 ps before the probe continuum. Then allow the excitation beam to reach the sample and show, as at the bottom of Fig. 8.38, that the blue–green part of the probe continuum has disappeared (absorption by the excited sample) and that at the same time the red part is extends further into the red (amplification by the excited sample). Then alternately block and allow through the excitation beam to demonstrate the role of the excitation beam.

(3) Move the optical delay line in both directions. Then two effects may be demonstrated. Reducing the delay between the excitation and probe beams will show that when the probe arrives at the sample just before the excitation, the absorption and gain disappear suddenly, for an optical delay range corresponding roughly to the pulse duration. Increasing the delay will show that the gain and absorption also progressively disappear (in this case for a total delay of the order of 500 ps); this corresponds roughly to the excited-state lifetime.

8.8.2 How to Observe Generation of a CARS Signal by Eye

The experimental setup used for this demonstration is shown in Fig. 8.39 and is a modified version of the setup described in Sect. 8.2.6.3.

The ω_1 pulse at 600 nm is focused on the sample by means of a convergent lens ($f = 20$ cm) with an incident energy of 3.8 μJ, and appears as a bright yellow spot on a screen placed after the sample. The ω_2 pulse, generated by amplification of a continuum in dye laser cells (DCM (4-dicyanomethylene-2-methyl-6-p-dimethyl-amino-styryl-4H-pyran) as a dye laser), extends spectrally from 630 nm to 700 nm, with an overall energy of 5.7 μJ, and appears as a bright red spot.

The sample is a glass cell (1 cm length) containing pure water. The angle between the ω_1 and ω_2 beams on the sample is of the order of 2–3° (adjusted by tuning the separation between the two beams just before the focusing lens) and ensures phase-matching conditions. The demonstration is then as follows.

(1) The delay between the ω_1 and ω_2 pulses is adjusted to zero by means of the optical delay line. When the two beams simultaneously cross the sample cell, observation of the screen shows the generation of a third pulse appearing as a green spot on the screen. This is the CARS signal generated at the anti-Stokes ($\omega_2 + \omega_V$) frequency corresponding to the water vibration ω_V near 1650 cm^{-1}. Point out that the direction of this generated pulse, relative to the yellow and red spots corresponds well to the phase-matching conditions.

(2) Block alternately the ω_1 and ω_2 pulses and show that the green spot disappears, showing that the CARS generation really is due to the interaction between the two pulses.

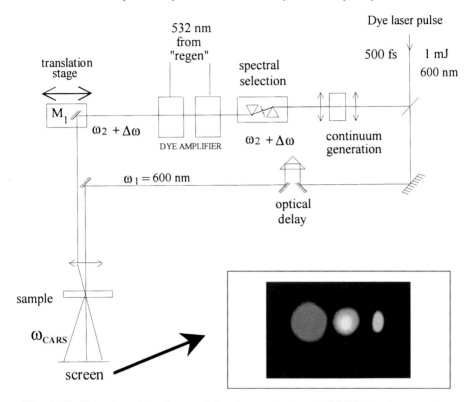

Fig. 8.39. Experimental setup used for demonstration of CARS signal generation. "regen": Nd^{3+}:YAG regenerative amplifier

(3) Change progressively the delay between the ω_1 and ω_2 pulses and show that the green spot disappears for delays of the order of the pulse durations. This demonstrates that the the two incident pulses must coincide in the sample to make CARS generation possible.

(4) Then, by means of the translation stage of mirror M_1 (Fig. 8.39), change the spatial separation between the ω_1 and ω_2 pulses, while keeping them parallel, in front of the convergent lens. Explain that, after focusing, this changes the angle between the two beams in the sample so that the phase-matching conditions are not strictly respected. Show that the green spot disappears or its intensity is progressively reduced when the translation stage is moved forwards and backwards. Comment on the ellipsoidal shape of the green spot in relation to the phase-matching conditions and the near-Gaussian shapes of the incident beams.

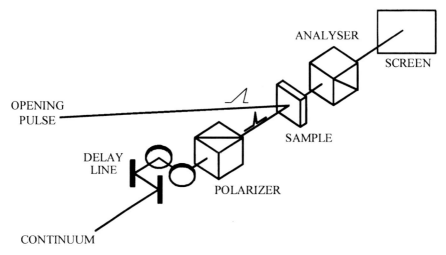

Fig. 8.40. Experimental setup used for demonstration of Kerr gate

8.8.3 How to Build a Kerr Shutter Easily for Demonstration

The experimental setup is shown in Fig. 8.40 and is of the type described in Sect. 8.7.1.

The sample cell is a cuvette (1 cm length) filled with a solution of cresyl violet in ethanol (10^{-3} M). This cell is placed between two crossed polarizers. The opening beam is a part of a dye-amplified laser pulse at 600 nm (energy $\approx 13\,\mu$J) as described in Sect. 8.8.1. The probe beam passed through the Kerr shutter is a continuum of light generated, as described above, in a quartz disk. The polarizations of the two beams are linear and at 45° to each other. The demonstration is as follows.

(1) When the optical delay line is adjusted to zero a bright red spot appears on a screen placed behind the analyzer. When the opening beam is blocked, no spot appears on the screen, demonstrating that the Kerr gate really is opened by the pulse at 600 nm. In fact, the bright red spot is a part of the continuum amplified by the transition gain of cresyl violet excited by the 600 nm pulse. It creates, at the same time, anisotropy in the sample [8.48], making transmission through the analyzer possible (see Sect. 8.7).

(2) Change the position of the delay line so that the opening pulse arrives later than the probe continuum. In this case no spot appears on the screen, demonstrating once again that the 600 nm pulse really opens the Kerr gate.

(3) Change progressively the adjustment of the delay line so that the probe pulse is more and more delayed with respect to the opening pulse. In this situation the intensity of the red spot decreases until it disappears. Measuring the optical delay gives an indication of the reorientation time of cresyl violet in ethanol.

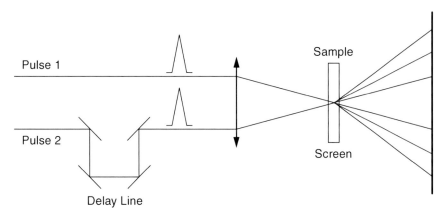

Fig. 8.41. Experimental setup for demonstration of DFWM technique

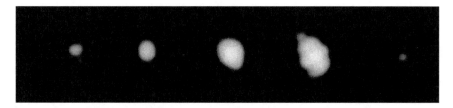

Fig. 8.42. Diffraction pattern observed on the screen in the setup shown in Fig. 8.41. *Left* to *right*: second order, first order, pulse 1, pulse 2, first order

8.8.4 How to Observe a DFWM Diffraction Pattern Directly

The experimental setup is shown in Fig. 8.41 and is very simple. Pulses from an amplified dye laser at 600 nm are divided into two parts, made parallel and focused by a convergent lens ($f = 20$ cm) so to interfere in a cell. The cell (400 μm thick) is filled with carbon disulfide (CS_2). This highly non-linear material produces a large self-diffraction efficiency from the induced grating, making it possible to observe several orders of diffraction, as shown in Fig. 8.42. The energies of the pulses are respectively 3.8 and 1.6 μJ. The demonstration is as follows.

(1) Adjust the optical delay line to ensure zero delay between the two pulses at the sample. Block alternately one pulse and the other to show that the diffraction pattern disappears.
(2) Adjust the delay line to show that when the pulses are not coincident in the sample the diffraction pattern also disappears.

References

[8.1]   R.R. Alfano, S.L. Shapiro: Phys. Rev. Lett. **24**, 592 (1970)

[8.2]   F. Salin, J. Watson, J.F. Cormier, P. Georges, A. Brun: *Ultrafast Phenomena VIII*, Springer Ser. Chem. Phys., Vol. 55 (Springer, Berlin, Heidelberg 1992) p. 306

[8.3]   N.P. Ernsting, M. Kaschke: Rev. Sci. Instrum. **62**, 600 (1991)

[8.4]   G.R. Fleming: *Chemical Applications of Ultrafast Spectroscopy* (Oxford University Press, New York 1986)

[8.5]   H.E. Von Lessing, A. Von Jena: Chem. Phys. Lett. **42**, 213 (1976)

[8.6]   J.F. Létard, R. Lapouyade, W. Rettig: Pure Appl. Chem. **65**, 1705 (1993)

[8.7]   J.F. Létard, P. Dumon, G. Jonusauskas, F. Dupuy, P. Pée, C. Rullière, R. Lapouyade: J. Phys. Chem. **98**, 10391 (1994)

[8.8]   J.N. Moore, P.A. Hansen, R.M. Hochstrasser: Chem. Phys. Lett. **138**, 110 (1987)

[8.9]   M. Dantus, M.J. Rosker, A.H. Zewail: J. Phys. Chem. **89**, 6128 (1988)

[8.10]   J. Manz, L. Wëste (eds.): *Femtosecond Chemistry* (VCH, Weinheim 1995)

[8.11]   P.J. Reid, S.D. Wickham, R. Mathies: J. Phys. Chem. **96**, 5720 (1992)

[8.12]   R.J.H. Clark, R.E. Hester (eds): *Time Resolved Spectroscopy*, Advances in Spectroscopy, Vol. 18 (Wiley, New York 1989). G.H. Atkinson (ed): *Time-Resolved Vibrational Spectroscopy: Proc. JSPS/NSF Symp.* (Honolulu, Hawaii, November 1985) (Gordon and Breach, Newark, NJ 1987)

[8.13]   H.G. Purucker, V. Tunkin, A. Laubereau: J. Raman Spectrosc. **24**, 453 (1993)

[8.14]   R. Brakel, V. Mudogo, F.W. Schneider: J. Chem. Phys. **84**, 2451 (1986)

[8.15]   H. Okamato, R. Inaba, K. Yoshihara, M. Tasumi: Chem. Phys. Lett. **202**, 161 (1993)

[8.16]   J. Oberlé, E. Abraham, A. Ivanov, G. Jonusauskas, C. Rullière: J. Phys. Chem. **100**, 10179 (1996)

[8.17]   G. Lucassen: *Polarization Sensitive Coherent Raman Spectroscopy on (Bio)Molecules in Solutions*, PhD thesis, University of Twente, Netherlands (1992)

[8.18]   T. Dudev, T. Kamisuki, N. Akamatsu, C. Hirose: J. Phys. Chem. **95**, 4999 (1991)

[8.19]   N.I. Koroteev, A.P. Shkurinov, B.N. Toleutaev: in *Coherent Raman Spectroscopy*, Springer Proceedings in Physics Vol. 63, eds. G. Marowsky and V.V. Smirnov (Springer, Berlin, Heidelberg 1992), p. 182

[8.20]   G. Mourou, W. Knox: Appl. Phys. Lett. **36**, 623 (1980)

[8.21]   M.A. Duguay: Prog. Opt. **14**, 161 (1976)

[8.22]   J. Shah: IEEE J. Quant. Electr. **24**, 276 (1988)

[8.23]   S. Vallogia: Thèse de doctorat, INSA, Toulouse, France (1988)

[8.24]   F. Zernike, J.E. Midwinter: *Applied Non Linear Optics* (Wiley, New York 1973)

[8.25]   T. Amand, X. Marie, P. Le Jeune, M. Brousseau, D. Robart, J. Barrau, R. Planel: Phys. Rev. Lett. **78**, 7 (1997)

[8.26]   D. Von der Linde, N. Fabricius, J. Kuhl, E. Rosengart: in *Picosecond Phenomena III*, Springer Series in Chemical Physics, Vol. 23, eds: K.B. Eisenthal, R.M. Hochstrasser, W. Kaiser, A. Laubereau (Springer, Berlin, Heidelberg 1982) p. 336

[8.27]   R. Christanell, R.A. Höpfel: J. Appl. Phys. **66**, 10 (1989)

[8.28] M.J. Rosker, F.W. Wise, C.L. Tang: Appl. Phys. Lett. **49**, 25 (1986)

[8.29] A.J. Taylor, D.J. Erskine, C.L. Tang: Appl. Phys. Lett. **43**, 11 (1983)

[8.30] H.J. Eichler, P. Günter, D.W. Pohl: in *Laser Induced Dynamic Gratings* (Springer, Berlin, Heidelberg 1985)

[8.31] J. Oberlé, G. Jonusauskas, E. Abraham, C. Rullière: Chem. Phys. Lett. **241**, 281 (1995)

[8.32] J. Oberlé, G. Jonusauskas, E. Abraham, C. Rullière: Opt. Commun. **124**, 616 (1996)

[8.33] D.W. Phillion, D.J. Kuizenga, A.E. Siegman: Appl. Phys. Lett. **27**, 85 (1975)

[8.34] T. Höfer, P. Kruck, W. Kaiser: Chem. Phys. Lett. **224**, 411 (1994)

[8.35] T. Höfer, P. Kruck, T. Elsaesser, W. Kaiser: J. Phys. Chem. **99**, 4380 (1995)

[8.36] G.P. Wiederrecht, W.A. Svec, M.P. Niemczyk, M. Wasielewski: J. Phys. Chem. **99**, 8918 (1995)

[8.37] P.P. Ho, R.R. Alfano: Phys. Rev. A **20**, 2170 (1979)

[8.38] M.E. Orczyk, J. Swiatkiewicz, G. Huang, P.N. Prasad: J. Phys. Chem. **98**, 7307 (1994)

[8.39] M.G. Giorgini, P. Foggi, R.S. Cataliotti, M.R. Distefano, A. Morresi, L. Mariani: J. Chem. Phys. **102**, 8763 (1995)

[8.40] M. Ricci, R. Torre, P. Foggi, V. Kamalov, R. Righini: J. Chem. Phys. **102**, 9537 (1995)

[8.41] H.P. Deuel, P. Cong, J.D. Simon: J. Phys. Chem. **98**, 12600 (1994)

[8.42] P. Foggi, V.F. Kamalov, R. Righini, R. Torre: Opt. Lett. **17**, 775 (1992)

[8.43] I. Santa, P. Foggi, R. Righini, J.H. Williams: J. Phys. Chem. **98**, 7692 (1994)

[8.44] N. Pfeffer, F. Charra, J.M. Nunzi: Opt. Lett. **16**, 1987 (1991)

[8.45] D.S. Alavi, R.S. Hartman, D.H. Waldeck: J. Chem. Phys. **92**, 4055 (1990)

[8.46] A.J. Cross, D.H. Waldeck, G.R. Fleming: J. Chem. Phys. **78**, 6455 (1983)

[8.47] E. Abraham, J. Oberlé, G. Jonusauskas, R. Lapouyade, C. Rullière: Chem. Phys. **214**, 409 (1997)

[8.48] G. Jonusauskas, J. Oberlé, E. Abraham, C. Rullière: Opt. Commun. **137**, 199 (1997)

Coherent Effects in Femtosecond Spectroscopy: A Simple Picture Using the Bloch Equation

M. Joffre

With 16 Figures

9.1 Introduction

Femtosecond spectroscopy aims at characterizing the dynamics of elementary excitations in material systems. One of the most common experimental techniques, spectrally resolved pump–probe spectroscopy, may in some cases present artifacts, the so-called *coherent effects*, which make data interpretation less straightforward than the incoherent picture would lead one to believe. It is therefore desirable to be able to rely on a theoretical model in order to assess the importance of coherent effects in these experiments. In this chapter, we will introduce such a formalism, which will also help in the discussion of other experimental techniques, such as photon echo, wave-packet excitation, and multidimensional spectroscopy.

We follow closely the standard nonlinear-optics textbooks, such as that by *Butcher* and *Cotter* [9.1], as well as the work of *Brito-Cruz* et al. [9.2] and the first chapters of the book by *Mukamel* [9.3] on nonlinear optical spectroscopy.

9.2 Theoretical Model

The number of degrees of freedom being usually far too large to allow an exact solution of the problem of light–matter interaction, it is usual to divide the system into two parts, a small quantum system containing only a few energy levels, and a macroscopic thermal bath, as shown in Fig. 9.1. Only the quantum system is coupled to light, but its phase coherence is strongly affected by its own coupling to the thermal bath. For example, for a dye molecule in a liquid, the quantum system consists of the electronic energy levels of the dye molecule, while the bath corresponds to all the degrees of freedom of the solvent. In a semiconductor, the quantum system could be the exciton, coupled to a large phonon bath.

The advantage of such an approach [9.3] is that the quantum system has a limited number of degrees of freedom, so that its reduced density matrix is

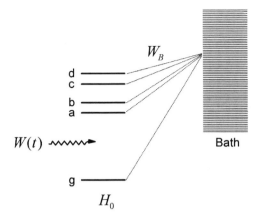

Fig. 9.1. Separation of the system into a quantum system, with only a few energy levels, and a macroscopic bath. H_0 is the Hamiltonian of the quantum system, $W(t)$ is the coupling to the electric field of the incident light and W_B is the coupling to the bath

of reasonable size. The coupling to the bath can be treated through approximations, or in a phenomenological way, so that the reduced density operator can be assumed to follow an equation whose solution is not beyond reach.

Among the various methods allowing one to take account of the coupling between the quantum system and the bath [9.3], we will discuss only the simplest one, which involves the Bloch equation. This is a very crude approximation, limiting relaxation to purely Markovian processes, which has been shown to fail in several cases, including liquids [9.3–5] and semiconductors [9.6, 7]. However, as the simplest model, it is very useful as an illustration and for understanding the basic physics of femtosecond dynamics, even though this is only an approximation, and a very drastic one.

9.2.1 Equation of Evolution

The quantum system is described by its density operator ρ, which is assumed to obey the Bloch equation; this is derived directly from the Schrödinger equation with an additional term accounting for relaxation due to the coupling to the bath:

$$i\hbar \frac{d\rho}{dt} = [H_0, \rho] + [W(t), \rho] + i\hbar \frac{\partial \rho}{\partial t}\bigg|_{\text{relax}}, \tag{9.1}$$

where H_0 is the reduced unperturbed Hamiltonian, $W(t) = -\mu E(t)$ is the dipolar interaction with the total electric field $E(t)$ of the incident femtosecond pulses, and μ is the electric-dipole operator. The eigenstates of H_0 are

$$H_0|n\rangle = \hbar\omega_n|n\rangle.$$

The last term in (9.1) corresponds to the interaction with the bath, leading to relaxation. In the Bloch model [9.1], this term is written in the form

$$\left.\frac{\partial \rho_{nm}}{\partial t}\right|_{\text{relax}} = -\Gamma_{nm}\left(\rho_{nm} - \rho_{nm}^{(0)}\right),$$

so that it leads to an exponential relaxation with a decay rate Γ_{nm}. Γ_{nm} corresponds to a dephasing rate when $n \neq m$ and to a population relaxation rate when $n = m$. $\rho^{(0)}$ is the density operator at thermal equilibrium, whose non-diagonal terms are all zero. In the following, we will assume that kT is much smaller than any transition energy from the ground state. As a consequence, only the ground state is populated, so that $\rho_{00}^{(0)} = 1$ and all other matrix elements are zero. Let us now take the matrix element of (9.1) between states $|n\rangle$ and $|m\rangle$. The matrix elements of the two commutators are

$$\langle n|[H_0, \rho]|m\rangle = \langle n|H_0\rho - \rho H_0|m\rangle = \langle n|\hbar\omega_n \rho - \rho\hbar\omega_m|m\rangle$$
$$= \hbar\omega_{nm}\rho_{nm},$$

with $\omega_{nm} = \omega_n - \omega_m$, and

$$\langle n|[W, \rho]|m\rangle = \langle n|W\rho - \rho W|m\rangle$$
$$= \sum_l (W_{nl}\rho_{lm} - \rho_{nl}W_{lm})$$
$$= -E(t)\sum_l (\mu_{nl}\rho_{lm} - \rho_{nl}\mu_{lm}).$$

The evolution equation of the density matrix elements can therefore be written as

$$\left(i\frac{d}{dt} - \omega_{nm} + i\Gamma_{nm}\right)\rho_{nm}(t) = -\frac{E(t)}{\hbar}\sum_l [\mu_{nl}\rho_{lm}(t) - \rho_{nl}(t)\mu_{lm}]. \quad (9.2)$$

In order to solve this equation, we define $G_{nm}(t)$, the zeroth-order Green function, which obeys a similar equation in which the right-hand side has been replaced with an impulsive source term:

$$\left(i\frac{d}{dt} - \omega_{nm} + i\Gamma_{nm}\right)G_{nm}(t) = -\frac{\delta(t)}{\hbar}, \quad (9.3)$$

where $\delta(t)$ is the Dirac distribution. This equation is easily solved, giving

$$G_{nm}(t) = \frac{i}{\hbar}\Theta(t)\exp(-i\omega_{nm}t - \Gamma_{nm}t),$$

where $\Theta(t)$ is the Heaviside function. The Fourier transform of (9.3) can be written as

$$G_{nm}(\omega) = \frac{-1/\hbar}{\omega - \omega_{nm} + i\Gamma_{nm}}.$$

Similarly, the Fourier transform of (9.2) reads

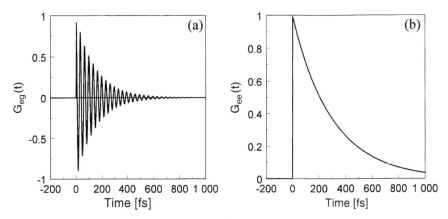

Fig. 9.2. Imaginary parts of the Green functions (**a**) $G_{eg}(t)$ and (**b**) $G_{ee}(t)$ in the case of a two-level system $\{|g\rangle, |e\rangle\}$

$$\rho_{nm}(\omega) = \frac{-1/\hbar}{\omega - \omega_{nm} + i\Gamma_{nm}} \, \text{F.T.} \left(E(t) \sum_l [\mu_{nl}\rho_{lm}(t) - \rho_{nl}(t)\mu_{lm}] \right)$$

$$= G_{nm}(\omega) \, \text{F.T.} \left(E(t) \sum_l [\mu_{nl}\rho_{lm}(t) - \rho_{nl}(t)\mu_{lm}] \right),$$

where "F.T." stands for Fourier transformation. After an inverse Fourier transform, we obtain the following equation:

$$\rho_{nm}(t) = G_{nm}(t) \otimes \left(E(t) \sum_l [\mu_{nl}\rho_{lm}(t) - \rho_{nl}(t)\mu_{lm}] \right). \tag{9.4}$$

The Green function can be interpreted as the impulsive response function of the material. The density operator is simply the convolution product of this response function and a source term (see the right-hand side of (9.2)), which is itself a product of the electric field and a term depending on the density operator. Figure 9.2 shows the Green functions for both the nondiagonal and the diagonal matrix elements.

9.2.2 Perturbation Theory

We now assume that the electric field is small enough that we can perform a perturbation expansion of the density operator [9.1, 2, 8, 9]

$$\rho(t) = \rho^{(1)}(t) + \rho^{(2)}(t) + \rho^{(3)}(t) + \dots,$$

where $\rho^{(p)}$ is of order p in the electric field. Equation (9.4) is particularly suited for a perturbation expansion, as it allows one to compute the term of

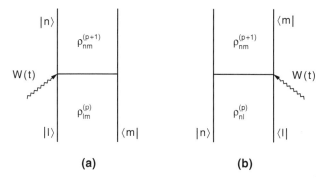

Fig. 9.3. Double-sided Feynman diagrams showing the action of the perturbation $W(t)$, either on the ket side (**a**), or on the bra side (**b**)

order $p+1$ from the term of order p. Indeed, if $\rho^{(p)}(t)$ is known, its product with the electric field is of order $p+1$. Using (9.4), we find

$$\rho_{nm}^{(p+1)}(t) = G_{nm}(t) \otimes \left(E(t) \sum_l \left[\mu_{nl}\rho_{lm}^{(p)}(t) - \rho_{nl}^{(p)}(t)\mu_{lm} \right] \right), \qquad (9.5)$$

which provides us with a sequence of equations allowing us to compute iteratively the density matrix up to any order in the electric field. This equation is often represented graphically using double-sided Feynman diagrams [9.3], as shown in Fig. 9.3.

We use the following conventions, as in [9.3]. A matrix element is represented as two vertical lines, with time going upwards. The left-hand line represents the ket $|n\rangle$, while the right-hand line stands for the bra $\langle m|$. The effect of each interaction appears as a vertex, with a wavy line representing the photon. For each interaction on the ket side, involving a transition from $|l\rangle$ to $|n\rangle$, a term $\mu_{nl}E(t)$ appears in the equation (Fig. 9.3a). This contribution is complex-conjugated and changes sign when the vertex is on the bra side (Fig. 9.3b). Finally, when the complex notation $E(t) = \mathrm{Re}\,\mathcal{E}(t)$ is used, the arrow points to the right, or the left when the interaction involves $\mathcal{E}(t)$, or $\mathcal{E}^*(t)$ respectively. The propagation of the matrix element $\rho_{nm}(t)$ to the next vertex is taken into account through a convolution product with the Green function $G_{nm}(t)$. Note that these conventions allow us to write down directly (9.5) from the two diagrams shown in Fig. 9.3. Indeed, using the above rules, we write down $G_{nm}(t) \otimes \left[\mu_{nl}E(t)\rho_{lm}^{(p)}(t) \right]$ for diagram (a) and $-G_{nm}(t) \otimes \left[\mu_{lm}E(t)\rho_{nl}^{(p)}(t) \right]$ for diagram (b). By adding these two terms, one recovers (9.5).

As an example, let us write down the second-order density operator, first by using only the equations, and second by drawing all possible diagrams and then translating them into equations.

We start from the density operator at zeroth order, which verifies that $\rho_{gg}^{(0)} = 1$. Using (9.5), we find, at first order,

$$\rho_{ng}^{(1)}(t) = G_{ng}(t) \otimes \left[E(t)\mu_{ng}\rho_{gg}^{(0)}\right] = \mu_{ng}G_{ng}(t) \otimes E(t),$$

$$\rho_{gn}^{(1)}(t) = G_{gn}(t) \otimes \left[-E(t)\rho_{gg}^{(0)}\mu_{gn}\right] = -\mu_{gn}G_{gn}(t) \otimes E(t) = \left[\rho_{ng}^{(1)}(t)\right]^*.$$

Then, at second order,

$$\rho_{nm}^{(2)} = G_{nm}(t) \otimes \left\{E(t)\left[\mu_{ng}\rho_{gm}^{(1)}(t) - \rho_{ng}^{(1)}(t)\mu_{gm}\right]\right\}$$

$$= \mu_{ng}\mu_{gm}G_{nm}(t) \otimes \left\{E(t)\left[(-G_{gm}(t) - G_{ng}(t)) \otimes E(t)\right]\right\}, \quad (9.6)$$

where we have again made use of (9.5). Using the complex notation for the electric field, $E(t) = \mathrm{Re}\mathcal{E}(t) = (\mathcal{E}(t) + \mathcal{E}^*(t))/2$, (9.6) can be expressed as a sum of eight terms. We now keep only the terms that are not too far off resonance, for which the ω_{ng} are of the same order of magnitude as the frequency of the optical wave. We can therefore make use of the standard rotating-wave approximation (RWA), which means we keep only resonant terms. For example, in $G_{ng}(t) \otimes E(t) = [G_{ng}(t) \otimes \mathcal{E}(t) + G_{ng}(t) \otimes \mathcal{E}^*(t)]/2$, $\mathcal{E}(t)$ is resonant while $\mathcal{E}^*(t)$ is not, as it is rotating the wrong way. This is also apparent in the frequency domain: $G_{ng}(\omega)\mathcal{E}(\omega)$ provides a much larger contribution than $G_{ng}(\omega)\mathcal{E}^*(-\omega)$, due to a much greater overlap in the spectra. According to the RWA, only two of the eight terms in (9.6) have a nonnegligible contribution. This yields

$$\rho_{nm}^{(2)}(t) = -\frac{\mu_{ng}\mu_{gm}}{4}G_{nm}(t)$$
$$\otimes \left\{\mathcal{E}(t)\left[G_{gm}(t) \otimes \mathcal{E}^*(t)\right] + \mathcal{E}^*(t)\left[G_{ng}(t) \otimes \mathcal{E}(t)\right]\right\}. \quad (9.7)$$

The second method of reaching this result consists of drawing the eight possible double-sided Feynman diagrams, shown in Fig. 9.4. It is easy to keep only the resonant diagrams, also called RWA diagrams, whose contributions are nonnegligible within the rotating-wave approximation. For this purpose, one makes use of the following rule, which is reminiscent of energy conservation. For example, in the case of diagram (a), the ket $|g\rangle$ "absorbs" a photon and goes into state $|n\rangle$, then the bra $\langle g|$ "absorbs" a photon and goes into state $\langle m|$. Energy conservation is roughly fulfilled so that this diagram satisfies the RWA. In contrast, in the case of diagram (c), the bra $\langle g|$ "emits" a photon and goes into state $\langle m|$. There is an energy mismatch of about twice the optical frequency, so that this diagram can be neglected. It can thus be shown easily that of all the

To summarize, the perturbative method allows one to write the density operator at any order as a succession of products and convolution products. Besides the advantage of this approach in terms of a physical picture, allowing one to draw the various terms as Feynman diagrams, it has also an advantage in terms of numerical computation. Indeed, (9.5) allows one to compute the

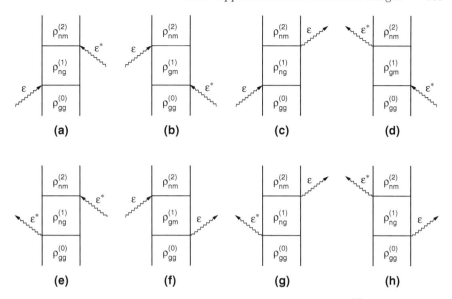

Fig. 9.4. The eight double-sided Feynman diagrams leading to $\rho_{nm}^{(2)}$. Only the first two (**a** and **b**) contribute within the rotating-wave approximation

density operator very easily, since convolution products can be evaluated very efficiently using fast-Fourier-transform routines. This method [9.2] is more stable than a direct integration of the Bloch equation in the time domain.

9.2.3 Two-Level Model

We consider here the important case of a two-level system, consisting of levels $|g\rangle$ and $|e\rangle$. Furthermore, we assume the system to have a center of inversion, so that μ_{eg} is the only non-zero matrix element of the electric-dipole operator: there are no permanent dipoles. Using (9.4), we have

$$\rho_{eg}(t) = \mu_{eg} G_{eg}(t) \otimes \{E(t)[\rho_{gg}(t) - \rho_{ee}(t)]\}, \tag{9.8}$$

which connects population terms to coherence terms, and

$$\rho_{ee}(t) = G_{ee}(t) \otimes \{E(t)[\mu_{eg}\rho_{ge}(t) - \rho_{eg}(t)\mu_{ge}]\}, \tag{9.9}$$

connecting coherence terms to population terms. The perturbation expansion therefore leads to an alternating development:

$$\rho_{gg}^{(0)} \rightarrow \rho_{eg}^{(1)} \rightarrow \rho_{ee}^{(2)} \rightarrow \rho_{eg}^{(3)} \rightarrow \dots.$$

Starting from the system at equilibrium at zero temperature ($\rho_{gg}^{(0)} = 1$), (9.8) yields the first-order coherence:

$$\rho_{\rm eg}^{(1)}(t) = \mu_{\rm eg} G_{\rm eg}(t) \otimes E(t),$$

which itself can be used, through (9.9), to compute the second-order population term:

$$\rho_{\rm ee}^{(2)}(t) = -|\mu_{\rm eg}|^2 G_{\rm ee}(t) \otimes (E(t)\,\{[G_{\rm eg}(t) + G_{\rm ge}(t)] \otimes E(t)\})\,. \qquad (9.10)$$

The second-order population difference is deduced easily from the fact that $\mathrm{Tr}\,\rho = 1$, so that $\mathrm{Tr}\,\rho^{(2)} = 0$. This yields $\rho_{\rm gg}^{(2)} - \rho_{\rm ee}^{(2)} = -2\rho_{\rm ee}^{(2)}$, which, when substituted in (9.8), provides the third-order coherence term:

$$\rho_{\rm eg}^{(3)}(t) = 2\mu_{\rm eg}\mu_{\rm ge}\mu_{\rm eg} G_{\rm eg}(t)$$
$$\otimes \{E(t)\,[G_{\rm ee}(t) \otimes (E(t)\,\{[G_{\rm eg}(t) + G_{\rm ge}(t)] \otimes E(t)\})]\}\,. \qquad (9.11)$$

Note that in a two-level system, only two relaxation rates appear: $\Gamma_{\rm eg} = 1/T_2$, where T_2 is the dephasing time, and $\Gamma_{\rm ee} = 1/T_1$, where T_1 is the population relaxation time.

9.2.4 Induced Polarization

The optical properties of the system can be deduced entirely from the polarization induced in the material. Indeed, it is the polarization oscillation which is responsible for the emission of the radiated electric field. Assuming that the material can be modeled as an assembly of N identical, independent systems per unit volume, the total polarization per unit volume reads

$$P(t) = N\langle\mu\rangle = N\,\mathrm{Tr}\,\mu\rho(t) = N \sum_{nm} \mu_{mn}\rho_{nm}. \qquad (9.12)$$

Corresponding to the perturbation expansion of the density operator, there is a similar expansion of the polarization

$$P(t) = P^{(1)}(t) + P^{(2)}(t) + P^{(3)}(t) + \dots.$$

The computation of the density operator allows one to compute the nonlinear susceptibility tensors [9.1], defined in the following equation:

$$P^{(p)}(t) = \varepsilon_0 \int_{-\infty}^{+\infty} \frac{\mathrm{d}\omega_1}{2\pi} \cdots \int_{-\infty}^{+\infty} \frac{\mathrm{d}\omega_p}{2\pi}$$
$$\times \chi^{(p)}(-\omega_\sigma; \omega_1, \dots, \omega_p) E(\omega_1) \dots E(\omega_p) \exp(-\mathrm{i}\omega_\sigma t),$$

where $\omega_\sigma = \omega_1 + \omega_2 + \dots + \omega_p$. The frequency dependence of the nonlinear susceptibility allows one to take into account memory effects in the material, and hence to compute the response of the material to any sequence of femtosecond pulses. However, in the following, we shall not use the susceptibility formalism; instead, we shall directly compute the relevant polarization terms, to show the physical picture more clearly.

9.3 Applications to Femtosecond Spectroscopy

We now discuss a few examples of optical responses, in order to illustrate a few of the most common femtosecond techniques. We will first consider the well-known linear response, and then proceed to the quadratic and third-order responses.

9.3.1 First Order

Let us compute the expression for the density operator to first order in the electric field, using (9.5):

$$\rho_{nm}^{(1)}(t) = G_{nm}(t) \otimes \left[E(t) \left(\mu_{nm}\rho_{mm}^{(0)} - \rho_{nn}^{(0)}\mu_{nm} \right) \right].$$

Through a Fourier transformation, we obtain

$$\rho_{nm}^{(1)}(\omega) = \mu_{nm}G_{nm}(\omega)E(\omega) \left(\rho_{mm}^{(0)} - \rho_{nn}^{(0)} \right),$$

from which we can deduce the induced polarization, using (9.12):

$$P(\omega) = N \sum_{nm} \mu_{mn}\mu_{nm}G_{nm}(\omega) \left(\rho_{mm}^{(0)} - \rho_{nn}^{(0)} \right) \mathcal{E}(\omega).$$

Using $P(\omega) = \varepsilon_0\chi^{(1)}(\omega)\mathcal{E}(\omega)$, we deduce the expression for the linear susceptibility:

$$\chi^{(1)}(\omega) = N \sum_{nm} \frac{-|\mu_{nm}|^2}{\hbar\varepsilon_0} \frac{\rho_{mm}^{(0)} - \rho_{nn}^{(0)}}{\omega - \omega_{nm} + i\Gamma_{nm}}.$$

The real part of the susceptibility yields the refractive index, while the imaginary part yields the absorption coefficient, which can be expressed in terms of the linear polarization:

$$\alpha(\omega) = \frac{n\omega}{c}\mathrm{Im}\chi^{(1)}(\omega) = \frac{n\omega}{\varepsilon_0 c}\mathrm{Im}\frac{P(\omega)}{\mathcal{E}(\omega)}. \tag{9.13}$$

Replacing the linear susceptibility with the expression obtained above, we obtain

$$\alpha(\omega) = N \sum_{nm} \frac{|\mu_{nm}|^2}{\hbar\varepsilon_0} \frac{\Gamma_{nm} \left(\rho_{mm}^{(0)} - \rho_{nn}^{(0)} \right)}{(\omega - \omega_{nm})^2 + \Gamma_{nm}^2}. \tag{9.14}$$

The absorption spectrum is therefore an assembly of Lorentzian lines, centered on the transition frequencies ω_{nm}, with full widths at half maximum $2\Gamma_{nm}$. Figure 9.5 shows the linear susceptibility in the case of a two-level system.

Note that (9.14) also includes stimulated emission, since the population difference occurs as a factor in the susceptibility. When the excited-state population is larger than the ground state population, the absorption coefficient is negative, which means that amplification can be observed. Figure 9.5 also

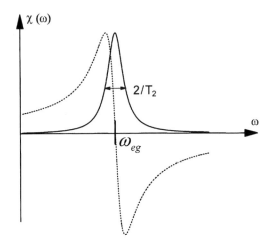

Fig. 9.5. Real part (*dashed line*) and imaginary part (*solid line*) of the linear susceptibility for a two-level model

shows that the refractive index is always increasing in regions of transparency (normal dispersion).

All the above results are well known in linear optics, and only show that the formalism yields reasonable results in this regime. Let us now consider the nonlinear terms, which are the only ones actually relevant to femtosecond spectroscopy.

9.3.2 Second Order

The second-order coherence terms of the density operator lead to a second-order polarization, responsible for sum- and difference-frequency mixing. Such processes occur only in noncentrosymmetric systems. The second-order density operator is also the lowest order in which a population appears. We will consider only two examples below.

9.3.2.1 Excited Population in a Two-Level System Let us consider a two-level system with an infinite population relaxation time. This means that any absorbed photon leaves the system in an excited state, with no relaxation path. The results obtained in this section will be relevant to the case where the pulse duration is much shorter than any of the population relaxation times. Applying (9.7) to the case of a two-level system, we obtain, within the rotating-wave approximation,

$$\rho_{ee}^{(2)}(t) = -\frac{|\mu_{eg}|^2}{4} G_{ee}(t) \otimes \left\{ \mathcal{E}^*(t) \left[G_{eg}(t) \otimes \mathcal{E}(t) \right] - \mathcal{E}(t) \left[G_{eg}^*(t) \otimes \mathcal{E}^*(t) \right] \right\}$$

$$= -\frac{\mathrm{i}|\mu_{eg}|^2}{2} G_{ee}(t) \otimes \mathrm{Im} \left\{ \mathcal{E}^*(t) \left[G_{eg}(t) \otimes \mathcal{E}(t) \right] \right\}.$$

Using the expression $G_{ee}(t) = i\Theta(t)/\hbar$, we find

$$\rho_{ee}^{(2)}(t) = \frac{|\mu_{eg}|^2}{2\hbar}\mathrm{Im}\int_{-\infty}^{t}\mathcal{E}^*(t')\left[G_{eg}(t')\otimes\mathcal{E}(t')\right]\mathrm{d}t'. \tag{9.15}$$

The energy absorbed per unit time is therefore not equal to the integral of the incident power, owing to the noninstantaneous nature of the medium response. Indeed, some frequency components will be favored because of the shape of the absorption spectrum. If we compute the excited population after the pulse is turned off, we obtain

$$\begin{aligned}\rho_{ee}^{(2)}(+\infty) &= \frac{|\mu_{eg}|^2}{2\hbar}\mathrm{Im}\int_{-\infty}^{+\infty}\mathcal{E}^*(t')\left[G_{eg}(t')\otimes\mathcal{E}(t')\right]\mathrm{d}t'\\&= \frac{|\mu_{eg}|^2}{2\hbar}\mathrm{Im}\int_{-\infty}^{+\infty}\mathcal{E}^*(\omega)G_{eg}(\omega)\mathcal{E}(\omega)\frac{\mathrm{d}\omega}{2\pi}\\&= \frac{|\mu_{eg}|^2}{2\hbar}\int_{-\infty}^{+\infty}|\mathcal{E}(\omega)|^2\mathrm{Im}G_{eg}(\omega)\frac{\mathrm{d}\omega}{2\pi},\end{aligned} \tag{9.16}$$

where we have made use of the Parseval–Plancherel theorem. $\mathrm{Im}G_{eg}(\omega)$ being closely connected to the absorption coefficient (see (9.14)), (9.16) can be interpreted in terms of energy conservation: the population excited by the short pulse is proportional to the absorbed energy, which is itself connected to the overlap between the pulse spectrum and the absorption spectrum.

We conclude that it is only after the pulse is turned off that the excited population assumes the simple form of (9.16). During the population buildup, (9.15) should be used instead, where the exact pulse profile appears in a more intricate form than simply its power spectrum.

9.3.2.2 Wave-Packet Excitation Let us now consider the more complex case where several energy levels are excited simultaneously by the incident pulse, as in the various examples shown in Fig. 9.6. This can involve nuclear vibrational levels, electronic levels in a Rydberg atom or electronic levels in a quantum well structure, to name a few. For the purposes of illustration, we shall consider only three levels here, as shown in Fig. 9.6c: the ground state $|0\rangle$, and two excited states, $|1\rangle$ and $|2\rangle$, both within the laser bandwidth. However, the calculation could be easily extended to take more states into account.

As shown in the diagrams of Fig. 9.7, it is possible to excite not only the population terms ρ_{11} and ρ_{22}, but also coherence terms, such as ρ_{21}. Similarly to (9.10), we obtain the following expression for the coherence term between two excited states:

$$\rho_{21}^{(2)}(t) = -\mu_{20}\mu_{01}G_{21}(t)\otimes\left(E(t)\left\{[G_{01}(t)+G_{20}(t)]\otimes E(t)\right\}\right),$$

from which it is possible to compute the polarization induced in the material:

$$P(t) = -\mu_{01}\mu_{12}\mu_{20}G_{21}(t)\otimes\left(E(t)\left\{[G_{01}(t)+G_{20}(t)]\otimes E(t)\right\}\right). \tag{9.17}$$

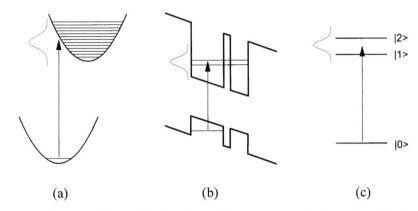

Fig. 9.6. Wave-packet oscillations from (**a**) coherent excitation of a vibrational manifold and (**b**) in the case of coupled electronic levels in a quantumwell structure. (**c**) Three-level model used in the calculation

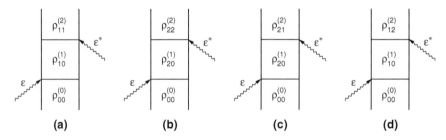

Fig. 9.7. Feynman diagrams associated with the excitation of (**a**) ρ_{11}, (**b**) ρ_{22}, (**c**) ρ_{21}, and (**d**) ρ_{12}

This polarization oscillates at frequency ω_{21} and decays with the dephasing rate Γ_{21}, as shown in Fig. 9.8. Unlike $G_{21}(t)$, this polarization exhibits a finite rise time corresponding to the duration of the exciting pulse. Note that the induced polarization vanishes in a centrosymmetric medium, where the product $\mu_{01}\mu_{12}\mu_{20}$ is zero as a result of the well-defined parity of the eigenstates. This is unsurprising since the induced polarization, oscillating at ω_{21}, can also be interpreted in terms of a resonant difference-frequency-mixing process, which can occur only in noncentrosymmetric systems, as for any quadratic optical nonlinearity.

The polarization is responsible for the radiation of an electromagnetic wave of frequency ω_{21}, typically in the infrared. This phenomenon has been reported by *Roskos* et al. [9.10] in the case of a far-infrared emission from a coupled quantum-well structure, and by *Bonvalet* et al. [9.11] in the case of a mid-infrared emission from an asymmetric quantum-well structure. We now shall discuss the latter case in more detail.

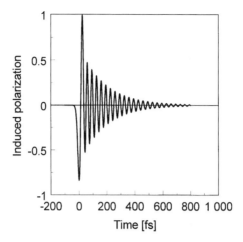

Fig. 9.8. Polarization induced in the system after wave-packet excitation. The two excited states are spaced by 30 THz and the pulse duration is 30 fs

Figure 9.9 shows the experimental setup used to monitor the electromagnetic wave emitted from a wave packet oscillating in an asymmetric quantum well. As shown in the inset, a 12 fs pulse initially creates a linear superposition of the first two electronic levels of the quantum well. The subsequent quantum beat between the two stationary states leads to an oscillation of the average position of the electron in the well. This charge oscillation results in the emission of an electromagnetic wave, which, considering the energy separation between the two states, lies in the mid-infrared domain, as is typical for intersubband transitions in GaAs quantum wells. The quantum-well emission is detected through linear interferometry, using an ultrashort, mid-infrared pulse as a reference. For this purpose, the beam delivered by a 12 fs Ti:sapphire oscillator is split into two parts. The first beam is simply focused on the quantum-well sample, tilted so that there is a nonvanishing component of the oscillating electric field perpendicular to the propagation axis. The mid-infrared emission is then collected using a gold-coated parabolic mirror and focused on a HgCdTe detector.

The other part of the Ti:sapphire beam first passes through a translation stage. It is then focused on a [1$\bar{1}$0] GaAs sample, where the 12 fs near-visible pulses undergo optical rectification. This results in the emission of nearly single-cycle mid-IR pulses, whose spectrum accordingly covers a large part of the whole mid-IR spectral domain [9.12]. The resulting reference beam is collected using a gold-coated parabolic mirror and transmitted through the quantum-well sample, where it undergoes no significant absorption. Finally, the reference beam interferes on the detector with the beam resulting from the wave-packet emission.

Figure 9.10 shows the experimental data obtained when the time delay is varied. It corresponds to the linear correlation between the reference pulse

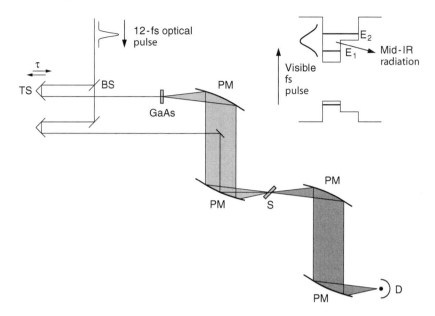

Fig. 9.9. Experimental setup used to observe the oscillation of electronic wave packets in asymmetric quantum wells. BS, beam-splitter; TS, translation stage; PM, gold-coated parabolic mirror; S, multiple-quantum-well sample; D, HgCdTe detector. The inset shows schematically the energy levels of the asymmetric quantum well. From [9.11]

and the quantum-well emission. Because the reference pulse is much shorter than the emission we want to characterize, the experimental data is closely related to the electric field emitted by the quantum well; this radiation itself is proportional to the first derivative of the second-order polarization. This explains the qualitative agreement between the data shown in Fig. 9.10 and the calculation shown in Fig. 9.8. The period of the wave-packet oscillation is observed to be about 33 fs, corresponding to a frequency of 30 THz, i.e. an emission wavelength of 10 μm. The oscillation decay results from dephasing, yielding a dephasing time $T_2 = 180$ fs. The inset shows the data in the frequency domain, exhibiting good agreement with a calculation based on (9.17).

The study of wave-packet oscillations thus provides spectroscopic data on the system under investigation, in this case relevant to intersubband transitions between *unpopulated* states. Wave-packet oscillations are also very useful in molecular physics, as a way to explore energy surfaces corresponding to nuclear motion in chemical reactions.

9.3.3 Third Order

In experiments aiming at measuring higher-order response functions, more than one beam is typically used to excite the sample, so that the additional

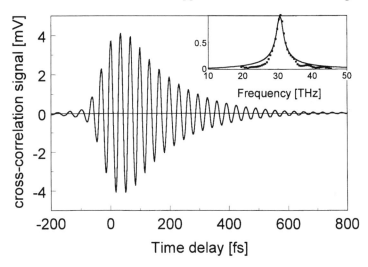

Fig. 9.10. Experimental observation of the electric field emitted by an electronic wave-packet oscillating in an asymmetric quantum well. The main group shows the linear correlation with an ultrashort, mid-IR reference pulse. The inset shows the magnitude of the Fourier transform of the data (*squares*). The *solid line* is a calculation based on (9.17), using an excitation pulse duration of 12 fs and a dephasing time of 180 fs. From [9.11]

parameters corresponding to the time delays between the various pulses allow better exploration of the response function. Figure 9.11 shows two common experimental arrangements, corresponding to the pump–probe and the four-wave-mixing (FWM) geometry. In these experiments, the total electric field applied to the material is therefore a sum of terms with different spatial dependences, so that the polarization induced in the material will include many different possibilities for spatial variation. It is by selecting the relevant terms in the polarization, as we will show below, that one is able to compute the electric field associated with the beam propagating in the direction of the detector. This is illustrated in the following for the case of a two-level system.

9.3.3.1 Pump–Probe Experiment We consider here the case of the pump–probe experiment, depicted in Fig. 9.11a. Two pulses are incident on the sample in a noncolinear geometry: the pump pulse, $\mathcal{E}_{\mathrm{P}}(\boldsymbol{r}, t) \propto \exp(i\boldsymbol{k}_{\mathrm{P}} \cdot \boldsymbol{r})$, propagating along the direction $\boldsymbol{k}_{\mathrm{P}}$, and the test pulse, $\mathcal{E}_{\mathrm{T}}(\boldsymbol{r}, t) \propto \exp(i\boldsymbol{k}_{\mathrm{T}} \cdot \boldsymbol{r})$, propagating in the direction $\boldsymbol{k}_{\mathrm{T}}$. The detector collects the beam propagating in the direction $\boldsymbol{k}_{\mathrm{T}}$. Assuming the sample is thin, the total electric field in the material is simply the sum of all incident electric fields, namely

$$E(\boldsymbol{r}, t) = \tfrac{1}{2}\left[\mathcal{E}_{\mathrm{P}}(\boldsymbol{r}, t) + \mathcal{E}_{\mathrm{T}}(\boldsymbol{r}, t) + \mathcal{E}_{\mathrm{P}}^{*}(\boldsymbol{r}, t) + \mathcal{E}_{\mathrm{T}}^{*}(\boldsymbol{r}, t)\right]. \qquad (9.18)$$

This total electric field should then be substituted into (9.11) in order to compute the third-order density operator and third-order polarization. As the

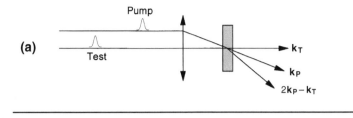

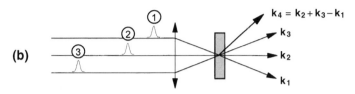

Fig. 9.11. Experimental geometries used to perform third-order nonlinear optical measurements. In (**a**) two excitation beams are used, referred to as the pump and the test (or probe). When the detector is placed in the direction of the test beam, this setup corresponds to the pump–probe geometry. The four-wave-mixing geometry is schematized in (**b**)

electric field appears three times in (9.11), many different combinations of the four terms of (9.18) will appear in the subsequent computation. However, few of these terms will be of relevance to the signal detected in a pump-probe experiment. For example, the term involving $\mathcal{E}_\mathrm{P}(\boldsymbol{r},t) \propto \exp(i\boldsymbol{k}_\mathrm{P}{\cdot}\boldsymbol{r})$ three times corresponds to frequency-tripling in the direction of the pump, of no interest here. The term involving $\mathcal{E}_\mathrm{P}(\boldsymbol{r},t)$ twice and $\mathcal{E}_\mathrm{P}^*(\boldsymbol{r},t)$ once corresponds to saturation absorption of the pump beam, again of no interest here. Of all the possible combinations, it is straightforward to show that only the one where $\mathcal{E}_\mathrm{P}(\boldsymbol{r},t)$, $\mathcal{E}_\mathrm{P}^*(\boldsymbol{r},t)$ and $\mathcal{E}_\mathrm{T}(\boldsymbol{r},t)$ all appear once should be considered. Indeed, this particular combination will exhibit a spatial dependence of the form $\exp(i\boldsymbol{k}_\mathrm{P}\boldsymbol{r})\exp(-i\boldsymbol{k}_\mathrm{P}\boldsymbol{r})\exp(i\boldsymbol{k}_\mathrm{T}\boldsymbol{r})$. The corresponding polarization will therefore radiate an electric field with a spatial phase dependence leading to its propagation in the direction $\boldsymbol{k}_\mathrm{P} - \boldsymbol{k}_\mathrm{P} + \boldsymbol{k}_\mathrm{T} = \boldsymbol{k}_\mathrm{T}$, i.e. in the direction of the probe beam and the detector.

The third-order polarization thus computed, $\mathcal{P}^{(3)}(t)$, can be shown to radiate an electric field $\mathcal{E}_\mathrm{R}(t)$, the expression for which reads

$$\mathcal{E}_\mathrm{R}(t) = -\frac{nL}{2\varepsilon_0 c}\frac{\partial \mathcal{P}^{(3)}}{\partial t},$$

where we have assumed a very thin sample (thickness L) and the slowly-varying-envelope approximation. The total electric field incident on the detector is thus $\mathcal{E}_\mathrm{T}'(t)+\mathcal{E}_\mathrm{R}(t)$, where $\mathcal{E}_\mathrm{T}'(t)$ is the linearly transmitted probe field. This leads to the following spectrum:

$$I(\omega) = |\mathcal{E}_\mathrm{T}'(\omega) + \mathcal{E}_\mathrm{R}(\omega)|^2 \approx |\mathcal{E}_\mathrm{T}'(\omega)|^2 + 2\mathrm{Re}\mathcal{E}_\mathrm{T}'^*(\omega)\mathcal{E}_\mathrm{R}(\omega),$$

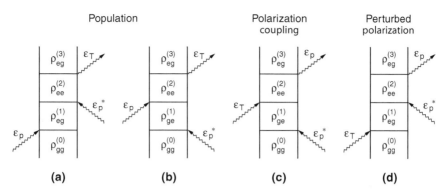

Fig. 9.12. RWA diagrams leading to the nonlinear polarization terms relevant into a pump–probe experiment

if we assume that the energy of the pump pulse is sufficiently small that the third-order radiated field is much smaller than the transmitted probe. In a spectrally resolved pump–probe experiment, one usually records the following signal:

$$\frac{\Delta I(\omega)}{I_0(\omega)} \approx \frac{2\mathrm{Re}\mathcal{E}_T'^*(\omega)\mathcal{E}_R(\omega)}{|\mathcal{E}_T'(\omega)|^2} = \frac{n\omega L}{\varepsilon_0 c}\mathrm{Im}\frac{\mathcal{P}^{(3)}(\omega)}{\mathcal{E}_T'(\omega)}, \tag{9.19}$$

where $I_0(\omega)$ is the unperturbed transmitted probe intensity. Equation (9.19) is easily evaluated once the density operator is known. Note that in some experiments, the term $\ln[I(\omega)/I_0(\omega)]$ is measured instead of the differential transmission signal, although both expressions are identical in the small-signal limit. Since $I(\omega) = I_0(\omega)\exp[-\Delta\alpha(\omega)L]$, the expression $\ln[I(\omega)/I_0(\omega)]$ is simply equal to $-\Delta\alpha(\omega)L$, so that the corresponding spectra are referred to as differential-absorption spectra. In the following calculations, we will assume the sample to be extremely thin so that $\mathcal{E}_T'(\omega)$ can be simply replaced with $\mathcal{E}_T(\omega)$ in (9.19).

We now proceed with the calculation of the density operator, keeping only those Feynman diagrams in which the probe appears once and the pump appears twice (once complex conjugated), which are drawn in Fig. 9.12. We follow closely an approach first proposed by *Brito-Cruz* et al. [9.2].

Three different contributions to the differential absorption can be identified, depending on the order of occurrence of the excitation fields in the diagrams. The first contribution corresponds to the case where the two pump fields are involved before the probe field (Fig. 9.12a and b). In this case the pump creates a second-order population term, $\rho_{ee}^{(2,\mathrm{PP})}$, which modifies the polarization induced afterwards by the probe. This contribution, called the *population term*, corresponds to the intuitive understanding of a pump–probe experiment: the pump creates a population which modifies the transmission of the probe pulse. Transcribing diagrams (a) and (b), the corresponding contribution to the density operator can be written as

$$\rho_{eg}^{(2,\text{TPP})}(t) = \frac{\mu_{eg}}{2} G_{eg}(t) \otimes \left\{ \mathcal{E}_T(t) \left[\rho_{gg}^{(2,\text{PP})}(t) - \rho_{ee}^{(2,\text{PP})}(t) \right] \right\}$$

$$= \frac{\mu_{eg}\mu_{ge}\mu_{eg}}{4} G_{eg}(t) \otimes \left[\mathcal{E}_T(t) \left(G_{ee}(t) \otimes \left\{ \mathcal{E}_P^*(t) \left[G_{eg}(t) \otimes \mathcal{E}_P(t) \right] \right. \right. \right.$$
$$\left. \left. \left. - \mathcal{E}_P(t) \left[G_{eg}^*(t) \otimes \mathcal{E}_P^*(t) \right] \right\} \right) \right].$$

Figure 9.13a shows the corresponding contribution in a typical differential-absorption spectrum. The signal appears for positive time delays, i.e. when the probe pulse follows the pump pulse. The bleaching in the absorption spectrum is proportional to the population in the excited state, which decays with the time constant T_1. This is the conventional interpretation of pump–probe spectra, in which coherent effects are not included. However, this is not the whole story. There is a second contribution to $\rho_{eg}^{(3)}$, called the *pump–polarization coupling*, associated with diagram (c) of Fig. 9.12. It involves first $\mathcal{E}_P^*$ then $\mathcal{E}_T$, and $\mathcal{E}_P$. As the pump must occur both before and after the probe, this term can only be present when the two pulses overlap, i.e. for time delays very close to zero, within the pulse duration. This is confirmed in the simulation shown in Fig. 9.13c. This term can be interpreted as the diffraction of the pump beam in the direction of the probe beam, after it has undergone a diffraction on an induced grating formed by the interference between the pump and the probe. This term also occurs in spectrally integrated pump–probe spectra, and is closely related to the so-called *coherent artifact* [9.13, 14].

Transcribing diagram (c) of Fig 9.12, we obtain

$$\rho_{eg}^{(3,\text{PTP})}(t) = \frac{\mu_{eg}}{2} G_{eg}(t) \otimes \left\{ \mathcal{E}_P(t) \left[\rho_{gg}^{(2,\text{TP})}(t) - \rho_{ee}^{(2,\text{TP})}(t) \right] \right\}$$

$$= -\frac{\mu_{eg}\mu_{ge}\mu_{eg}}{4} G_{eg}(t)$$
$$\otimes \left[\mathcal{E}_P(t) \left(G_{ee}(t) \otimes \left\{ \mathcal{E}_T(t) \left[G_{eg}^*(t) \otimes \mathcal{E}_P^*(t) \right] \right\} \right) \right].$$

Finally, the last diagram, (d), contributes to the signal when the probe pulse *precedes* the pump pulse, i.e. for negative time delays. It can be interpreted in terms of a perturbation of the probe-induced polarization by the pump. It is called *perturbed polarization decay* [9.2, 15] and is written as

$$\rho_{eg}^{(3,\text{PPT})}(t) = -\frac{\mu_{eg}\mu_{ge}\mu_{eg}}{4} G_{eg}(t)$$
$$\otimes \left[\mathcal{E}_P(t) \left(G_{ee}(t) \otimes \left\{ \mathcal{E}_P^*(t) \left[G_{eg}(t) \otimes \mathcal{E}_T(t) \right] \right\} \right) \right].$$

As shown by (9.19), the detected signal corresponds to the Fourier transform of the induced polarization, so that an abrupt perturbation of the polarization results in spectral oscillations whose period is inversely proportional to the time delay between the two pulses, as observed in Fig. 9.13b.

Figure 9.14 shows a series of differential absorption spectra recorded in an optical-Stark-effect experiment performed on a bulk GaAs sample. In this case, the probe induces an excitonic polarization in the sample which lasts for about 1 or 2 ps. The pump pulse later shifts the exciton frequency, which

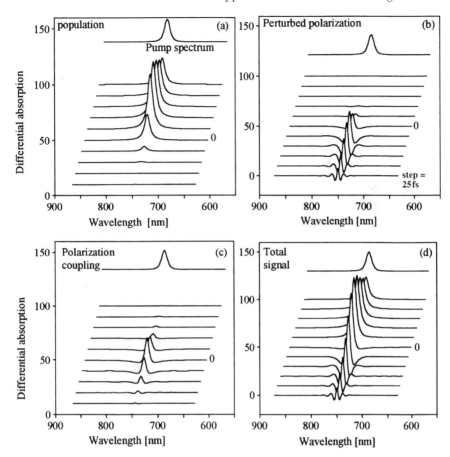

Fig. 9.13. Contribution to the pump–probe differential absorption signal from (**a**) population, (**b**) perturbed polarization and (**c**) polarization coupling; (**d**) is the total signal. The model parameters are $T_1 = 80\,\mathrm{fs}$ and $T_2 = 50\,\mathrm{fs}$. From [9.8]. See also [9.2]

is also the oscillation frequency of the decaying polarization. This induces a phase shift in the time-dependent polarization, which appears as spectral oscillations in the recorded differential-absorption signal [9.16]. Note that the experimental conditions have been optimized in order to enhance the importance of coherence effects, by choosing a relatively spectrally-narrow probe pulse [9.17]. As can be seen, the results are entirely different from what would be expected from a simple ultrafast shift of the exciton line, in the absence of coherent effects.

   To summarize, coherent effects play a major role in spectrally resolved pump–probe experiments when the pulse duration is shorter than the dephasing times of the medium. An intrinsic difficulty in these experiments is the

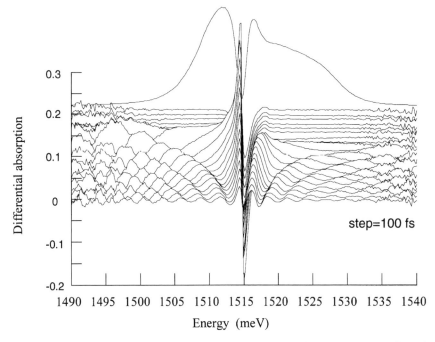

Fig. 9.14. Coherent oscillations observed in a pump–probe experiment, when the probe spectrum is chosen so as to enhance coherent effects. From [9.17]

simultaneous use of spectral and temporal resolution, which is the very source of the observed artifacts, owing the time–frequency uncertainty principle. We will now discuss another spectroscopic technique which does not suffer from these drawbacks.

9.3.3.2 Photon Echo Figure 9.11b shows the experimental setup corresponding to four-wave mixing, or photon echo, experiments. Three beams, propagating along wave vectors k_1, k_2 and k_3, are focused on the sample, and the detection scheme is arranged so as to collect only the beam propagating along $k_4 = k_2 + k_3 - k_1$. The relevant term in the nonlinear polarization is therefore proportional to $\mathcal{E}_1^* \mathcal{E}_2 \mathcal{E}_3$, according to the discussion at the beginning Sect. 9.3.3.1. Assuming the three pulses are well separated and incident in the order 1, 2, 3, there is only one Feynman diagram contributing to the nonlinear polarization, which is drawn in Fig. 9.15a. The first two pulses (the writing pulses) write a grating in the material, while the third pulse, the reading pulse, is diffracted off this grating in the direction $k_4 = k_2 + k_3 - k_1$ [9.4]. The experimental geometry shown in Fig. 9.11b corresponds to the most general setup, also referred to as *three-pulse photon echo*. Another, simplified geometry is available, where only two beams are used, pulses 2 and 3 then being identical. This is the so-called *two-pulse photon echo* geometry. It is actually identical

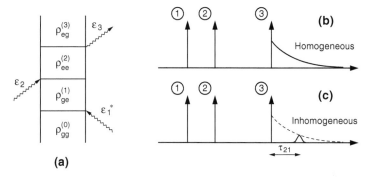

Fig. 9.15. (a) RWA diagram relevant to photon echo experiments. (b) Induced polarization in a homogeneous system, to be compared with an inhomogeneous system (c), where the polarization looks like an echo of the reading pulse

to a pump–probe excitation scheme, as shown in Fig. 9.11a, except that the detector now collects the beam emitted in the direction $k_4 = 2k_P - k_T$.

In the general case, the contribution to the density operator can be directly written from the diagram shown in Fig. 9.15a. We find

$$\rho_{eg}^{(3)} = -\frac{|\mu_{eg}|^2 \mu_{eg}}{4} G_{eg}(t) \otimes \left[\mathcal{E}_3(t) \left(G_{ee}(t) \otimes \left\{ \mathcal{E}_2(t) \left[G_{eg}^*(t) \otimes \mathcal{E}_1^*(t) \right] \right\} \right) \right].$$

If we now assume that the excitation pulses are infinitely short, i.e. $\mathcal{E}_\alpha(t) = \mathcal{E}_\alpha \delta(t - \tau_\alpha)$, where $\alpha = 1, 2, 3$, we obtain

$$\rho_{eg}^{(3)}(t) = -i \frac{|\mu_{eg}|^2 \mu_{eg}}{4\hbar^3} \Theta(t - \tau_3) \Theta(\tau_{21}) e^{-\tau_{32}/T_1} e^{-(t - \tau_3 + \tau_{21})/T_2}$$
$$\times \mathcal{E}_1^* \mathcal{E}_2 \mathcal{E}_3 \exp\left[-i\omega_{eg}(t - \tau_3 - \tau_{21}) \right],$$

which results in a polarization occurring at time $t = \tau_3$, just after the reading pulse, and decaying with the dephasing rate of the material, as shown in Fig. 9.15b. The initial amplitude of this polarization is $\exp(-\tau_{21}/T_2)$, so that an integrating detector will measure a signal proportional to $\exp(-2\tau_{21}/T_2)$. The FWM signal will then exhibit an exponential dependence on time delay, with a time constant $T_2/2$.

Let us now consider the case of an inhomogeneously broadened system, modeled as an assembly of independent two-level systems whose transition frequencies ω_{eg} are distributed according to a Gaussian distribution $g(\omega_{eg})$. The total polarization induced in such a system is simply the sum of the individual polarizations, weighted with the function $g(\omega_{eg})$. The averaged density operator then reads

$$\left\langle \rho_{eg}^{(3)}(t) \right\rangle = -i \frac{|\mu_{eg}|^2 \mu_{eg}}{4\hbar^3} \Theta(t - \tau_3) \Theta(\tau_{32}) \Theta(\tau_{21}) e^{-\tau_{32}/T_1} e^{-(t - \tau_3 + \tau_{21})/T_2}$$
$$\times \mathcal{E}_1^* \mathcal{E}_2 \mathcal{E}_3 \int_{-\infty}^{+\infty} g(\omega_{eg}) \exp\left[-i\omega_{eg}(t - \tau_3 - \tau_{21}) \right] d\omega_{eg}$$

$$= -\mathrm{i}\frac{|\mu_{\mathrm{eg}}|^2\mu_{\mathrm{eg}}}{4\hbar^3}\Theta(t-\tau_3)\Theta(\tau_{32})\,\mathrm{e}^{-\tau_{32}/T_1}\,\mathrm{e}^{-(t-\tau_3+\tau_{21})/T_2}$$
$$\times\,\mathcal{E}_1^*\mathcal{E}_2\mathcal{E}_3\,g(t-\tau_3-\tau_{21}),$$

where $g(t)$ is the inverse Fourier transform of the frequency distribution $g(\omega_{\mathrm{eg}})$. Note that the phase accumulated in the induced polarization between pulses 1 and 2 is $+\omega_{\mathrm{eg}}\tau_{21}$, while the phase accumulated after the reading pulse is $-\omega_{\mathrm{eg}}(t-\tau_3)$. The change in sign, corresponding to a change in the rotation direction of the density operator, is analogous to what occurs in the spin-echo process in nuclear magnetic resonance. It results in a rephasing at time $t=\tau_3+\tau_{21}$, when the phase is always zero for any values of the transition frequencies. This creates a rephasing between all the inhomogeneous components of the polarization, hence the *echo* observed at a time τ_{21} after the reading pulse [9.18]. The time dependence of the corresponding emission is governed by $g(t)$, so that its duration is the inverse of the inhomogeneous linewidth, i.e. the inhomogeneous dephasing time. In the extreme inhomogeneous limit, the echo amplitude is proportional to $\exp(-2\tau_{21}/T_2)$, since the polarization has decayed once in the time interval $[0,\tau_{21}]$ and a second time in the time interval $[\tau_3,\tau_3+\tau_{21}]$. In a time-integrated measurement of the echo, the signal will then decay with a time constant $T_2/4$. Studying the echo signal therefore provides information about homogeneous dephasing processes in the system, independently of the inhomogeneous broadening.

However, in many experiments, the echo signal is more carefully analyzed using either up-conversion [9.5–7] or interferometric techniques [9.19–21], as shown in Fig. 9.16. Besides, a drawback of measuring only the echo energy is that an exponential decay of the echo energy as a function of time delay is no proof of the validity of the model. In a number of experiments, it has been shown that it is only by recording the echo as a function of both time delay and time that valid experimental results can be obtained [9.5–7]. The Bloch model has thus been shown to fail in a number of cases. The alternate theories required to properly describe the experimental system are out oft he scope of this book, but we refer the reader to [9.3] and [9.22] for more information on these emerging topics.

9.4 Multidimensional Spectroscopy

As mentioned in the previous section, pump-probe and time-integrated photon-echo signals result from the third-order nonlinear response of the sample. However, such experiments provide only a partial measurement of the nonlinear response function, and it would be clearly desirable to directly measure the response function itself. This is the aim of multidimensional spectroscopy, a new approach to femtosecond spectroscopy that has been developed in recent years [9.23–25].

In order to define the nonlinear response, we assume that we apply an arbitrary electric field $E(t)$ to our sample and that the radiated electric field

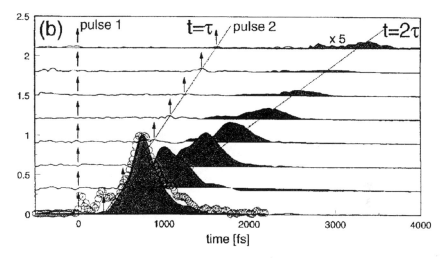

Fig. 9.16. Time-resolved photon-echo amplitude obtained in a two-pulse photon-echo experiment for different values of the time decay τ between the two exciting pulses (shown with vertical arrows). The sample is a multiple–quantum well structure presenting a strong inhomogeneous broadening of the exciton absorption line, hence the observed behavior of the echo occurring at a time delay τ after the second pulse. The echo field has been measured using Fourier-transform spectral interferometry (from [9.19])

$E_R(t)$ can be developed in a power series of the exciting field, similarly to the expansion of the polarization in Sect. 9.2.4:

$$E_R(t) = E^{(1)}(t) + E^{(2)}(t) + E^{(3)}(t) + \dots . \qquad (9.20)$$

With the assumption of time invariance of the sample response, the radiated field of order n, $E^{(n)}(t)$ can be written as

$$E^{(n)}(t) = \int \Xi^{(n)}(\omega_1, \dots, \omega_n) E(\omega_1) E(\omega_2) \dots E(\omega_n)$$
$$\times \exp(-i(\omega_1 + \omega_2 + \dots + \omega_n)t) \frac{d\omega_1}{2\pi} \frac{d\omega_2}{2\pi} \dots \frac{d\omega_n}{2\pi},$$

where $\Xi^{(n)}(\omega_1, \omega_2, \dots, \omega_n)$ is, by definition, the nth-order nonlinear response of the sample. This multidimensional response can be measured using an approach related to Fourier-transform nuclear magnetic resonance [9.26–27]: The sample is excited by a sequence of n ultrashort pulses, $E_0(t)$, $E_0(t - \tau_2), \dots, E_0(t - \tau_n)$. While scanning the $n-1$ time delays, we record the radiated electric field resulting from the cross interaction between the n incident pulses:

$$E^{(n)}_{\tau_2, \dots, \tau_n}(t) = \int \Xi^{(n)}(\omega_1, \dots, \omega_n) E_0(\omega_1) E_0(\omega_2) e^{i\omega_2 \tau_2} \dots E_0(\omega_n) e^{i\omega_n \tau_n}$$
$$\times \exp(-i(\omega_1 + \omega_2 + \dots + \omega_n)t) \frac{d\omega_1}{2\pi} \dots \frac{d\omega_n}{2\pi}.$$

This cross term can be selected from the entire nonlinear emission by either using a noncollinear geometry or performing a series of measurements while applying appropriate phase shifts on the different exciting pulses. Note that it is important to perform a complete measurement of the emitted electric field, using either time-domain interferometry [9.24] or frequency-domain interferometry [9.23,25]. Finally, a Fourier transform of $E^{(n)}_{\tau_2,\ldots,\tau_n}(t)$ with respect to the n time variables $t, \tau_2, \ldots, \tau_n$ yields the quantity $\Xi^{(n)}(\omega_1,\ldots,\omega_n)$, $E_0(\omega_1)E_0(\omega_2)\ldots E_0(\omega_n)$, from which one can compute the multidimensional nonlinear response $\Xi^{(n)}(\omega_1,\ldots,\omega_n)$ assuming the exciting field $R_0(\omega)$ is known in both amplitude and phase.

It is noteworthy that $\Xi^{(n)}$ contains all the information on the nth-order response of the sample to an arbitrary electric field. For example, the knowledge of $\Xi^{(3)}(\omega_1,\omega_2,\omega_3)$ for all frequency values makes possible the prediction of any third-order experiment such as a pump-probe or a photon-echo measurement (in the low-intensity regime). However, actual experiments are usually limited to two dimensions, yielding either the whole response in the case of second-order processes [9.23] or a 2D projection of $\Xi^{(3)}(\omega_1,\omega_2,\omega_3)$ in the case of third-order processes [9.24–25]. In all cases, a 2D measurements directly provides very valuable information on the material, such as the coupling between electronic excited state or between different vibrational modes of a molecule.

9.5 Conclusion

We have applied perturbation theory for a system described by the Bloch equation when excited by a sequence of femtosecond pulses. This approach allows us to interpret many experimental techniques in femtosecond spectroscopy and is a useful step toward a simulation of femtosecond experiments. However, many experimental results show that the Bloch equation is but an approximation, as many processes, involving many-body interactions or non-Markovian dynamics cannot be described in such a simple framework. The investigation of such complex coherent processes should greatly benefit from new experimental techniques such as multidimensional spectroscopy.

Acknowledgment

I gratefully acknowledge Adeline Bonvalet, Jean-Paul Foing, Danièle Hulin, Jean-Pierre Likforman, Jean-Louis Martin and Arnold Migus for their contributions to the work summarized in this chapter.

9.6 Problems

1. (a) A two-level system is resonantly excited by a sequence of two short, identical pulses. The dephasing time is assumed to be much longer than

the time delay between the two pulses, itself assumed to be much longer than the pulse duration. Using the time-domain expression of $\rho_{ee}^{(2)}(t)$, show that the population excited in the system is the sum of three terms: the population created by the first pulse, the population created by the second pulse, and a cross-term which involves the interaction between the field of the second pulse and the coherent polarization induced by the first pulse.
(b) Using simple arguments and approximations when needed, show that for some values of the phase $\omega_{eg}\tau$, the coherent term can be as large in magnitude as twice the population created by a single pulse. What about its sign?
(c) Make a schematic plot of the population in the system as a function of time when:
(i) $\omega_{eg}\tau = 2n\pi$,
(ii) $\omega_{eg}\tau = (2n+1)\pi$;
where n is an integer. Interpret this effect in terms of coherent population control. What happens when the time delay between the two pulses is comparable to or larger than the dephasing time? Why?
(d) Plot the power spectrum of the sequence of the two pulses (do not forget interferences!) in the above two cases. Compare this with the absorption spectrum of the two-level system. Can you suggest an alternative explanation for the effect interpreted above in the time domain?
2. (a) Compute numerically the photon-echo signal for the case of a Gaussian distribution of two- level systems, assuming infinitely short pulses.
(b) Compute the decay time of the integrated echo signal and plot the time constant as a function of the ratio between the inhomogeneous and homogeneous widths. Do you recover the limits of $T_2/4$ and $T_2/2$ when the distribution is purely inhomogeneous and purely homogeneous respectively?

References

[9.1]   P.N. Butcher, D. Cotter: *The Elements of Nonlinear Optics*, Cambridge Studies in Modern Optics (Cambridge University Press, Cambridge 1990)
[9.2]   C.H. Brito-Cruz, J.P. Gordon, P.C. Becker, R.L. Fork, C.V. Shank: IEEE J. Quant. Electr.**QE 24**, 261 (1988)
[9.3]   S. Mukamel: *Principles of Nonlinear Optical Spectroscopy* (Oxford University Press, Oxford 1995)
[9.4]   E.T.J. Nibbering: *Femtosecond Optical Dynamics in Liquids*, PhD thesis, Groningen University (1993)
[9.5]   M.S. Pshenichnikov, K. Duppen, D.A. Wiersma: Phys. Rev. Lett. **74**, 674 (1995)
[9.6]   S. Weiss, M.-A. Mycek, J.-Y. Bigot, S. Schmitt-Rink, D.S. Chemla: Phys. Rev. Lett. **69**, 2685 (1992)
[9.7]   D.-S. Kim, J. Shah, T.C. Damen, W. Schäfer, F. Jahnke, S. Schmitt-Rink, K. Köhler: Phys. Rev. Lett. **69**, 2725 (1992)

[9.8]   J.-P. Foing: *Etude d'une distribution de porteurs hors d'équilibre dans l'arséniure de gallium par spectroscopie femtoseconde*, PhD thesis, Ecole Polytechnique (1991)

[9.9]   J.-P. Likforman: *Saturation sélective d'excitons et effets de cohérence dans des puits quantiques en arséniure de gallium*, PhD thesis, Ecole Polytechnique (1994)

[9.10] H.G. Roskos, M.C. Nuss, J. Shah, K. Leo, D.A.B. Miller, A.M. Fox, S. Schmitt-Rink, K. Köhler: Phys. Rev. Lett. **68**, 2216 (1992)

[9.11] A. Bonvalet, J. Nagle, V. Berger, A. Migus, J.-L. Martin, M. Joffre: Phys. Rev. Lett. **76**, 4392 (1996)

[9.12] A. Bonvalet, M. Joffre, J.-L. Martin, A. Migus: Appl. Phys. Lett. **67**, 2907 (1995)

[9.13] Z. Vardeny, J. Tauc: Opt. Commun. **39**, 396 (1981)

[9.14] H.J. Eichler, P.D. Langhans, F. Massmann: Opt. Commun. **50**, 117 (1984)

[9.15] M. Joffre, D. Hulin, A. Migus, A. Antonetti, C. Benoità la Guillaume, N. Peyghambarian, S.W. Koch: Opt. Lett. **13**, 276 (1988)

[9.16] M. Joffre, D. Hulin, J.-P. Foing, J.-P. Chambaret, A. Migus, A. Antonetti: IEEE J. Quant. Electr. **25**, 2505 (1989)

[9.17] J.-P. Likforman, M. Joffre, G. Chériaux, D. Hulin: Opt. Lett. **20**, 2006 (1995)

[9.18] T. Yajima, T. Taira: J. Phys. Soc. Japan **47**, 1620 (1979)

[9.19] J.-P. Likforman, M. Joffre, V. Thierry-Mieg: Opt. Lett. **22**, 1104 (1997)

[9.20] M.F. Emde, W.P. de Boeij, M.S. Pshenichnikov, D.A. Wiersma: Opt. Lett. **22**, 1338 (1997)

[9.21] S.M. Gallagher, A.W. Albrecht, J.D. Hybl, B.L. Landin, B. Ranaram, D.M. Jonas: J. Opt. Soc. Am. B **15**, 2338 (1998)

[9.22] H. Haug, S. Koch: *Quantum Theory of the Optical and Electronic Properties of Semiconductors*, World Scientific (1990)

[9.23] L. Lepetit, M. Joffre: Opt. Lett. **21**, 564 (1996)

[9.24] M.C. Asplund, M.T. Zanni, R.M. Hochstrasser: Proc. Nat. Acad. Sci. **97**, 8219 (2000)

[9.25] J.D. Hybl, A.A. Ferro, D.M. Jonas: J. Chem. Phys. **115**, 6606 (2001)

[9.26] L. Müller, A. Kumar, R.R. Ernst: J. Chem. Phys. **63**, 5490 (1975)

[9.27] R.R. Ernst, G. Bodenhausen, A. Wokaun: *Principles of Nuclear Magnetic Resonance on One and Two Dimensions*, Oxford (1987)

10

Terahertz Femtosecond Pulses

A. Bonvalet and M. Joffre

With 20 Figures

10.1 Introduction

As was shown in previous chapters, the large peak intensities associated with femtosecond laser pulses make them well suited to nonlinear wave-mixing processes, allowing the generation of new colors. Such processes include second-harmonic generation, sum-frequency generation, parametric oscillation and amplification, and continuum generation. However, there is yet another route for generating new frequencies, which relies on a quite different approach. It is actually based on a very old technique, first developed by Hertz in the last century: a transient polarization or current surge, occurring for example in a spark, will act as a source term in the Maxwell equations and radiate a pulsed electromagnetic wave. If the polarization transient does not exhibit any well-defined oscillatory feature but is, rather, a rapid change such as a step or a pulse, then the radiated wave has no well-defined frequency. Its spectrum will therefore be extremely broad and will peak at a frequency inversely proportional to the timescale of the transient. Because all these frequency components are emitted in phase, this technique is very well suited to the generation of broadband coherent radiation at the lower end of the electromagnetic spectrum.

In Hertz's time, transients with nanosecond features could be produced, so that this technique was suitable for the generation of Hertzian or radio waves. However, with the advent of picosecond and femtosecond lasers, the very same technique has been pushed to much higher frequencies (or shorter wavelengths), as shown in Fig. 10.1. Typically, a transient of about 100 fs corresponds to an emitted central frequency of about 3 THz, which corresponds to a wavelength of 100 μm. This is in the far-infrared domain. This chapter is devoted to the generation of such terahertz pulses, which constitutes a very active field of femtosecond laser science.

Other broadband sources in the far-infrared domain are not entirely satisfactory. Indeed, this spectral domain is in the low-frequency tail of black-body radiation, so that the brightness of black-body sources is rather small and the

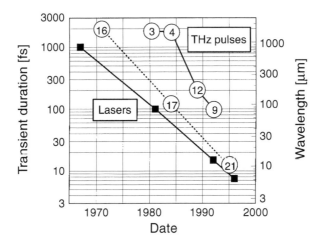

Fig. 10.1. Progress in the generation of laser pulses (*squares*) and terahertz pulses obtained from laser-induced polarization transients (*circle with numbers*). As described in this chapter, two different techniques have been used: photoconductive switching (*solid line*) and optical rectification (*dashed line*). The numbers in the circles refer to the references cited in this chapter

achieved signal-to-noise ratio is quite limited. The other competing source, the free-electron laser, requires a large and heavy installation. In many respects, femtosecond terahertz sources are nearly ideal far-infrared coherent light sources.

In the following, we shall review the various techniques available for the generation of a fast polarization transient from an ultrashort pulse. Then we shall describe several detection techniques allowing the measurement of these terahertz electromagnetic pulses. Finally, we shall show some experimental results and briefly review an application to terahertz spectroscopy.

10.2 Generation of Terahertz Pulses

The emission of terahertz femtosecond pulses relies on the generation of a time-dependent polarization $P(t)$, varying on a short timescale Δt. If we consider the case of a point source, i.e. of size much smaller than the generated wavelength, it is straightforward to show [10.1] that, far from the source, the emitted electric field is proportional to the second derivative of the induced polarization:

$$E(t) \propto \frac{\partial^2 P}{\partial t^2}. \tag{10.1}$$

Consider for example the case shown in Fig. 10.2a, where the induced polarization exhibits a step-like behavior, with a rise time Δt. Its second derivative

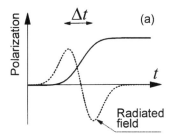
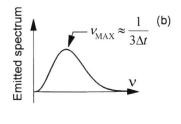

Fig. 10.2. (**a**) Step-like polarization transient. In the case of a point source, the radiated electric field is the second derivative, shown as a dashed line. (**b**) Power spectrum of the emitted radiation

then corresponds to a single cycle of radiation, whose spectrum is shown in Fig. 10.2b.

The shape of the emitted electric field depends greatly on the geometry of the source. In the case of an extended source, of size much larger than the radiated wavelength, all the point sources of the emitting area interfere so that the field is now proportional to the *first* derivative of the polarization. This is analogous to the propagation of an electromagnetic wave in a dielectric medium: in the case of a plane wave, the radiated field is proportional to the first derivative of the polarization:

$$E(t) \propto \frac{\partial P}{\partial t}. \tag{10.2}$$

In the case of a step-like polarization, the radiated field should then be half a cycle. Let us now consider the various techniques allowing one to generate the polarization transient, which is the only ingredient needed for the generation of terahertz radiation.

10.2.1 Photoconductive Switching

The generation of a transient polarization through photoconductive switching consists of focusing an ultrashort light pulse on a DC-biased, optoelectronic, ultrafast switch. When the device switches from the "off" to the "on" state, the applied voltage is reduced and a transient polarization is produced. The required ultrashort switching time can be achieved through the creation of electron–hole pairs in a semiconductor, as shown in Fig. 10.3 and first demonstrated by *Auston* [10.2] in 1975.

Using such an Auston switch triggered by a subpicosecond laser pulse, Mourou et al. observed the generation of freely propagating picosecond microwave pulses [10.3]. Through the use of more elaborate generation and detection schemes, this method was considerably improved [10.4–6], and eventually led to the generation of subpicosecond electromagnetic pulses with frequencies extending up to a few terahertz [10.7–9].

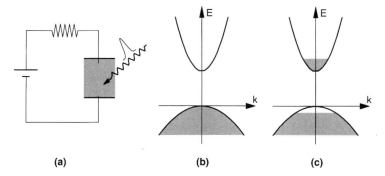

Fig. 10.3. (a) A DC voltage is applied to a semiconducting material, which is irradiated by an ultrashort light pulse. (b) Before irradiation, virtually all the carriers are in the valence band. The semiconductor is insulating. (c) Upon absorption of the light pulse, electron–hole pairs are created in the semiconductor, which thus becomes conducting. The voltage across the semiconductor drops

The exact shape of the polarization transient thus produced depends on the characteristics of both the semiconducting material and the device itself. In the case of a standard semiconductor sample, the carrier lifetime is typically much longer than the timescale of interest, so that upon excitation the current will flow continuously in the device until the capacitance is fully discharged. Then, after the carriers recombine, the DC source charges the capacitance again and the voltage slowly recovers its initial value; the device is ready for the next laser shot. However, this recovery takes a very long time, so that the polarization can be approximated here by a step-like function, with a rise time determined by the device capacitance. Using this technique, subpicosecond polarization transients have been achieved. Yet, even better performances are achieved when the current is allowed to flow for a reduced amount of time, so that the rise time of the polarization transient is shorter and leads to greater generated frequencies. Several techniques have been demonstrated for reducing the carrier lifetime to a few hundreds of femtoseconds, using for example radiation-damaged silicon [10.10] or low-temperature-grown GaAs. It is also possible to reduce the rise time of the polarization transient by using particular excitation geometries [10.11]. Either of these techniques allows the generation of polarization transients with durations of a few hundreds of femtoseconds.

Figure 10.4 shows a now standard terahertz generation scheme, as developed by Fattinger and Grischkowsky [10.12]. The laser beam is focused onto radiation-damaged silicon on sapphire, in the gap between two aluminum electrodes spaced by a few microns. Since the size of the emitting dipole is much smaller than the generated wavelength, this device can be considered as a point source, radiating in all directions. A spherical sapphire lens on the back of the sapphire substrate allows one to collect as much of the emitted radia-

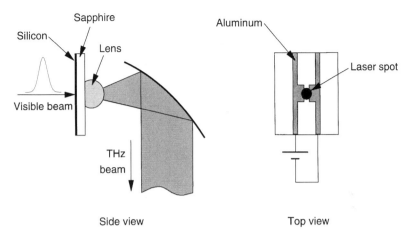

Fig. 10.4. Typical experimental arrangement for the generation of terahertz pulses from a point source excited by a femtosecond visible laser beam (adapted from [10.7, 12])

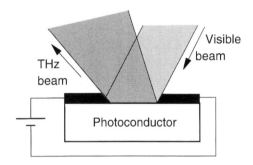

Fig. 10.5. Typical experimental arrangement for the generation of terahertz pulses from an extended source (adapted from [10.13])

tion as possible. The divergent beam can then be collimated by a parabolic mirror, and a parallel beam is produced [10.7].

Another terahertz generation scheme, demonstrated by Hu et al. [10.13], is shown in Fig. 10.5. In this case, the spacing between the two electrodes is much greater than in the arrangement above, and is typically of the order of 1 cm, i.e. much larger than the far-infrared wavelength. As a result, the divergence of the terahertz beam is greatly reduced and the far-infrared emission occurs only in the transmission and specular-reflection directions. The main advantage of such an extended source is that the applied voltage and the excitation pulse energy can be scaled up, allowing the generation of greater energies. Using an amplified laser source, *You* et al. [10.14] have demonstrated the generation of terahertz pulses of nearly $1\,\mu$J energy.

10.2.2 Optical Rectification in a Nonlinear Medium

As have been shown above, fast polarization transients can be efficiently generated through photoconductive switching. However, the same purpose can also be achieved more directly in nonlinear optical materials. Indeed, it was shown in Chap. 2 that the polarization induced in a material can be written as a Taylor expansion of the incident electric field. Keeping only the first two terms in (2.1), we write

$$P(t) = \chi^{(1)} E(t) + \chi^{(2)} E(t)^2. \tag{10.3}$$

Let us consider the case, illustrated in Fig. 10.6a, of excitation by an electric field associated with a short laser pulse,

$$E(t) = A(t) \cos \omega_0 t, \tag{10.4}$$

where $A(t)$ is the pulse envelope and ω_0 is the carrier frequency. The induced polarization is shown in Fig. 10.6b. According to (10.3), the polarization will not be equal for the two opposite directions of the applied electric field. This statement obviously holds only in the case of a noncentrosymmetric medium and hence a nonvanishing $\chi^{(2)}$. The nonlinear polarization, $P^{(2)}(t)$, can then be decomposed into two terms, according to the following expression

$$
\begin{aligned}
P^{(2)}(t) &= \chi^{(2)} E(t)^2 \\
&= \frac{\chi^{(2)} A(t)^2}{2} + \frac{\chi^{(2)} A(t)^2}{2} \cos 2\omega_0 t.
\end{aligned} \tag{10.5}
$$

The second term is associated with a carrier frequency $2\omega_0$ and is responsible for the process of second-harmonic generation, already discussed in detail in Chap. 2. The first term, shown as a thick line in Fig. 10.6b, is directly proportional to the pulse intensity. This term is also the average of the induced polarization over one optical cycle. Owing to the noncentrosymmetric nature of the medium, this period-averaged value is nonzero, so that this term of the induced polarization always points in the same direction despite an oscillating excitation. By analogy with the corresponding electronic device, the medium is said to act as an optical rectifier for the electric field of the light pulse and this nonlinear effect is called *optical rectification* [10.15]; it is also referred to as the inverse electro-optic effect. In the case of CW excitation, optical rectification simply results in the occurrence of a DC voltage across the crystal, which has no practical application. However, in the case of a short laser pulse, optical rectification results in the generation of a transient polarization which follows directly the time-dependent laser intensity. This is exactly what is needed for the generation of terahertz pulses.

Figure 10.6c shows the frequency-domain picture, as obtained from a Fourier transform of the nonlinear polarization. The two terms of (10.5) yield two spectra, centered on zero and the doubled frequency respectively.

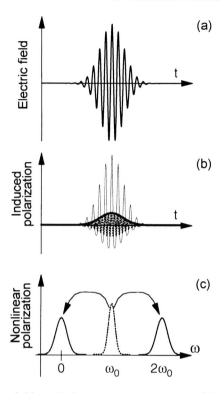

Fig. 10.6. (a) Electric field applied to a nonlinear material. (b) Induced polarization in the material (*thin line*). The *dotted line* shows the contribution from frequency-doubling and the *thick line* the contribution from optical rectification. (c) Optical rectification and second-harmonic generation in the frequency domain. The *dotted line* shows the spectrum of the incident laser pulse

Frequency-doubling of a short laser pulse can also be viewed as all the possible sum-frequency mixings between its different spectral components. Similarly, the spectrum generated through optical rectification is nothing else than all the possible difference-frequency mixings among the incident spectral components. The difference-frequency generation picture is more appropriate when the incident spectrum consists of two different spectral components. In contrast, the optical rectification picture is more appropriate when a very broad spectrum is generated from a broad incident spectrum.

Optical rectification was actually the very first process put to use for the generation of microwave pulses from picosecond laser pulses. Indeed, *Yang* et al. [10.16] demonstrated in 1971 the generation of far-infrared radiation in LiNbO$_3$. In 1984, *Auston* et al. [10.17] demonstrated the high-bandwidth capabilities of optical rectification by the generation of frequencies up to 4 THz in lithium tantalate. The generation of terahertz radiation through optical rectification has subsequently been demonstrated in a variety of materials, such as lithium tantalate (LiTaO₃) [10.13], semiconductors [10.18, 19] and organic crystals [10.20].

As compared with photoconductive switching, optical rectification is more direct since it simply requires one to focus the laser beam on a nonlinear material. The transient polarization is generated at a microscopic level, directly through the material response, unlike that in photoconductive switching, which involves the realization of an optoelectronic device. The advantage of photoconductive switching is obviously its much greater efficiency, as very large voltages can be switched by the laser pulse, unlike optical rectification, which must rely on existing nonlinear coefficients. On the other hand, optical rectification is truly instantaneous. As a result, while photoconductive switching is limited to a few terahertz, optical rectification has been demonstrated to allow the generation of frequencies up to 50 THz [10.21], well into the mid-infrared spectral domain. Finally, note that the low efficiency of optical rectification can be greatly improved through the use of phase-matched materials, which make possible the generation of higher energy pulses at the cost of a reduced spectral bandwidth [10.22–24].

10.3 Measurement of Terahertz Pulses

As compared to the visible, the far-infrared spectral domain is clearly towards the lower end of the electromagnetic spectrum. Specific detection schemes must therefore be designed for this new spectral range, the main difficulty being that the spectra generated using the methods discussed in Sect. 10.2 are extremely broad, so that the detection bandwidth must be equally broad. The first approach consists of extending Fourier transform spectroscopy, a reliable and extensively used technique for CW infrared radiation. This means using an infrared detector, such as a bolometer, placed after a Michelson interferometer. Another approach, specific to femtosecond terahertz pulses, consists of using a gated detection, either through the electro-optic effect or through photoconductive switching. We will now discuss these techniques in more detail, and point out their respective advantages.

10.3.1 Fourier Transform Spectroscopy

Fourier transform spectroscopy is clearly the most efficient technique for achieving the measurement of the power spectrum of infrared or far-infrared radiation [10.25]. It relies on the acquisition of the linear autocorrelation of the incident beam in a Michelson interferometer, followed by the numerical computation of its Fourier transform. As was explained in Chap. 7, this yields the square of the modulus of the electric field in the frequency domain, i.e. the power spectrum, $|E(\omega)|^2$. As demonstrated by *Greene* et al. [10.8], Fourier transform spectroscopy can be successfully adapted to the measurement of terahertz pulses, using the setup shown in Fig. 10.7.

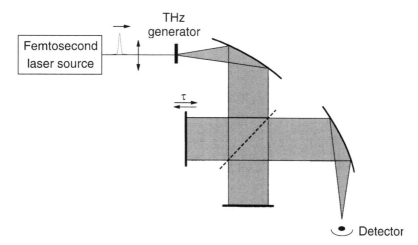

Fig. 10.7. Michelson interferometer for the measurement of the power spectrum of terahertz pulses using Fourier transform spectroscopy. The terahertz generator can be a photoconductive switch or a quadratic nonlinear material, as discussed in Sect. 10.2

Figure 10.8 shows another experimental implementation of Fourier transform spectroscopy, as demonstrated by *Ralph* and *Grischkowsky* [10.9], in which there are no moving parts in the infrared part of the experimental setup. The technique consists of using two terahertz generators, which are excited by two different visible pulses. By varying the time delay between the two visible pulses, one is able to generate two infrared pulses in sequence, thus performing the same function as the infrared Michelson interferometer shown in Fig. 10.7.

The setup shown in Fig. 10.9 relies on the same idea, namely generating the terahertz pulses from two visible pulses. However, a single terahertz generator is used here, which makes the setup much less bulky. This technique is suitable when there are no saturation effects in the terahertz generator, which is often the case when optical rectification is used as the infrared generation process. In order to avoid the detection of any nonlinear mixing between the two visible pulses when they overlap, a noncolinear geometry must be used. Owing to the greater diffraction in the infrared, the generated infrared beams are still colinear, so that this setup, called a *diffracting Fourier transform spectrometer*, achieves the same purpose as an infrared Michelson interferometer [10.26]. The advantage is a simplified setup in the infrared domain, since no broad band beam-splitter is needed. Furthermore, a sample and a detector can be placed in the near-field region just after the generator, avoiding any optical elements in the infrared and thus providing an extremely compact setup.

The Fourier transform of the data acquired using any of the above experimental setups actually provides $R(\omega)|E(\omega)|^2$, where $R(\omega)$ is the spectral response of the detector [10.25]. In this respect, a Fourier transform spectrom-

Page 318

A. Bonvalet and M. Joffre

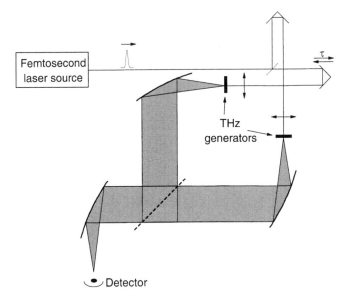

Fig. 10.8. Modified Fourier transform spectrometer [10.9]. Two terahertz generators are used, thus avoiding any moving part in the infrared part of the setup

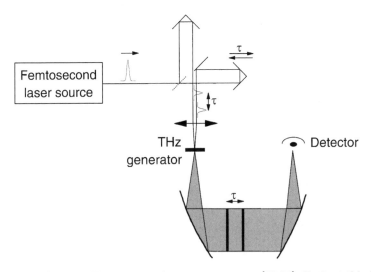

Fig. 10.9. Diffracting Fourier transform spectrometer [10.26]. Both visible beams are focused on the same terahertz generator, thus transferring most of the optics to the visible part of the setup

eter is no different from a grating spectrometer: only the photons to which the detector is sensitive matter. By choosing the right detector, such as a bolometer, the spectral detection bandwidth can be extremely large, which is the main advantage of Fourier transform spectroscopy. This feature alone has made possible the detection of very high-frequency components [10.8], up to the mid-infrared domain [10.21, 26]. However, the main drawback of measuring only the power spectrum $|E(\omega)|^2$ is that no information is available on the spectral phase. Any chirp present in the terahertz pulses will therefore remain undetected. To obtain phase information, other techniques such as those described below must be used. However, variants of the above setups can be used to perform linear cross-correlations between a terahertz reference pulse and an unknown pulse [10.27]. Similarly, phase information on materials in the terahertz domain can be readily obtained by placing the sample in one arm of the Michelson interferometer [10.8, 25].

10.3.2 Photoconductive Sampling

Photoconductive sampling relies on the idea that a duplicate of the very same Auston switch used for the generation of an electric transient can be used just as easily as a gate for detection purposes [10.2]. With this technique, the electric field associated with terahertz pulses can be recorded as a function of time [10.5, 6, 28]. The corresponding experimental setup for the generation and detection of terahertz pulses is shown in Fig. 10.10 [10.7]. A first laser pulse triggers the terahertz generator, as described in Sect. 10.2.1. The radiated terahertz beam is collimated and focused onto the detector. This latter chip is identical to the generator, except that the DC voltage source generator is replaced with a current meter. The voltage between the two aluminum electrodes of the detector is then directly proportional to the electric field of the focused terahertz pulse. A current proportional to this voltage will flow through the meter, but only when the photoconductive switch is made conductive by the second visible-laser pulse (the detection pulse). By varying the time delay between the excitation and the detection pulses using a standard optical delay line, it is possible to record the electric field in the terahertz pulse as a function of time. The time resolution is that of the photoconductive switch, i.e. a few hundreds of femtoseconds.

The great advantage of this gated-detection technique is that it is directly sensitive to the electric field of the terahertz pulse, which is thus completely characterized, in contrast to a measurement of the power spectrum only. However, owing to the finite response time of photoconductive switches, the maximum frequency which can be detected using this technique is a few terahertz only.

10.3.3 Free-Space Electro-Optic Sampling

Electro-optic sampling takes advantage of the Pockels effect [10.15], which, like optical rectification, is a second-order nonlinear effect. Indeed, if we consider a crystal placed in both a static electric field E_0 and the electric field $E(t)$ of a laser pulse, the polarization oscillating at the laser frequency is

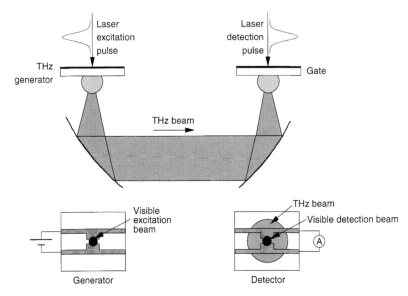

Fig. 10.10. Experimental setup for the generation and detection of terahertz pulses. The detector is a switch identical to the generator, shown in Fig. 10.4. The current measured at the detector is proportional to the electric field present in the terahertz beam when the laser detection pulse hits the chip

$$P(t) = \chi^{(1)} E(t) + \chi^{(2)} E_0 E(t)$$
$$= \left(\chi^{(1)} + \chi^{(2)} E_0 \right) E(t), \tag{10.6}$$

which results in a variation of the refractive index proportional to the static field. This is the electro-optic, or Pockels, effect. Under some conditions, this change in refractive index appears as a rotation of the polarization of the laser beam, which can be easily measured using balanced detection. In the case of a slowly varying electric field $E_0(t)$, the detected signal is proportional to the value taken by the electric field when the visible laser pulse passes through the electro-optic crystal. Using a correlation technique similar to that of Fig. 10.10, it is then possible to sample $E_0(t)$, as long as the time scale of variation of $E_0(t)$ is slow compared to the duration of the laser pulse. The oscillation period of the measured field must be longer than the duration of the optical pulse, an additional limit resulting from group velocity dispersion in the electro-optic crystal. This approach has been successfully demonstrated, first in the case of electric pulses propagating on transmission lines [10.29, 30] and then as a free-space detection of terahertz beams [10.31], up to oscillation frequencies of 40 THz [10.24, 32].

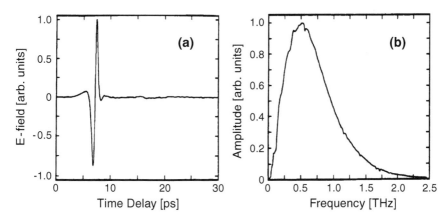

Fig. 10.11. (a) Measured electrical pulse of a freely propagating terahertz beam. (b) Amplitude spectrum of (a). From [10.33]

10.4 Some Experimental Results

A great variety of terahertz-generation experiments have been performed in the last ten years, so that only a few can be mentioned here. Figure 10.11 shows one of the most striking results, obtained by *Grischkowsky* et al. [10.33], using photoconductive switching and the experimental setup shown in Fig. 10.10. Since the authors use a point source, of size much smaller than the radiated wavelength, the emitted electric field is expected to resemble the second derivative of the (step-like) transient polarization, i.e. a single cycle. This is exactly what is observed in the experiment. The spectrum extends from 0 to about 2 THz.

Using the extended photoconducting source of Fig. 10.5, *You* et al. [10.14] have obtained half-cycle pulses, shown in Fig. 10.12. Indeed, in the case of an extended source, the field is only the first derivative of the step-like polarization, hence the half cycle. Such pulses have many specific applications, owing to their unipolar nature. As an example, they have been used for the ionization of Rydberg wave packets [10.34].

As mentioned in Sect. 10.2.1, in photoconductive switching, the radiated frequencies are limited to below a few terahertz. In contrast, as illustrated in Fig. 10.1, optical rectification does not have such limitations. Through the use of a Ti:sapphire laser delivering 12 fs pulses, pulses with frequencies extending up to 50 THz have been produced [10.21]. This corresponds to a wavelength of $6\,\mu m$, i.e. the mid-infrared spectral domain. The experimental data, shown in Fig. 10.13, were recorded using the setup shown in Fig. 10.9.

Figure 10.14 shows the variation of the infrared emitted power as a function of the azimuthal angle in GaAs. Such a measurement in agreement with theory allows one to assign the physical origin of the emission here to optical

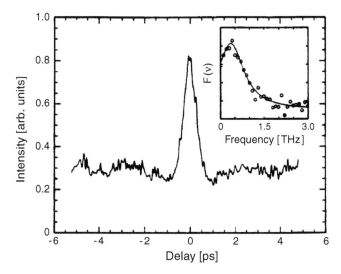

Fig. 10.12. Interferogram obtained in a Michelson interferometer from an extended terahertz source. The inset shows the power spectrum as obtained from a Fourier transform of the interferogram. From [10.14]

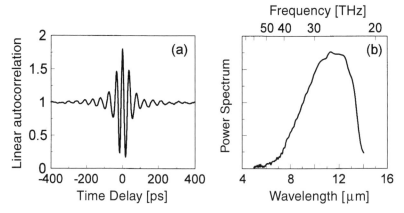

Fig. 10.13. Terahertz pulses produced through optical rectification of 12 fs laser pulses in bulk GaAs. (**a**) shows the linear autocorrelation and (**b**) its Fourier transform. The cutoff on the low-frequency side results from the spectral bandwidth of the HgCdTe detector

rectification, ruling out in this case other possible processes such as the acceleration of electrons in the depletion layer of the semiconductor [10.19, 35–37].

Another interesting development of terahertz generation is the possibility of shaping the emitted pulses [10.38–41]. For example, Fig. 10.15 shows an experimental setup allowing one to produce tunable terahertz radiation, as suggested by *Weling* et al. [10.38]. A sequence of two colinear, linearly

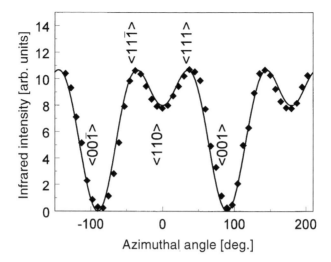

Fig. 10.14. Infrared emission from a GaAs substrate as a function of the azimuthal angle between the crystal axis and the polarization of the incident light [10.21]. *Solid line*, calculation based on the assumption that the signal results from optical rectification

chirped laser pulses is sent to the terahertz generator. Owing to the linear chirp, the frequency varies linearly as a function of time, but the frequency difference between the two interfering pulses is constant, so that beating occurs in the total incident intensity at this frequency difference. As a result, the periodically excited terahertz generator emits not a single cycle, but a sinusoidal wave whose envelope is the total duration of the chirped pulse, as shown in Fig. 10.15. The system thus produces narrow-band terahertz radiation, whose frequency can be tuned simply by varying the time delay between the two laser pulses.

Figure 10.16 shows the corresponding experimental data obtained in a Michelson interferometer for different values of the time delay between the two excitation pulses. As expected, the oscillation frequency is shown to be directly proportional to the time delay [10.38]. The same technique has been extended to the mid-infrared domain where shaped pulses have been obtained, for example, through difference-frequency mixing between two chirped phase-locked pulses, as shown in Fig. 10.17. In contrast with the experiment of *Weling et al.* [10.38], the two incident pulses now have different chirps so that the beating frequency is no longer constant with time, resulting in the generation of a chirped infrared pulse. Such experiments, in both the far- and the mid-infrared domain, demonstrate that by shaping the intensity of the incident light pulse, it is possible to generate any arbitrary electric field in the infrared spectral region.

To summarize, terahertz pulses have been generated and detected using many different experimental configurations, including photoconductive

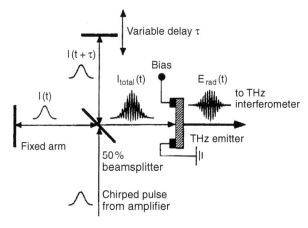

Fig. 10.15. Experimental setup allowing generation of tunable terahertz radiation from the overlap of two chirped laser pulses [10.38]

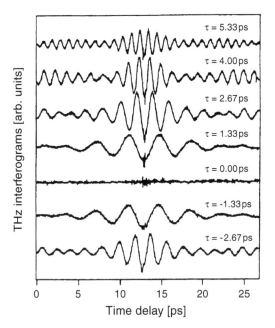

Fig. 10.16. Terahertz interferograms from beating of 22.5 ps chirped pulses at various delays [10.38]

switching and optical rectification for the generation, and photoconductive or electro-optic sampling as well as Fourier transform spectroscopy for the detection. A great versatility has thus been achieved in source characteristics such as maximum frequency, pulse energy and pulse duration. These new tech-

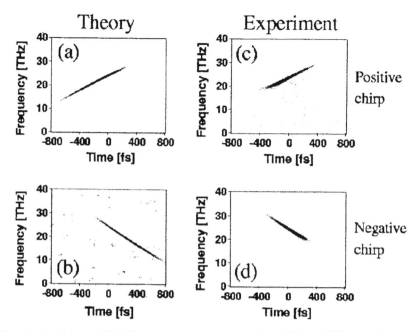

Fig. 10.17. (a) and (b), Wigner functions expected; and (c) and (d) experimentally measured, corresponding to the mid-infrared electric field emitted by optical rectification of a superposition of two phase-locked chirped pulses. Depending on the time order between the two exciting pulses, and infrared pulse of either positive (c) or negative (d) chirp can be generated. From [10.41]

niques make possible many applications, some of which are briefly mentioned below.

10.5 Time-Domain Terahertz Spectroscopy

The ability to measure the electric field as a function of time makes the approach of time-domain spectroscopy [10.33] natural. As shown in Fig. 10.18, a short terahertz pulse is simply sent through the material under investigation. An analysis of the transmitted field as a function of time allows one to deduce the linear response function of the material in the time domain. A simple Fourier transformation then yields the absorption coefficient and refractive index of the material.

Time-domain far-infrared spectroscopy has been applied to many different systems, including sapphire, quartz, GaAs, Ge [10.33], a two-dimensional electron gas in GaAs [10.42], and superconductors [10.43]. In the following, we consider the example of water vapor [10.44].

Figure 10.19 shows the shape of a terahertz pulse obtained after transmission through water vapor. Decaying oscillations are apparent in the trailing

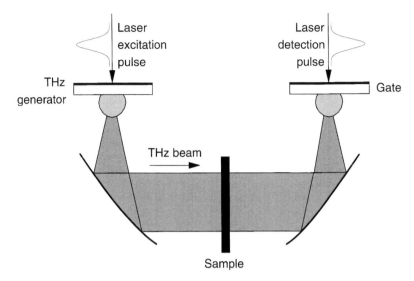

Fig. 10.18. Experimental setup used for time-domain far-infrared spectroscopy

edge of the transmitted pulse, as a consequence of free induction decay in water. Figure 10.19 shows the corresponding frequency-domain data, obtained from a Fourier transformation of Fig. 10.19. The data provide both the absorption coefficient and the refractive index with great accuracy.

Although the same kind of information could be obtained in principle by putting the sample in one arm of a far-infrared Fourier transform spectrometer [10.25], the gated detection scheme and the high brightness of the source provide time-domain terahertz spectroscopy with a much better signal-to-noise ratio.

Finally, the fact that different chemical elements exhibit very specific absorption lines can be put to use in imaging applications; these have been demonstrated recently as an interesting development of terahertz spectroscopy [10.31, 45].

10.6 Conclusion

To summarize, new techniques have been developed in order to generate and detect low-frequency electromagnetic pulses in the far-infrared and mid-infrared spectral domains. These pulses are assuredly the shortest possible electromagnetic pulses when expressed in units of optical cycles, since single-and even half-cycle pulses have been obtained.

Surprisingly, many of the applications of these terahertz femtosecond pulses do not take advantage of the time resolution thus achieved. Indeed, it is already very interesting to be able to make use of the broad frequency

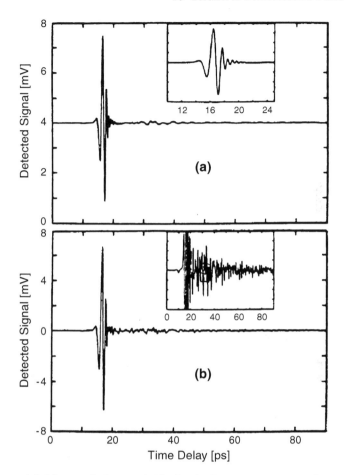

Fig. 10.19. (a) Measured electric field of a freely propagating terahertz beam in
pure nitrogen. The inset shows the pulse on an expanded timescale. (b) Measured
electric field with 1.5 torr of water vapor in the enclosure. The inset shows the pulse
on a 20 times expanded vertical scale. From [10.44]

spectrum obtained in the unusual spectral range of the far infrared. However, there is the additional possibility to design experiments where both
far-infrared and visible pulses are exploited, although these have not been
discussed in this chapter. For example, wave-packet oscillations have been observed in atoms [10.37, 46], as well as in semiconductors [10.27, 47, 48]. More
recently, the buildup of Coulomb screening in semiconductors has been measured by recording the infrared field transmitted through a GaAs sample for
different time delays with respect to a pump pulse at 800 nm [10.49].

Terahertz femtosecond pulses have many applications, a very few of which
have been mentioned in this chapter. Furthermore, with the availability of

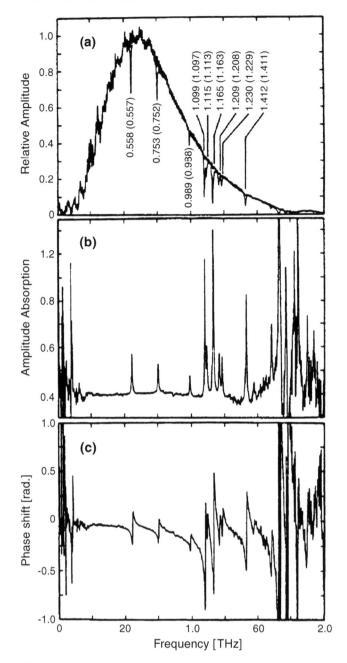

Fig. 10.20. (a) Amplitude spectra of data shown in Fig. 10.18a, b. (b) Absorption coefficient of water obtained from (a). (c) Relative phase. From [10.44]

diode-pumped femtosecond laser systems, terahertz pulses may soon find industrial applications.

10.7 Problems

1. (a) A Gaussian pulse induces a polarization transient whose time dependence is proportional to the integral of the temporal profile of the pulse. What is the temporal profile of the emitted electric field in the case of a point source?
   (b) Compute the power spectrum and determine the frequency of its maximum. Compare with Fig. 10.2. What is the numerical value of this frequency in THz, and the corresponding wavelength in μm, for an excitation pulse with an FWHM of 100 fs? what are the corresponding values for an FWHM of 10 fs?
2. The figure below shows the linear autocorrelation of a terahertz pulse as measured in a Michelson interferometer.

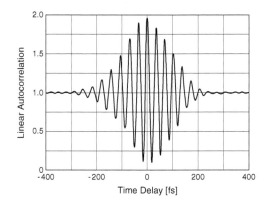

(a) What is the central frequency ν_0 (in THz), the spectral width and the central wavelength (in μm) of the pulse thus characterized. Can you say anything about the duration of the pulse?
(b) A linear material having a strong positive group velocity dispersion is placed in the terahertz beam before it enters the detection apparatus. Describe the data then obtained.
(c) What would be the answer to the above question if the apparatus was an electro-optic sampling setup instead of a Michelson interferometer?
(d) Using again the Michelson interferometer, what would be the data obtained if the above linear material was placed in one arm of the interferometer?

Acknowledgments

Some of the experimental results presented in this chapter were obtained in collaboration with Nadia Belabas, Jean-Pierre Likforman, Jean-Louis Martin and Arnold Migus, whose input is gratefully acknowledged. We are also grateful to Phil Bucksbaum for very helpful discussions on the generation of terahertz pulses from photoconductive switches.

References

[10.1]   J. D. Jackson: *Classical Electrodynamics* (Wiley, New York 1975)

[10.2]   D.H. Auston: Appl. Phys. Lett. **26**, 101 (1975)

[10.3]   G. Mourou, C.V. Stancampiano, A. Antonetti, A. Orszag: Appl. Phys. Lett. **39**, 295 (1981)

[10.4]   D.H. Auston, K.P. Cheung, P.R. Smith: Appl. Phys. Lett. **45**, 284 (1984)

[10.5]   A.P. DeFonzo, M. Jarwala, C. Lutz: Appl. Phys. Lett. **50**, 1155 (1987)

[10.6]   C. Fattinger, D. Grischkowsky: Appl. Phys. Lett. **53**, 1480 (1988)

[10.7]   N. Katzenellenbogen, D. Grischkowsky: Appl. Phys. Lett. **58**, 222 (1991)

[10.8]   B.I. Greene, J.F. Federici, D.R. Dykaar, R.R. Jones, P.H. Bucksbaum: Appl. Phys. Lett. **59**, 893 (1991)

[10.9]   S.E. Ralph, D. Grischkowsky: Appl. Phys. Lett. **60**, 1070 (1992)

[10.10] M.B. Ketchen, D. Grischkowsky, T.C. Chen, C.-C. Chi, I.N. Duling III, N.J. Halas, J.-M. Halbout, J.A. Kash, G.P. Li: Appl. Phys. Lett. **48**, 751 (1986)

[10.11] D. Krökel, D. Grischkowsky, M.B. Ketchen: Appl. Phys. Lett. **54**, 1046 (1989)

[10.12] C. Fattinger, D. Grischkowsky: Appl. Phys. Lett. **54**, 490 (1989)

[10.13] B.B. Hu, J.T. Darrow, X.-C. Zhang, D.H. Auston, P.R. Smith: Appl. Phys. Lett. **56**, 886 (1990)

[10.14] D. You, R.R. Jones, P.H. Bucksbaum, D.R. Dykaar: Opt. Lett. **18**, 290 (1993)

[10.15] P.N. Butcher, D. Cotter: *The Elements of Nonlinear Optics*, Cambridge Studies in Modern Optics (Cambridge University Press, Cambridge 1990)

[10.16] K.H. Yang, P.L. Richards, Y.R. Shen: Appl. Phys. Lett. **19**, 320 (1971)

[10.17] D.H. Auston, K.P. Cheung, J.A. Valdmanis, D.A. Kleinman: Phys. Rev. Lett. **53**, 1555 (1984)

[10.18] X.-C. Zhang, Y. Jin, K. Yang, L.J. Schowalter: Phys. Rev. Lett. **69**, 2303 (1992)

[10.19] A. Rice, Y. Jin, X.F. Ma, X.-C. Zhang, D. Bliss, J. Larkin, M. Alexander: Appl. Phys. Lett. **64**, 1324 (1994)

[10.20] X.-C. Zhang, X.F. Ma, Y. Jin, T.-M. Lu, E.P. Boden, P.D. Phelps, K.R. Stewart, C.P. Yakymyshyn: Appl. Phys. Lett. **61**, 3080 (1992)

[10.21] A. Bonvalet, M. Joffre, J.-L. Martin, A. Migus: Appl. Phys. Lett. **67**, 2907 (1995)

[10.22] M. Joffre, A. Bonvalet, J.-L. Martin, A. Migus: Ultrafast Phenomena X, Springer Series in Chemical Physics **62**, Springer, Berlin (1996)

[10.23] R.A. Kaindl, D.C. Smith, M. Joschko, M.P. Hasselbeck, M. Woerner, T. Elsaesser: Opt. Lett. **23**, 861 (1998)

[10.24] R. Huber, A. Brodschelm, F. Tauser, A. Leitenstorfer: Appl. Phys. Lett. **76**, 3191 (2000)

[10.25] P.R. Griffiths, J.A. de Haseth: *Fourier-Transform Infrared Spectrometry*, Chemical Analysis, Vol. 83 (Wiley, New York 1986)

[10.26] M. Joffre, A. Bonvalet, A. Migus, J.-L. Martin: Opt. Lett. **21**, 964 (1996)

[10.27] A. Bonvalet, J. Nagle, V. Berger, A. Migus, J.-L. Martin, M. Joffre: Phys. Rev. Lett. **76**, 4392 (1996)

[10.28] P.R. Smith, D.H. Auston, M.C. Nuss: IEEE J. Quant. Electr. **24**, 255 (1988)

[10.29] J.A. Valdmanis, G.A. Mourou, C.W. Gabel: IEEE J. Quant. Electr. **19**, 664 (1983)

[10.30] G.A. Mourou, K.E. Meyer: Appl. Phys. Lett. **45**, 492 (1984)

[10.31] Q. Wu, T.D. Hewitt, X.-C. Zhang: Appl. Phys. Lett. **69**, 1026 (1996)

[10.32] Q. Wu, X.-C. Zhang: Appl. Phys. Lett. **71** 1285 (1997)

[10.33] D. Grischkowsky, S. Keiding, M. van Exter, C. Fattinger: J. Opt. Soc. Am. B **7**, 2006 (1990)

[10.34] C. Raman, C.W.S. Conover, C.I. Sukenik, P.H. Bucksbaum: Phys. Rev. Lett. **76**, 2436 (1996)

[10.35] X.-C. Zhang, B.B. Hu, J.T. Darrow, D.H. Auston: Appl. Phys. Lett. **56**, 1011 (1990)

[10.36] S.L. Chuang, S. Schmitt-Rink, B.I. Greene, P.N. Saeta, A.F.J. Levi: Phys. Rev. Lett. **68**, 102 (1992)

[10.37] P.N. Saeta, B.I. Greene, S.L. Chuang: Appl. Phys. Lett. **63**, 3482 (1993)

[10.38] A.S. Weling, B.B. Hu, N.M. Froberg, D.H. Auston: Appl. Phys. Lett. **64**, 137 (1994)

[10.39] Y. Liu, S.G. Park, A.M. Weiner: Opt. Lett. **21**, 1762 (1996)

[10.40] F. Eickemeyer, R.A. Kaindl, M. Woerner, A.M. Weiner: Opt. Lett. **25**, 1472 (2000)

[10.41] N. Belabas, J.-P. Likforman, L. Canioni, B. Bousquet, M. Joffre: Opt. Lett. **26**, 743 (2001)

[10.42] W.J. Walecki, D. Some, V.G. Kozlov, A.V. Nurmikko: Appl. Phys. Lett. **63**, 1809 (1993)

[10.43] M.C. Nuss, K.W. Goosen: IEEE J. Quant. Electr. **QE 25**, 2596 (1989)

[10.44] M. van Exter, C. Fattinger, D. Grischkowsky: Opt. Lett. **14**, 1128 (1989)

[10.45] B.B. Hu, M.C. Nuss: Opt. Lett. **20**, 1716 (1995)

[10.46] R.R. Jones: Phys. Rev. Lett. **76**, 3927 (1996)

[10.47] H.G. Roskos, M.C. Nuss, J. Shah, K. Leo, D.A.B. Miller, A.M. Fox, S. Schmitt-Rink, K. Köhler: Phys. Rev. Lett. **68**, 2216 (1992)

[10.48] C. Waschke, H.G. Roskos, R. Schwedler, K. Leo, H. Kurz, K. Köhler: Phys. Rev. Lett. **70**, 3319 (1993)

[10.49] R. Huber, F. Tauser, A. Brodschelm, M. Bichler, G. Abstreiter, A. Leitenstorfer: Nature **414**, 286 (2001)

Coherent Control in Atoms, Molecules and Solids

T. Amand, V. Blanchet, B. Girard and X. Marie

With 35 figures

11.1 Introduction

The idea of controlling optical, physical and chemical processes in matter using the coherence properties of light appeared in the sixties in the scientific community, since the early proposal of Manykin and Alfanasev [11.1] to control the absorption of an atomic medium by adjusting the phases of two or more incident beams. It relies on the principle that interferences between different quantum mechanical pathways connecting the same initial and final states of a given system control the electron transfer between these states.

After a latent period of about two decades, the interest in such experiments was renewed, principally due to the advent of stable, flexible ultrafast laser sources. The broad spectral width of these pulses offers a wide range of possibilities. Coherent control has now been demonstrated in various fields, from atoms and molecules to solid-state quantum structures, in chemical physics and biology. Many different schemes have been proposed and demonstrated, and extensive reviews can be found [11.2–7]. We illustrate here the basic principles of coherent control, with typical examples taken in different fields. This chapter does not intend to give an exhaustive review of all the contributions made to the field. For instance, an approach to coherent control sharing strong similarities with the pump-probe scheme (see Chap. 8) will not be described in more detail here. After an initial theoretical proposal [11.8,9], several experimental demonstration were achieved [11.10,11].

In this book dedicated to ultrashort pulses, this chapter will focus on coherent control schemes involving such pulses. Therefore, except for the example detailed in section 11.2 for historical purposes, we will ignore a large amount of work involving adiabatic transfers or Rabi transfers performed with CW lasers or long (i.e., nanosecond) pulses. STIRAP (stimulated raman adiabatic passage) [11.12–14] is one striking example among others. We will review here the different approaches for coherent control, in their successive historical appearance. In the initially investigated scheme, a one-photon transition interferes with a multiphoton transition: the coherent control is

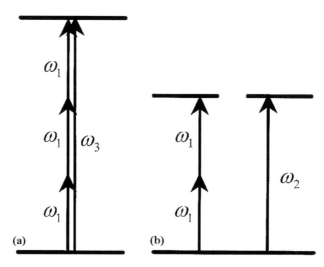

Fig. 11.1. (a) $3\omega_0/\omega_3$ scheme: both paths reach the same final state, a total cross section can be controlled; (b) $2\omega_0/\omega_2$ scheme, the final states have opposite parities; a differential cross section can be controlled

performed in the frequency domain (section 11.2). Then we will move to temporal coherent control, where the two excitation paths are now separated in time (section 11.3). These experiments can be generalized by using chirped or more generally shaped pulses, thus leading to the notion of optimal control (section 11.4). Strong field effects will be described in section 11.5, and finally, some applications and other examples will be described in the conclusion (section 11.6).

11.2 Coherent Control in the Frequency Domain

Historically, the first proposed scheme of coherent control was based on interferences between several excitation paths leading to the same final state [11.2,15]. One path is excited by a multiphoton transition, and the other one is excited by a one-photon transition. These latter photons are produced by harmonic generation from the first ones. This ensures that the final states reached by both paths have the same energy. Moreover the coherence relationship between the electric fields is well defined. Although this scheme does not involve ultrashort pulses, it is an excellent illustration of some basic principles of coherent control. Figure 11.1 presents the two most common cases: a three-photon transition and a two-photon transition each interfering with a one-photon transition.

In the $3\omega_0/\omega_3$ scheme (Fig. 11.1a), the large photons are produced by the third harmonic generation from the small ones: $\omega_3 = 3\omega_0$. The number of photons in each path is odd. This means that the excited state can be the same, with a parity opposite to the initial state's. Therefore, the total excitation cross section can be controlled. If several excited states are accessible, then it becomes possible to control the product ratio between these channels. This requires that different values of the control parameter (the relative phase between the laser fields) maximize the various final states. The excitation probability toward the final channel (q)

$$P^{(q)} = P_1^{(q)} + P_3^{(q)} + P_{1,3}^{(q)} \qquad (11.1)$$

results from contributions of each excitation path separately, $P_1^{(q)}$ and $P_3^{(q)}$, and from an interfering contribution

$$P_{1,3}^{(q)} \propto \cos\left(3\theta_1 - \theta_3 + \delta_{1,3}^{(q)}\right), \qquad (11.2)$$

which depends on the relative phase between the laser fields $\Delta\theta = 3\theta_1 - \theta_3$ and on a molecular phase $\delta_{1,3}^{(q)}$ related to the specific channel (q). The relative laser phase $\Delta\theta$ is in general stable for most experimental conditions. It can be scanned by propagating the beams through a dispersive medium of variable optical thickness. This is usually achieved with a buffer gas of variable pressure. The first experimental demonstration of this effect was achieved in mercury atoms [11.16] as depicted in Fig. 11.2. Subsequent observations were obtained in molecules such as HCl [11.17,18], HI [11.19–21] or CH_3I [11.22,23] and even larger molecules [11.24]. In most of these examples, either only one final channel is obtained or several channels (such as dissociation or ionisation) are open but they vary in the same way with the control phase. This means that one single excitation path is sensitive to the control parameter and leads to the different channels. One spectacular result was obtained in HI (and DI) where the product ratio between parent HI^+ (DI^+) and fragment ions I^+ could be controlled with a high contrast as displayed in Fig. 11.3 [11.19–21]. This behavior results from interferences between direct ionization and autoionization resonances [11.21,25].

In the $2\omega_0/\omega_2$ scheme (Fig. 11.1b), the second harmonic generation is used: $\omega_2 = 2\omega_0$. The final states have opposite parities. A coherent superposition of states with opposite parities is therefore created. This corresponds to an anisotropic wave function, so that the angular distribution of a fragmentation product or a photoelectron can be controlled.

Many theoretical as well as experimental studies have applied this scheme to control various systems: the angular distribution of photoelectrons produced in atomic photoionization [11.26,27], from photocathodes [11.28]; the asymmetric angular distributions of H^+/D^+ from HD^+ photodissociation [11.29,30]; The direction of photocurrents produced from bulk semiconductors [11.31,32].

As a particular case to the two main cases described so far, the total cross section can also be controlled with the $2\omega_0/\omega_2$ scheme in the saturation regime. Here, the saturation corresponds to several absorption/emission cycles and is

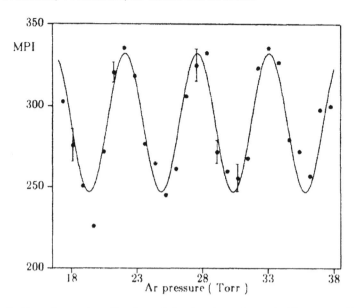

Fig. 11.2. Experimental realization of the $3\omega_0/\omega_3$ scheme in Hg atoms ($6s^1S_0 \rightarrow 6p^1P_1$ transition); the $6p^1P_1$ population is probed with multiphoton ionization; ion rate as a function of the pressure of the Ar buffer gas [11.16]

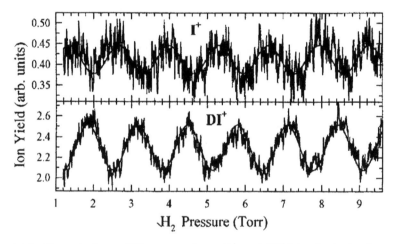

Fig. 11.3. Control of two different products (I^+ and DI^+) in the ionization of DI in the $3\omega_0/\omega_3$ scheme; the oscillations of the two products are phase-shifted by 150° [11.19–21]

thus equivalent to higher-order photon processes. Several studies have dealt with the photodissociation of molecules [11.33,34]. It was also suggested that the symmetry-breaking properties of this scheme could be used to produce chiral molecules from achiral precursors.

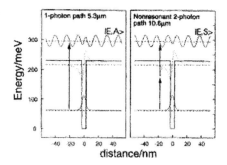

Fig. 11.4. Conduction electron energy band diagram of a 55-Å GaAs/Al$_{0.26}$Ga$_{0.74}$As QW and wave functions of the states implied in 5.3-μm single-photon and a 10.6-μm two-photon pathways; the QW ground state is populated by electrons from the doping; the spatial coordinate is taken along the QW [001] growth axis [11.31]

In condensed matter, early experiments aimed at controlling the photoemission from Stilbium-Cesium photocathodes [11.28] were achieved using one- and two-photon transitions toward electron vacuum states. The possibility to induce a photocurrent within a semiconductor using a similar control scheme, as initially suggested by Kurizki et al. [11.35], is far from obvious, due to the fast dephasing processes experienced by the carriers in delocalized ("free") semiconductor states. It was achieved first on nanostructures by Dupont, Corkum et al. [11.31]. These authors used, in order to stay close to the atomic physics analogy, N-doped AlGaAs/GaAs quantum well (QW) superlattices at 82 K. The energy band diagram of one spatial period of the structure is presented in Fig. 11.4.

The electrons are excited from the symmetrical ground level of the QW to a highly energetic free conduction state $|E\rangle$ using two quantum pathways: a two-photon intersubband absorption at 10.6 μm and a single-photon process at 5.3 μm (Fig. 11.4). The associated electric fields must have a nonzero projection on the QW growth axis to couple the optical fields with the intersubband dipoles. This is obtained by processing 45° edge facets. The phase relation between the two beams can be tuned. Figure 11.5 shows the resulting photocurrent, the direction of which can be coherently controlled as a function of the phase difference, just as in the rubidium experiment [11.27].

The observation of a coherently controlled photocurrent in unbiased bulk GaAs semiconductors was later observed by Haché et al. at room temperature [11.36]. These authors used a metal-semiconductor-metal structure (MSM), and phase-related 1 ps (or 175 fs) pulses at 0.775 μm and 1.55 μm that excite GaAs across the band gap (E$_g$ ≈ 1.42 eV) with one- and two-photon processes, respectively. The photocurrent oscillates with a similar law with the phase of the two beams, its direction being simply determined by the electric field polarization. Another example of coherent control in the frequency domain

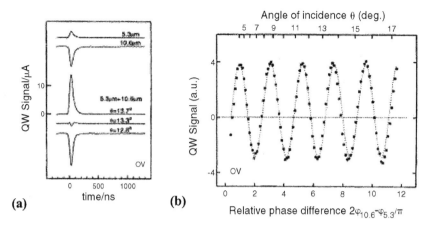

Fig. 11.5. (a) Quantum well photo-current pulse response [11.31]. The residual response with only one beam (the first two curves) at 5.3 μm and 10.6 μm. The three lower traces represent the QW response when both beams illuminate the sample. An NaCl dispersive plate allows to modify the phase relation between the two beams; the extreme angles $\theta = 13.7°$ and $\theta = 12.8°$ correspond to a phase shift of π; (b) integrated QW response versus the relation phase shift; dashed line: sinusoidal fit

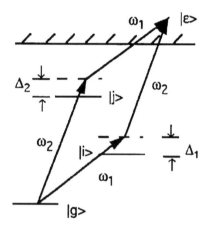

Fig. 11.6. Levels and excitation paths in the $\omega_1 + \omega_2/\omega_2 + \omega_1$ scheme [11.37]

consists in creating interferences between two sequential excitations $|g\rangle \rightarrow |i\rangle \rightarrow |e\rangle$ and $|g\rangle \rightarrow |j\rangle \rightarrow |e\rangle$ as depicted in Fig. 11.6 [11.37]. Two laser frequencies, $\omega_1 \simeq \omega_{ig}$ and $\omega_2 \simeq \omega_{jg}$, are chosen close to resonance with the first transition of each path. The same final state is chosen to be in the continuum, at an energy $\omega_{eg} = \omega_1 + \omega_2$, and can be reached with both paths. Changing the sign of the detuning ($\Delta_1 = \omega_1 - \omega_{ig}$ or $\Delta_2 = \omega_2 - \omega_{jg}$) of one path can change the interference from destructive to constructive or vice versa. This provides an efficient way to control these interferences. However, the only available experiments (in Ba) have been realized in the strong field regime where adiabatic transfers take place [11.38]. Although the results look like those predicted by the interference scheme, they correspond to total transfer through one path or the other, with an efficiency varying with detuning.

11.3 Temporal Coherent Control

In temporal coherent control, the two paths are excited by two ultrashort pulses separated by some time delay, with the same central frequency. This simple scheme shares many features with Ramsey fringes experiments and Fourier transform spectroscopy. It allows creation of an excited state wave packet by the first laser pulse. This wave packet can be destroyed or enhanced by interference with a second identical wave packet created by the second laser pulse. In the most common case (one-photon transitions in the linear regime), these experiments can also be interpreted in terms of optical interferences. We show here how the introduction of nonlinearities in the excitation or detection step precludes the interpretation with optical interferences. This is demonstrated in two-photon transitions and in one-photon transitions in the saturation regime (see section 11.3.1). Another example is obtained in the coherent control of exciton orientation in semiconductor quantum wells (11.3.2.2).

Two coherent control schemes can be used: in the first, the two pulses of the sequence have the same polarization; this most commonly used scheme leads to the control of the populations of the excited states. In the second, the two pulses have orthogonal polarization (circular or linear); this leads to the control of the coherences between the different excited states addressed by the pulse sequence.

11.3.1 Principles of Temporal Coherent Control

11.3.1.1 Introduction to the Theory of Temporal Coherent Control in One-Photon Transitions. Weak field interaction between an ultrashort laser pulse and a quantum system can be adequately described by perturbation theory. For one-photon transitions, this means that the interaction varies linearly with the excitation field. Thus, the response of a quantum system to a two laser pulse sequence can be described by summing the contribution of each laser pulse or by considering directly the effect of the total laser field. Two different physical interpretations result, as discussed here.

We consider a quantum system that consists of a ground state $|g\rangle$ (used as an energy reference), and of a set of two excited states $|e_k\rangle$ ($k = 1, 2$) of energies $\hbar\omega_k$, which can be excited from the ground state through a dipole-allowed transition as shown in Fig. 11.7 [11.39]. The extension to the case of several excited states, as encountered with molecular vibrational states, is straightforward [11.40–42].

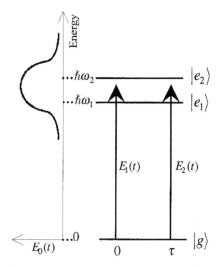

Fig. 11.7. Principle of the experiment: A ground state $|g\rangle$ is excited to a set of two excited states $|e_k\rangle$ (energy $\hbar\omega_k$) by a two-pulse sequence; the spectral width of the two pulses is broad enough to simultaneously excite both levels (see left-hand side)

The interaction is induced by a sequence of two identical short pulses (pulse duration τ_L, same polarization, spectral width ($\Delta\omega_L \simeq \tau_L^{-1}$) large enough to populate both excited states $|\omega_L - \omega_k| \leq \Delta\omega_L$).

The total laser field on the polarization direction $\varepsilon = \varepsilon_1 = \varepsilon_2$ can be expressed as:

$$E_T(t) = E_1(t) + E_2(t) \tag{11.3}$$

with

$$E_2(t) = E_1(t - \tau), \tag{11.4}$$

and assuming Fourier transform limited pulses

$$E_1(t) = E_0(t)e^{-i\omega_L t}, \tag{11.5}$$

where $E_0(t)$ represents the single pulse envelope and τ_0, ω_L the pulse duration and central frequency, respectively. τ is the time delay between the two pulses. In the low field regime, starting from the time-dependent Schrödinger equation, and using the rotating wave approximation (RWA), the system wave function can be written as:

$$|\psi(t)\rangle \approx |g\rangle + \sum_k \sum_{p=1,2} \frac{i}{\hbar}\mu_{kg} \int_{-\infty}^t E_p(t')e^{i\omega_k(t'-t)}dt' |e_k\rangle = |g\rangle + |\psi_e(t)\rangle, \tag{11.6}$$

where μ_{kg} is the transition dipole moment from state $|g\rangle$ toward the excited state $|e_k\rangle$, $\mu_{kg} = \langle k|q\mathbf{r}.\varepsilon|g\rangle$. We introduce the transition probability amplitude

$a_k^{(p)}$ from the ground state $|g\rangle$ toward the excited state $|e_k\rangle$ after interaction $(t \to +\infty)$ with the single pulse p $(p = 1, 2)$:

$$a_k^{(p)} = \frac{i}{\hbar} \mu_{kg} \tilde{E}_p(\omega_k), \qquad (11.7)$$

which is proportional to the Fourier transform $\tilde{E}_p(\omega_k)$ of the p pulse electric field at the transition energy. Note that

$$\tilde{E}_1(\omega_k) = \tilde{E}_0(\omega_k - \omega_L) \qquad (11.8)$$
$$\tilde{E}_2(\omega_k) = \tilde{E}_0(\omega_k - \omega_L)e^{i\omega_k\tau}. \qquad (11.9)$$

For $t \gg \tau_0$, τ and $|a_k^{(p)}| \ll 1$, expression (11.6) takes the simpler form:

$$|\psi_e(t)\rangle = \sum_k \sum_{p=1,2} e^{-i\omega_k t} a_k^{(p)} |e_k\rangle. \qquad (11.10)$$

Several interpretations of the result (11.10) can be given, writing it in several different forms.

(a) Interpretation in Terms of Quantum Beats

The total amplitude of the excited state $|e_k\rangle$ is obtained by summing the partial amplitudes corresponding to the different quantum paths followed by the system

$$|\psi_e(t)\rangle = \sum_k e^{-i\omega_k t} b_k(\tau) |e_k\rangle, \qquad (11.11)$$

where, using Eqs. (11.7), (11.8) and (11.9),

$$b_k(\tau) = a_k^{(1)} + a_k^{(2)}(\tau) = a_k^{(1)}(1 + e^{i\omega_k\tau}).$$

For a *one-photon transition*, there are only two quantum paths, corresponding to the excitation of the system by the first or second optical pulse. The probability amplitude $b_k(\tau)$ to find the system in the excited state $|e_k\rangle$ oscillates at the pulsation ω_k as a function of τ as a result of quantum interferences between these excitation paths. Defining the population of the excited state $|e_k\rangle$ by

$$n_k(t) = |\langle e_k|\psi_e(t)\rangle|^2. \qquad (11.12)$$

The total excited states population

$$n(\tau) = \sum_k n_k(t) = \sum_k |b_k(\tau)|^2 \qquad (11.13)$$

shows rapid oscillations at the mean pulsation $\langle \omega_k \rangle = (\omega_1 + \omega_2)/2$, and slower beats at the difference pulsation $\Delta\omega = \omega_1 - \omega_2$. This explains the quantum beat interpretation. In usual experiments, a large number of microscopic systems are simultaneously excited. The observed oscillations can be damped by interaction processes (such as lattice vibrations in solid-state systems) or by averaging over several systems having slightly different excitation transitions (this is equivalent to inhomogeneous broadening). These behaviors are equivalent to the appearance of an additional random phase factor $e^{i\phi}$ in the terms $p = 1$ of expression (11.10) for microscopic systems having experienced a diffusion process when the second pulse arrives. Averaging on the ensemble results in a damping of the oscillations with a characteristic time called the optical dephasing time T_2. So the condition required to observe oscillations with respect to the time delay τ is:

$$\tau \leq T_2. \tag{11.14}$$

(b) Spectral Interpretation

Due to the linearity of the interaction in the low field regime, it is possible to evaluate the global effect of the total field on the system from Eq. (11.10):

$$|\psi_e(t)\rangle = \sum_k \frac{1}{i\hbar} \mu_{kg} \tilde{E}_T(\omega_k) e^{-i\omega_k t} |e_k\rangle. \tag{11.15}$$

This equality simply shows that the system absorbs at its eigenfrequencies among the frequencies of the total field. The corresponding population is:

$$n(\tau) = \sum_k \left(\frac{\mu_{kg}}{\hbar}\right)^2 S(\omega_k), \tag{11.16}$$

where

$$S(\omega) = |\tilde{E}_T(\omega)|^2 = |\tilde{E}(\omega)|^2 |1 + e^{i\omega\tau}|^2 \tag{11.17}$$

represents the intensity spectrum of the total laser field. This spectrum has the same envelope as the spectrum of a single pulse but is modulated due to optical interferences. The fringe spacing is $2\pi/\tau$ (see Fig. 11.8). It should be noted that this modulated spectrum is an intrinsic property of the two pulse sequences. It originates in the interference of the laser optical modes belonging to two different pulses with same frequencies. These fringes damp with the coherence time of one laser *single* mode, which is usually orders of magnitude greater than T_2. They can be observed in a monochromator when the resolution is better than the fringe spacing. Here, the quantum system behaves as a spectrometer with several slits, each placed at one absorption frequency ω_k. The first pulse excites a microscopic dipole that keeps the memory of the optical phase. The second pulse interacts with this oscillating dipole, and interferences result. The main difference with the spectrometer is that the system induces interferences only at its absorption frequencies, whereas the interferences in the spectrometer are present at all the frequencies contained in the spectrum of the pulse. The resolution is limited in general by the system line width $\Gamma = 2\hbar/T_2$, which is equivalent to the condition (11.14) expressed in the frequency domain. In this description, everything occurs as if the system

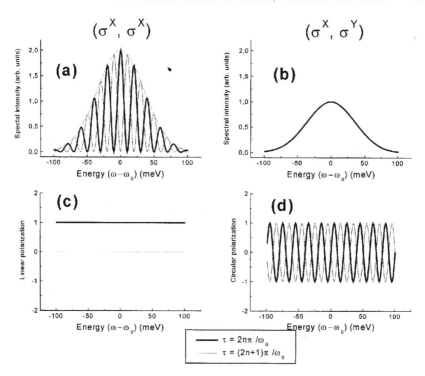

Fig. 11.8. Spectral intensity of the total electric field for; (a) a copolarized sequence (σ^X, σ^X); (b) a cross-polarized sequence (σ^Y, σ^Y); the respective (c) linear and (d) circular polarization degrees are presented. Here ω_0 is the central laser frequency, the two pulses have the same intensity, and their duration is about 20 fs; two time delays τ between the pulses are presented

excitation resulted from the optical interferences between the two pulses. We shall see (section 11.3.1.2) the limits of such an approach, when two-photon transitions are used to excite the system.

(c) Wave Packets Interferences

For laser pulses short enough, the wave function of the excited state (eq.(11.10)) can be written as the sum of two wave packets created by each laser pulse p:

$$|\psi_e(t)\rangle = \sum_{p=1,2} |\psi_p(t)\rangle = \sum_{p=1,2}\left(\sum_k a_k^{(p)}(\tau)e^{-\omega_k t}|e_k\rangle\right). \qquad (11.18)$$

Each pulse creates a wave packet $|\psi_p(t)\rangle$ as a superposition of stationary states. The wave packet $|\psi_1(t)\rangle$ evolves freely between the two laser pulses and then interferes with the second wave packet $|\psi_2(t)\rangle$. The wave packet created by the first pulse oscillates between two orthogonal limit states: the

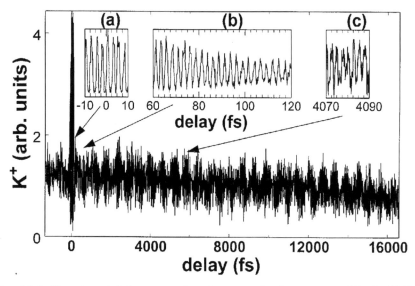

Fig. 11.9. Experimental demonstration of temporal coherent control in the K (3p) fine structure doublet state [11.39]. The excited state population is probed through two-photon ionization. The envelope of the fast oscillations reflects the wave packet oscillations. The insets show expanded views at higher resolution (0.3 fs). (a) Optical interferences when the pulses overlap; (b) transition between optical and quantum interferences when the pulses overlap partially; (c) quantum interferences for larger delay.

"bright" state $|\psi_b(t)\rangle \propto (a_\alpha^{(1)}|e_\alpha\rangle + a_\beta^{(1)}|e_\beta\rangle)$, coupled to the ground state by dipolar interaction, and the "dark" state $|\psi_d(t)\rangle \propto (a_\beta^{(1)}|e_\alpha\rangle + a_\alpha^{(1)}|e_\beta\rangle)$, which is uncoupled to the ground state due to destructive interferences in the probability amplitude. The period of the oscillations is $2\pi/\Delta\omega$. The second wave packet, created in the bright state by the second pulse (with a well-defined phase relatively to the first one) is coherent with the first one and thus can interfere with it. The interference contrast is proportional to the overlap between these two wave packets. The oscillations are thus modulated at the pulsation $\Delta\omega$ of the beating between the bright and dark states.

It is important to note that these three interpretations are only valid in the weak field regime where the interaction is linear with respect to the electric field, i.e., for unsaturated one-photon transitions. We shall see later that for multiphoton transitions [11.39,43] or saturated one-photon transitions [11.44,45], only the last interpretation is relevant.

An example illustrating this technique is shown in Fig. 11.9. The two fine structure states of potassium atoms (K $^2P_{1/2}$, $^2P_{3/2}$) are excited by a one-photon transition. The interference envelope reveals the wave packet oscillation between the bright state and the dark state. This motion corresponds to the precession of the spin and orbital angular momentum around the total angular momentum [11.39]. The insets display enlargements at different delay times, showing the interference period of 2.7 fs, associated to the frequency of the transition.

We turn now to the case of a sequence of two orthogonally polarized pulses, which excite two orthogonally excited states of the system, labeled: $|e_k\rangle$, $k = 1, 2$, with $\langle e_k | e_{k'} \rangle = \delta_{k,k'}$. The total electric field is thus here:

$$E_T = E_1(t)\varepsilon_1 + E_2(t)\varepsilon_2. \tag{11.19}$$

Relations (11.3), (11.4) and (11.5) are still valid, but we now have $\varepsilon_1 \perp \varepsilon_2$. We define the dipole moments: $\mu_{kg} = \langle e_k | qr.\varepsilon_k | g \rangle$. Due to assumed symmetry properties of the system, we have $\mu_{1g} = \mu_{2g}$, and $\langle e_1 | r.\varepsilon_2 | g \rangle = \langle e_2 | r.\varepsilon_1 | g \rangle = 0$ so that $a_2^{(1)} = a_1^{(2)} = 0$. The excited state time evolution can thus be written:

$$|\psi_e(t)\rangle = e^{-i\omega_1 t} a_1^{(1)} |e_1\rangle + e^{-i\omega_2 t} a_2^{(2)}(\tau) |e_2\rangle. \tag{11.20}$$

A coherence is thus excited between the states $|e_1\rangle$ and $|e_2\rangle$. This coherence can be written:

$$\rho_{21}(t) = \langle e_2 | \psi(t) \rangle \langle \psi(t) | e_1 \rangle. \tag{11.21}$$

It oscillates with time at the pulsation $\Delta\omega$ and depends on the time delay τ between the two pulses. For the sake of simplicity, we restrict to the simple case where the two eigenstates are degenerate, so that $\omega_1 = \omega_2 = \omega_{eg}$. Rewriting Eq. (11.20) in the stationary basis

$$|e_+\rangle = \frac{|e_1\rangle + i|e_2\rangle}{\sqrt{2}}, \quad |e_-\rangle = \frac{|e_1\rangle - i|e_2\rangle}{\sqrt{2}}, \tag{11.22}$$

one obtains

$$|\psi_e(t)\rangle = e^{-i\omega_{eg}t} \frac{a_1^{(1)}}{\sqrt{2}} \left\{ (1 - ie^{i\omega_{eg}\tau})|e_+\rangle + (1 + ie^{i\omega_{eg}\tau})|e_-\rangle \right\}, \tag{11.23}$$

where we have (from Eq. (11.7)), $a_2^{(2)} = a_1^{(1)} e^{i\omega_{eg}\tau}$. Note that $\mu_{\pm g} \equiv \langle e_\pm | qr.\varepsilon_\pm | g \rangle = \mu_{kg}$, and $\langle e_\pm | r.\varepsilon_\pm | g \rangle = 0$, where $\varepsilon_\pm \equiv \varepsilon_1 \pm i\varepsilon_2/\sqrt{2}$. The population in the $|e_\pm\rangle$ state, can be simply written in the new basis:

$$n_\pm = \langle e_\pm \psi(t) \rangle \langle \psi(t) | e_\pm \rangle = \frac{1}{2} |1 \mp ie^{i\omega_{eg}\tau}|^2 |a_1^{(1)}|^2. \tag{11.24}$$

Assuming that the two pulses of the sequence are linearly polarized (e.g., $\varepsilon_1 = \varepsilon_x$, $\varepsilon_2 = \varepsilon_y$) and that the photon emission rate with $\sigma^\pm$ polarization is proportional to the $n_\pm$ population, the rate of circular polarization of the detected light is simply given by the expression:

$$P_c = \frac{n_+ - n_-}{n_+ + n_-} = \sin(\omega_{eg}\tau). \tag{11.25}$$

Note that we have also $P_c = (2\mathrm{Im}\rho_{yx})/(n_x + n_y)$, so that the rate of circular polarization is also a measure of the quantum coherence ρ_{yx} between the $|e_x\rangle$ and the $|e_y\rangle$ states. The use of a (σ^x, σ^y) optical sequence allows us to excite this coherence. P_c oscillates with the time delay τ between the two pulses with the period $2\pi/\omega_{eg}$ and the two pulses must be in phase quadrature to obtain circular light. This description is analogous to the quantum beat description (see section 11.3.1.1(a)) for a copolarized optical pulse sequence.

It is indeed also possible, due to the system linear response to the electric field, to give a spectral interpretation of the experiment with cross-polarized pulses, which is the counterpart of section 11.3.1.1(b). The global effect of the total field on the system is evaluated as

$$|\psi_e(t)\rangle = \left(\sum_{k=1,2} \frac{1}{i\hbar} \mu_{kg}\varepsilon_k|e_k\rangle \right) \tilde{E}_T(\omega)e^{-i\omega_{eg}t}, \qquad (11.26)$$

where

$$\tilde{E}_T(\omega_{eg}) = \tilde{E}_1(\omega_{eg})(\varepsilon_1 + e^{i\omega_{eg}\tau}\varepsilon_2). \qquad (11.27)$$

This equation shows that the system absorbs its eigenfrequencies among the frequencies of the total electric field, with the ellipticity determined by the time delay τ. Here, as illustrated in Fig. 11.8, the intensity spectrum of the laser field is independent of the frequency, while the circular polarization degree oscillates with the time delay τ at the pulsation ω_{eg}. Defining the circular components of the electric field by $E_T \equiv E_+\varepsilon_+ + E_-\varepsilon_-$, we obtain:

$$P_c(\omega_{eg}) = \frac{|E_+|^2 - |E_-|^2}{|E_+|^2 + |E_-|^2} = \sin(\omega_{eg}\tau). \qquad (11.28)$$

Expression (11.28) coincides with (11.26), which shows that the system polarization is the same as the circular polarization of the excitation field at its eigenfrequency ω_{eg} (in this description, coherence relaxation processes are neglected).

An experimental illustration of the use of a cross-polarization sequence is given in section 11.3.2.2 for semiconductor quantum well excitons, showing the true coherent superposition achieved in the quantum system by the excitation sequence.

Observation of these wave packet interferences requires an interferometric stability of the delay lines. For measuring only the wave packet motion, it is sufficient to measure the interference envelope. Two schemes have been used: either the delay oscillates at a chosen frequency (with a piezo plate or other mechanical device) and the envelope is measured with a lock-in amplifier [11.46], or the delay fluctuates randomly and the amplitude of the noise is directly related to the envelope [11.47].

11.3.1.2 Temporal Coherent Control in the Nonlinear Regime. In the perturbative regime, the linearity of the interaction with respect to the

exciting field validates the various interpretations given in the previous section. As soon as the interaction is nonlinear, only the temporal interpretation remains valid.

We first consider the case of a one-photon transition in the saturated regime [11.44]. A wave packet is created by the first laser pulse. This wave packet evolves freely until the second pulse reaches the sample. Two possibilities can occur. Either the wave packet reaches a state where it cannot interact with the second laser pulse, which creates then a second wave packet in a state orthogonal to the first wave packet. Both wave packets add incoherently. If the wave packet is in a state coupled to the ground state, then it can be pumped down to the ground state or amplified through pumping of new population from the ground state. This depends on the evolution of the relative phase (at the transition frequency) between the ground state and the excited state. An example is depicted in Fig 11.10 on the same transition as studied in the low field regime. Saturation induces the disappearance of the wave packet interferences when a complete population transfer is induced by the first pulse. A slow oscillation corresponding to the envelope of the previous beats remains. This corresponds to the oscillation between the bright state (from which the wave packet can be pumped down to the ground state) and the dark state, which is inactive when the second pulse arrives. The saturation induces a coupling between the two excitation lines (toward the two fine structure states). It allows then to discriminate between the mixture of two independent two-level systems and a single "V" system (a common initial state for two excitation transitions).

In the case of multiphoton transitions, the spectral interpretation fails; however the other two interpretations are still valid. Besides the wave packet interferences, the quantum beats picture is still appropriate. The interference period corresponds to the frequency of the transition, twice the optical frequency for a two-photon transition. When the pump-probe delay is short enough so that the two pulses overlap, optical interferences (at the optical period) dominate the quantum interferences. The smooth transition between these two regimes can be observed when the delay increases gradually [11.43]. This case is exactly equivalent to replacing the two-photon transition by a frequency doubling crystal followed by a one-photon transition [11.48].

11.3.2 Temporal Coherent Control in Solid State Physics

We describe a few examples typical of the use of temporal coherent control in solid-state physics.

11.3.2.1 Coherent Control of Exciton Population in Semiconductor Quantum Wells. In solids, and more particularly in semiconductors, the optical dephasing times are usually much shorter (in the picosecond domain typically) than in atomic systems. The temporal coherent control experiments

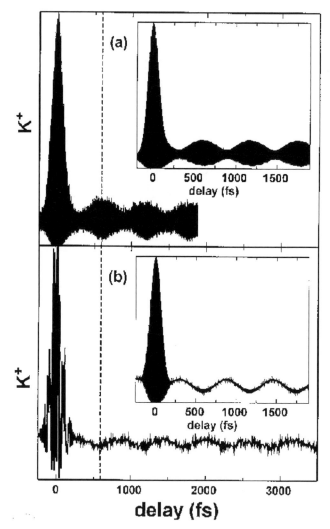

Fig. 11.10. Comparison between wave packet interferences in the (a) perturbative and (b) saturated regime [11.44]. When the field is strong enough so that the first laser pulse excites totally the system, the interferences disappear. The remaining modulation reflects the wave packet oscillation. When the wave packet is back to the bright state, it can be pumped down by the second laser pulse. This reduces the ionization probability (insets: model of the ion signal in the two regimes)

in solids thus require ultrafast laser sources. For instance, Heberle et al. have used 100-fs laser pulses to demonstrate the coherent destruction of excitons (electron-hole pair bound by Coulomb interaction) in GaAs/AlGaAs semiconductor quantum wells (for a description of these structures, see Chap. 5 [11.49]. In these experiments, summarized below, a sequence of two phase-

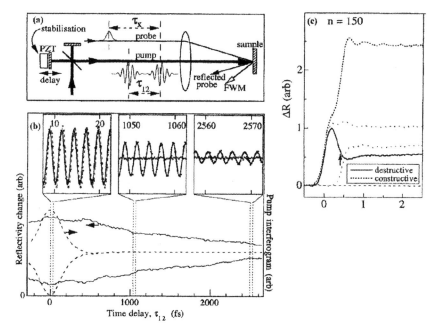

Fig. 11.11. Pump-probe coherent control of excitons in semiconductor quantum wells [11.49]: (a) Experimental setup; (b) maximum and minimum pump-induced reflectivity changes for a probe delay $\tau_x = 10$ ps (solid) and pump-pulse interferogram (dashed) versus the time delay τ_{12} between the two pump pulses (insets: expanded interference oscillations); (c) time-resolved enhancement (dashed, $\tau_{12} = nT_{HH}$) and destruction (solid, $\tau_{12} = (n + 0.5)T_{HH}$) of the pump-induced reflection change. The dotted lines record the effects of each pump pulse alone

locked pulses "creates" excitons within the quantum well and then "removes" them by destructive interference. In other words, the first pulse induces in the material a coherent exciton polarization that can interfere constructively or destructively with the electric field of the second pulse, as long as the time delay between the two pulses is shorter than the optical dephasing time T_2. Fig. 11.11(a) depicts the experimental setup used by Heberle et al. An actively stabilized Michelson interferometer produces two phase-locked pump pulses of temporal separation τ_{12} with a remarkable phase stability of $\lambda/100$. The Ti:Sa laser pulse energy is resonant with the heavy hole exciton absorption energy E_{HH} in the quantum well. The coherent manipulation of the exciton polarization is analyzed by a probe pulse (delayed by a time τ_x) through the pump-induced reflectivity change (ΔR), which measures directly the exciton population photogenerated by the pump pulses.

Fig. 11.11(b) displays the maximum and minimum pump-induced reflectivity changes at a probe delay $\tau_x = 10$ ps as a function of the delay τ_{12} between the two pump pulses. The insets display the expanded interference

oscillations with sinusoidal fits. These results show that the exciton population oscillates with a period $T_{HH} = h/E_{HH}$, as a result of the interference
between the second pump pulse and the coherent exciton polarization sur-
viving from the first pump pulse. The time-resolved evolution of the exciton
population is displayed in Fig. 11.11(c) for the fixed delays $\tau_{12} = 402$ and
403.34 fs, equivalent to 150 and 150.5 exciton period T_{HH}. The dotted curves
show the effect of each pump pulse alone. The enhancement (dashed line) or
destruction (solid line) of the exciton polarization is clearly observed. In particular, if the second pulse arrives $150T_{HH}$ later (dashed line), more excitons
are generated than from the sum of the two pump pulses individually.

The coherent manipulation of excitons in semiconductor quantum wells
can also be probed through the resonant secondary emission (SE) [11.50]. This
SE signal corresponds to resonant Rayleigh scattering due to static disorder in
the sample [11.51]. As in atomic systems [11.43], the coherent control of two-
photon transitions has also been demonstrated for the 2p-heavy hole exciton
transition in GaAs/AlGaAs quantum wells [11.52]. In that case, the temporal
evolution of the amplitude of coherent control, monitored in the SE, shows
two different contributions: one at the frequency of the exciting optical field
and a second at twice that frequency, from the interference of the excitonic
polarization.

11.3.2.2 Coherent Control of the Optical Orientation of Excitons in Semiconductor Quantum Well.

The exciton spin can also be coher-
ently controlled in semiconductor quantum wells. This coherent manipulation
can be monitored in either pump-probe reflectivity [11.53] or SE experiments
[11.54].

Fig. 11.12(a) presents the exciton SE dynamics following the excitation of
two picosecond pulses with orthogonal polarization (σ^X, σ^Y), and a time separation $t_1 = 4$ ps such that the second pulse is in phase quadrature with respect
to the first one. The excitation with the second laser pulse results in a sharp
rise of the circular polarization of the exciton SE. This circular polarization
originates from the interaction of the second pulse with the coherent excitonic
polarization created in the crystal by the first pulse. The SE circular polarization degree P_c measured after the second pulse exhibits clear oscillations
as a function of the temporal separation between the two excitation pulses
(inset in Fig. 11.12(a)), with the period T_{HH}. These oscillations reflect the
rotation of the orientation of the exciton average spin $\boldsymbol{S}$ in a plane containing
a direction normal to the quantum well, P_c being simply proportional to the
$\boldsymbol{S}$ projection on this direction. The amplitude of the oscillations is directly
proportional to the fraction of linear excitons created by the first pulse that
still oscillates in phase with their photogenerating optical field when the second pulse excites the sample. The decay of the oscillation's visibility (Fig.
11.12(b)) thus probes the decay of the coherent exciton polarization created
by the first pulse. It gives some estimate of the exciton optical dephasing time
T_2 (here about 6 ps). As can be seen in Fig. 11.12, the circular polarization

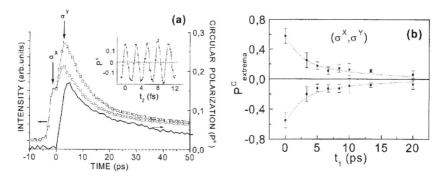

Fig. 11.12. Coherent control of the exciton circular polarization in semiconductor quantum wells [11.54]. The first (second) pulse is σ^X (σ^Y) polarized, the time separation is $t_1 = 4$ ps: (a) The time evolution of the circularly polarized I^+ ( ), and I^- (O) components of the emission and the circular polarization P^c (full line). Inset: The circular polarization P^c measured 4 ps after the second excitation pulse as a function of the fine temporal tuning t_2 between the two excitation pulses, separated by the time delay $t_1 + t_2$. (b) The maxima and minima of the linear polarization oscillations as a function of t_1 (the dotted line is a guide for the eyes)

decay time (about 35 ps here) is much longer than the optical dephasing time, which supports the fact that a true coherent superposition of linear exciton quantum states is achieved after the pulse sequence.

It has been shown similarly that the exciton alignment can be controlled using a sequence of cross-circularly polarized pulses (σ^+, σ^-) [11.54]. In that case, the orientation of the polarization field of the photogenerated linear excitons (coherent superposition of circular excitons states) in the quantum well plane is determined by the pulse time separation.

Finally, temporal coherent control has also been achieved in the context of strong coupling of excitons and photons in semiconductor microcavities, where exciton-polariton population and spin manipulations have been demonstrated [11.55].

11.3.2.3 Coherent Control of Exciton Charge Oscillation in a Semiconductor Quantum Well.

Coherent control experiments have also been used to enhance, weaken and phase-shift Terahertz (THz) radiation emitted by optically excited quantum beats in a semiconductor coupled quantum well (CQW). Planken et al. have nicely demonstrated this type of coherent manipulation in a sample consisting of 10 pairs of 14.5-nm GaAs wide well (WW) with a 10-nm GaAs narrow well (NW) separated by 2.5-nm AlGaAs barrier (see Fig. 11.13(a)) [11.56]. By adjusting the applied electric field in the CQW, the electron levels in the NW and the WW can be aligned. The new excitonic eigenstates of the coupled system consist of a bonding state and an antibonding states, labeled $|1\rangle$ and $|2\rangle$, respectively, with an energy separation measured to be 6 meV. The phase-locked pulses, tuned to the NW interband

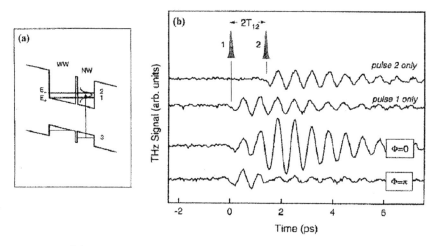

Fig. 11.13. (a) Energy diagram of the coupled quantum well consisting of a wide well (WW) and a narrow well (NW); 1 and 2 are the bonding and antibonding electron states of the coupled system and 3 is the localized hole state. (b) Measured THz waveforms. The upper two curves are generated by each pulse individually. The lower two are generated by the phase-locked pulse pair for a $\Phi = 0$ and $\Phi = \pi$ relative dephasing [11.56]

transition, are focused onto the quantum well sample. The bandwidth of the pulsed excitation laser (80 fs Ti:Sa pulses) is larger than the energy splitting between the bonding and antibonding exciton eigenstates of the CQW, resulting in a coherent excitation of both transitions. As the dipole matrix element between states $|1\rangle$ and $|2\rangle$ is nonzero, the beating between the two exciton states results in emission in the THz frequency range. The generated THz signal is detected with a photoconductive dipole antenna. The measured THz waveforms for different values of the phase difference between the two pulses are plotted in Fig. 11.13(b). The upper two traces are the individual THz waveforms generated by the second and first optical pulses, respectively. The other traces result when both pulses excite the sample. For a phase difference of $\phi = 0$ the THz radiation is clearly enhanced, while for a phase difference of $\phi = \pi$ the THz radiation is strongly weakened. These coherent manipulations result from the interference of the excitonic wave functions excited by the two phase-locked pulses.

11.3.2.4 Coherent Control of Electron-Phonon Scattering in Semiconductor. We have presented coherent manipulations of electron-photon interactions in different systems (semiconductors, polymers, metals). In these temporal coherent control applications, the interference between the coherent polarizations induced by two phase-locked excitation pulses can reverse or enhance the uncompleted absorption process, as long as the phase of the

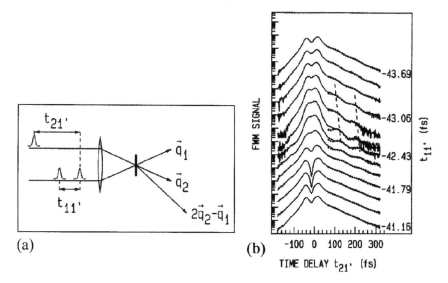

Fig. 11.14. (a) Schematic of the four wave mixing coherent control experiment of electron-phonon scattering in bulk GaAs (see text); (b) four-wave mixing signal from GaAs as a function of the time delay $\tau_{11'}$ in steps of 0.21 fs ($T = 77$ K) [11.57,58]

associated polarization has not been destroyed. Since the scattering processes among elementary excitations in condensed matter (for instance, electron-phonon) are not instantaneous in time, for short times they have not yet become irreversible. It is thus possible in principle to reverse or enhance the interaction process after it has already started. Wehner et al. demonstrated that such a coherent control of scattering processes is feasible [11.57,58]. These authors considered electron-LO phonon scattering in bulk GaAs. The system consists in electrons in conduction and valence bands coupled to the lattice vibration (considered as a thermal bath), which leads to oscillations of the polarization with a frequency close to the LO phonon one ($\approx \nu_{LO}$).

Fig. 11.14(a) depicts the coherent control experimental setup. It consists in two phase-locked 15-fs pulses 1 and 1′ separated by a time delay $t_{11'}$, propagating in the direction $\mathbf{q}_1$. They interact with another pulse propagating along $\mathbf{q}_2$ with time delay $t_{21'}$ in a GaAs sample ($T = 77$ K). The intensity of the four wave mixing signal in the direction $2\mathbf{q}_1 - \mathbf{q}_2$ is plotted in Fig. 11.14(b). We observe that the phonon oscillations seen in the time delay domain $t_{21'}$ can almost disappear around certain values (-43.69 or -41.16 fs, for instance) while pronounced oscillations occur at low absolute signal levels (-42.43 fs). These results demonstrate the coherent control of the phonon oscillations visibility and more generally the fact that an interaction process can be manipulated even after it has already started. Coherent optical control of acoustic phonon oscillations in InGaN semiconductor quantum wells has also been achieved

[11.59,60] and complete cancellation of generated acoustic phonons was, in particular, clearly demonstrated.

11.3.2.5 Coherent Control in Conjugated Polymers. The coherent control of optical emission from a conjugated polymer PPV (Poly-p-Phenylene Vinylene) has been reported [11.61]. These polymers are organic systems with extended electronic states, which are very promising for "plastic" electronic devices. These systems exhibit a well-defined energy gap analogous to that in inorganic semiconductors. However the stronger spatial localization of electronic and vibrational states on the polymer chain yields a greater coupling between electronic and vibrational properties. This leads to rapid loss of phase coherence (in the subpicosecond range typically) of the quantum states generated by optical excitation. Fig. 11.15(a) displays the time-integrated optical emission spectra of PPV for four different energies of excitation (marked by arrows). The excitation pulse ($\sim$ 100 fs) results from frequency mixing between Ti:sapphire mode-locked laser pulses and pulses from a synchronously pumped optical parametric oscillator. For the excitation energy (A), we observe a great enhancement of the resonant fluorescence by more than a factor of 100 as the excitation and detection energies coincide. This effect corresponds to the resonant enhancement of Rayleigh scattering. The results of temporal coherent control experiments in these conditions are displayed in Fig. 11.15(b). The PPV sample is excited by two replicas of the laser pulses, with the second pulse delayed interferometrically with an accuracy of 50 attoseconds ($\lambda/40$). Fig. 11.15(b) displays two interesting features:

1. The density plot shows that the interference persists beyond the time for which the pulses overlap. This demonstrates the continuance of a coherent material polarization after the excitation by optical pulses.
2. The far-field radiation pattern exhibits speckles which can be viewed semiclassically as the results of emission from spatially distributed coherent dipoles. In the case of the conjugated polymer PPV, the short laser pulse excites excitons distributed across the 20 meV pulse bandwidth, which therefore produces a variation of the emission phase. However, the emission phase in a given direction is still largely geometrically controlled, yielding the observation of speckles.

11.3.2.6 Coherent Control of Electrons Dynamics in Metals. Phase relaxation times in metals are usually very short due to ultrafast charge carrier scattering. Line width analysis of photoemission spectra of bulk bands in metals yields dephasing times shorter than 10 fs. Longer dephasing times are expected near the Fermi surface of metals and in the surface states that are weakly coupled to the bulk bands. In particular time-resolved photoemission studies of Cu(111) demonstrated that the dephasing times of electrons at metal surfaces can exceed 20 fs [11.62]. The control of the electron distribu-

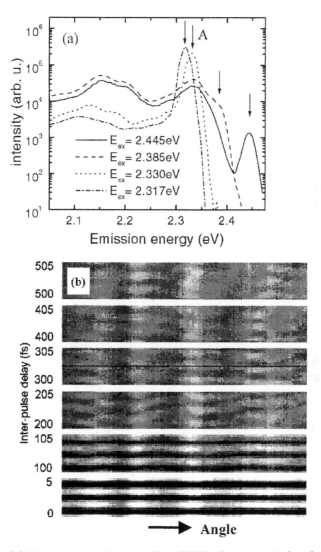

Fig. 11.15. (a) Time-integrated spectra from PPV polymer excited at four different photon energies (marked by arrows); (b) coherent control of speckle, angle-resolved PPV polymer intensity variation as a function of interpulse delay [11.61]

tion excited in the metal through the optical phase of the excitation is thus possible, as demonstrated by Petek et al. [11.63].

Fig. 11.16 displays the experimental setup [11.63]. The second harmonic light (3.1 eV) of a Ti:sapphire laser with 15 fs pulses excites two-photon photoemission (2PP) from the crystal surface of Cu(111) under ultrahigh vacuum ($< 10^{-10}$ Torr). Photoemission is measured for electrons with specific

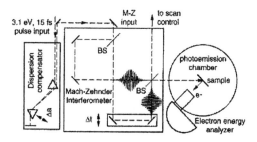

Fig. 11.16. Experimental setup for coherent control of two photon photoemission (see text) [11.63]

energy and momentum by an electron energy analyser. A feedback controlled Mach-Zender interferometer generates phase-locked pump-probe pulse pairs with a delay accuracy of $\lambda/50$. The band structure of Cu(111) and the excitation scheme for the 2PP spectra are shown in the inset of Fig. 11.17(a). The $L_{2'} - L_1$ band gap, which extends from -0.85 to 4.3 eV (for $k_{//} = 0$), supports two surface states; the occupied Shockley-type surface state (SS) at -0.39 eV and an image potential state series starting with $n = 1$ state (IP) at 4.1 eV. The 2PP photoemission signal as a function of the delay between the two excitation pulses is displayed in Fig. 11.17(a). The oscillations arise from single- and two-photon coherences induced in the sample; the decay is consistent with dephasing due to hole scattering on a 20-fs time scale. Petek et al. demonstrated that the corresponding photoemission spectra depend not only on the frequency, as in conventional spectroscopy, but also on the phase of the excitation light. Fig. 11.17(b) shows that the intensity and energy distribution in the 2PP spectra can be manipulated through the optical phase of the excitation light.

11.4 Coherent Control with Shaped Laser Pulses

The ability to change the temporal envelope (shape) of ultrashort laser pulses has opened a wide range of possibilities in controlling the dynamics of quantum systems. Pulse shapes of various complexities can be generated. Simple linearly chirp pulses depend on a single parameter. More complex shapes depending on hundreds or thousands of parameters can also be generated using passive spectral filters. We present in this section results obtained with these two shaping techniques.

In simple systems, an analytical solution of the interaction between the laser pulse and the quantum system allows a direct determination of the required shape. For more complex systems, a feedback loop, together with an optimization or learning algorithm are used to reach the desired goal. This scheme is often called "closed-loop optimal control" [11.64,65] whereas the first scheme can be named "open-loop coherent control." We first describe

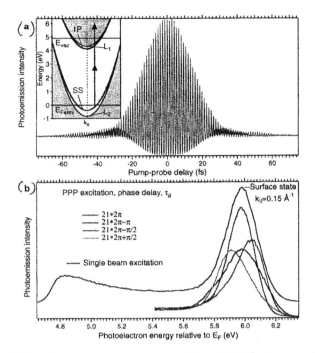

Fig. 11.17. (a) Interferometric two-pulse correlation (12PC) of photoemission measured at the SS resonance in Cu(111), inset: excitation scheme of the SS photoemission; (b) demonstration of coherent control of two-photon photoemission (see text) [11.63]

the shaping techniques and then give examples of applications of these techniques.

11.4.1 Generation of Chirped or Shaped Laser Pulses

Propagation of a Fourier-transform-limited ultrashort pulse in a dispersive medium leads to lengthening of its duration. This property is often a serious limitation to the manipulation of ultrashort pulses. Besides, it can be used to generate simply shaped pulses. The dispersion is characterized by a variation of the index of refraction with wavelength, $n(\lambda)$ or pulsation $n(\omega)$. This leads to a nonlinear variation of the spectral phase

$$\phi(\omega) = \phi_0 + \phi'(\omega - \omega_0) + \frac{1}{2}\phi''(\omega - \omega_0)^2 + \frac{1}{3}\phi'''(\omega - \omega_0)^3 + \dots. \quad (11.29)$$

The first-order term $\phi' = \partial\phi/\partial\omega(\omega_0)$ corresponds to a delay of the pulse envelope and has no effect on the shape. The second-order term $\phi'' = \partial^2\phi/\partial\omega^2(\omega_0)$ produces a linear variation of the instantaneous frequency within the temporal envelope of the pulse. In most cases this contribution (named "chirp") is

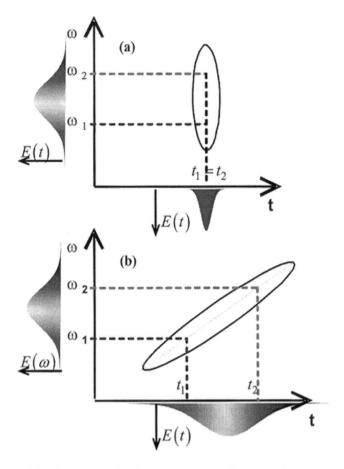

Fig. 11.18. (a) Wigner plot of a Fourier transform limited pulse and (b) the corresponding pulse after chirping. The spectrum remains unaffected, but the frequencies are staggered; the pulse is also stretched

dominant. Higher-order terms correspond to nonlinearities in this frequency variation. For Gaussian laser pulses and second-order term only in Eq. (11.29), one gets $\omega(t) = \omega_0 + 2\alpha t$ with $\alpha = 2\phi''/[\tau_0^4 + 4\phi''^2]$ where τ_0 is the transform-limited pulse duration. As illustrated in Fig. (11.18), the chirp does not modify the spectral content of the pulses. The various frequency components have the same weight. Only their relative phase is affected. The pulse duration can, however, be strongly lengthened to:

$$\tau_c = \tau_0 \sqrt{1 + \frac{4\phi''^2}{\tau_0^4}}. \tag{11.30}$$

For large chirp rates ($\phi'' \gg \tau_0^2$) the pulse duration is approximately $\tau_c \simeq 2\phi''/\tau_0$ and $\alpha \simeq 1/2\phi''$. Chirp laser pulses can be used in at least two different ways. The pulse lengthening described by Eq. (11.30) implies a reduction of the peak intensity, while keeping the same pulse energy. In applications where the spectral content of the pulse is not critical, such as high intensity experiments with complex systems or nonresonant experiments, the chirp can be used to study peak intensity effects. In other cases, the frequency variation is critical and can be used for adiabatic transfer or interference effects when a resonance frequency lies within the spectral range.

Since the typical pulse duration lies in the femtosecond to picosecond range, direct shaping of the pulse in the time domain is not possible. The complex temporal shaping is then realized in the Fourier domain by modulating the phase and amplitude of each frequency component of the broadband laser pulse [11.66]. This is achieved by applying a passive spectral filter. This filter can also affect the polarization of each spectral component [11.67]. The spectral width $\Delta\omega$ of the input pulse gives the smallest temporal resolution $\delta t \simeq 1/\Delta\omega$, which can be achieved for the shaped pulse. Its duration can be stretched up to $\Delta t \simeq 1/\delta\omega$ where $\delta\omega$ is the spectral resolution of the shaping device. The most common devices for both high fidelity and wide flexibility of shapes involve an optical Fourier plane where a liquid crystal spatial light modulator (SLM) [11.66], acousto-optic modulator (AOM) [11.68] or deformable mirror [11.69] modulates the spectral phase and/or amplitude. Shaping without Fourier transform optics is also possible by using an actively controlled acousto-optic programmable filter (DAZZLER) [11.70]. In this case the resolution is mainly limited by the laser bandwidth and the length of the crystal.

Pixellated SLM offers the highest number of independent parameters. Up to 128 pixels for phase and amplitude devices or 640 pixels for phase only (or amplitude only) [11.71] shaping are currently available. The modulator is placed in the Fourier plane of a 4f-nondispersive setup (Fig. (11.19)). The 4f line must have low geometrical aberrations and the diffraction limited focal spot associated to each frequency component should be roughly adjusted to the pixel size [11.66]. Larger focal spots reduce the number of independent parameters. Smaller focal spots induce undesired replica of the shaped pulse. This finite spot size smooths the phase and amplitude variations that can be applied between adjacent pixels. Moreover, a phase step results also in a reduction of the spectral amplitude [11.98].

Characterization of the pulse shape depends on its complexity. For simple shapes, self-referenced methods can be used, such as frequency resolved optical shaping (FROG) [11.72] or spectral phase interferometry for direct electric-field reconstruction (SPIDER) [11.73] (see chap. 7). More complex shapes can be analyzed by comparison with a reference pulse with spectral interferometry [11.74]. Experimentally, some artefacts can be introduced as spatio-temporal shaping side effects [11.75]. This spatio-temporal shaping (2D shaping [11.76])

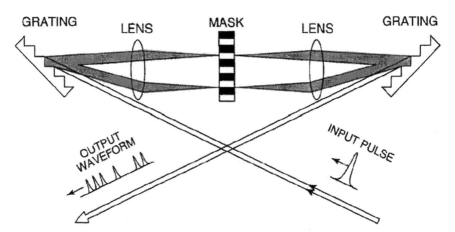

Fig. 11.19. Pulse shaper based on a spatial light modulator (MASK) placed in a 4f line

can be used to control crystal lattice vibrations, THz radiation or polaritons dynamics [11.77–79].

11.4.2 Coherent Control with Chirped Laser Pulses

A large fraction of experiments carried on molecules involve the creation of a vibrational wave packet. The typical time scale of the wave packet evolution is comparable to or larger than the pulse duration. Chirped pulses have been used, for instance, to prepare a wave packet with a precompensation of the spreading result of the dispersion of matter wave. This wave packet can therefore be focused at a desired position, as first demonstrated in iodine [11.80]. In other examples, the chirp of the pulse is used, in the low field regime, to compensate the variation of the absorption frequency resulting from the evolution of the wave packet. This can be used either to create a ground state vibrational wave packet in a Raman scheme [11.81] or to perform a stepwise excitation as in Na_2 [11.82]. For larger intensities, adiabatic transfer following the dressed states can be achieved. Examples have been demonstrated in I_2 [11.83] and Rb [11.84].

In the liquid phase, strong chirp effects have also been observed [11.85–87]. Negative chirps can again produce vibrational wave packets in the ground electronic state through a Raman cycle. Positive chirps have been shown to enhance the population of the excited electronic state. A strong sensitivity on the chirp rate to the chemical environment has been observed [11.88].

Subtle interference effects can be achieved with chirped pulses. In ladder climbing, a series of nearly equidistant eigenstates are excited with an ultra-short pulse. These states can belong, for instance, to a vibrational ladder. The spectral width is broad enough so that direct multiphoton transition as well as sequential transitions can be excited. Due to the chirp, the steps are climbed

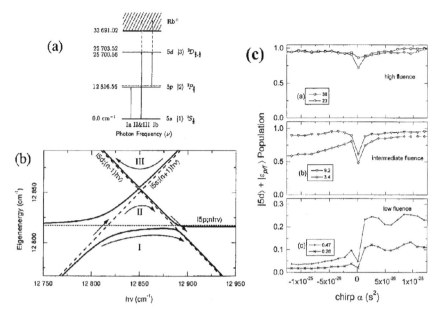

Fig. 11.20. Ladder climbing in Rb atoms [11.84,89,90]: (a) Levels involved (ground state 5s, excited states 5p, $^2P_{3/2}$ and 5d); (b) dressed state picture. With positive chirp, the system can follow either the sequential path (I), or the direct two-photon excitation (II). With negative chirp, only the direct two-photon excitation (III) is possible. (c) 5d (+ ionization continuum) population as a function of chirp (α and ϕ'') for various laser fluences as indicated (in mJ/cm^2)

at different times, so that the direct and sequential paths encounter a phase difference proportional to the chirp rate. A first demonstration was achieved in Rb atoms and is illustrated in Fig. (11.20) [11.84,89,90]. Figure 11.20(a) presents the three bare states involved, whereas Fig. 11.20(b) presents the same states in the dressed state picture, as a function of laser frequency. The intermediate state is taken as energy reference. Three excitation paths can be distinguished. Path I corresponds to a sequential excitation of the ladder. This can be achieved only with positive chirp (increasing laser frequency with time). Paths II and III are direct two-photon transitions excited with positive or negative chirp respectively. Figure 11.20(c) displays the result obtained for various laser fluences. At low fluence, a strong asymmetry between positive and negative chirps is observed. For negative chirp, only the direct two-photon path (III) is excited. Its probability is low and decreases when $|\phi''|$ increases. For positive chirp, the direct two-photon path (now II) is present, with the same behavior (of the probability amplitude) with $|\phi''|$ as for negative chirp. In addition, the sequential path (I) is allowed. This is the dominant contribution. The corresponding excitation probability increases strongly between slightly negative ($\phi'' \simeq -\tau_0^2$) and slightly positive chirp ($\phi'' \simeq \tau_0^2$). It reaches a constant value (as a function of chirp rate ϕ'') as soon as the frequencies of the two sequential transitions are separate in time ($\phi'' \geq \tau_0^2$). Since paths I and II connect the same final state from a given initial state, interferences are possible. The interference phase is equal to the product of the enclosed area between the two paths with the chirp rate. Indeed, the corresponding oscillations as a function of chirp can be seen for low fluence. At higher fluence, one sees a progressive transition toward adiabatic transfer (for nonzero chirps). The three paths have a probability high enough to result in a total transfer. However, the wave packet follows the adiabatic states (solid lines on Fig. 11.20(b)) corresponding to path I for positive chirp and path III for negative chirp, with almost equal probability. Interferences have therefore disappeared. A minimum remains for unchirped pulses. This is due to Rabi oscillations, which should result in an average (over the beam transverse profile) probability of 0.5. However, the leak toward the ionization continuum increases this value, and this minimum disappears at the highest fluences.

A second example of interferences in quantum ladder climbing has been observed in Na atoms [11.91]. This atom presents two major differences with the textbook case of Rb. The two fine structure states of the intermediate state can be simultaneously excited. Therefore two sequential paths can be excited, and three interference patterns associated to the three possible pairs of paths are present. Besides, the balance between dipole moments provides a high interference contrast of 0.95. Figure 11.21 presents the level scheme (bare state Fig. 11.21(a), dressed states Fig. 11.21(b)) and the resulting excitation probability as a function of chirp rate (Fig. 11.21(c)). Again, the excitation probability is very small for positive chirp for which only a direct excitation is possible (path IV). For negative chirp, one sees a strong increase of the population, followed by rapid and slow oscillations. The rapid oscillations result from the interference between both sequential paths and the direct process. Their slight frequency difference produces beats at the same frequency as the slow oscillation. Moreover, the amplitude of oscillation decreases with $|\phi''|$ because the direct path probability decreases when the range of frequencies simultaneously present decreases. The slow oscillation corresponds to interferences between both sequential paths (I and II). Its contrast is about 0.75, independent of chirp.

In all the previous examples, the effect of the chirp pulse was observed after the end of the pulse. In time-resolved experiments, coherent transients (CT) result from beats between the atomic dipole excited during the passage through resonance and the instantaneous laser frequency [11.92] (see in Fig. 11.22 an example in Rb). The excited state probability amplitude at time t is given by

$$a_e(t) \propto \int_{-\infty}^{t} e^{-(t'/\tau_c)2} e^{-i\frac{(t'-t_0)^2+t_0^2}{2\phi''}} dt'. \tag{11.31}$$

The pulse duration is broadened to τ_c and the quadratic phase factor $[(t' - t_0)^2 + t_0^2]/2\phi''$ in Eq. (11.31) (where t_0 is the time of passage through reso-

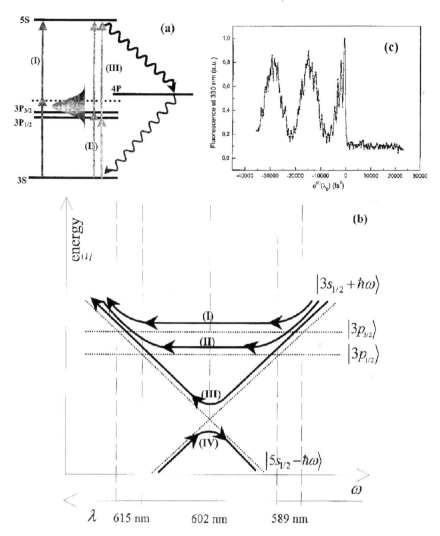

Fig. 11.21. Multiple interferences between sequential and direct two-photon transitions in quantum ladder climbing in Na [11.91]: (a) Levels involved, excitation and detection scheme; (b) dressed state picture with the relevant quantum paths; sequential paths for negative chirp ((I) and (II)), direct path for negative (III) or positive (IV) chirp; (c) experimental result of the excited state (5s) population recorded through the (4p → 3s) fluorescence as a function of chirp

nance) reflects the two main features observed. The transferred population is negligible before resonance. During the passage of the instantaneous frequency through resonance, the population increases (on a time interval of the order of $2\sqrt{2|\phi''|}$, when the phase factor remains less than 1) up to a maximum approximately 1.3 times the asymptotic value. The width of this transition

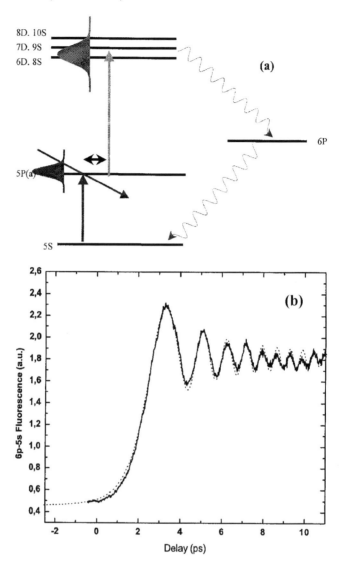

Fig. 11.22. Coherent transients observed in the excitation of the Rb(5s-5p$_{1/2}$) transition with a negatively chirped pulse (τ_c = 20 ps, λ = 795 nm) [11.87]; (a) Levels scheme: The 4p state is probed with an ultrashort transform limited pulse (30 fs) toward the 8s–10s, 6d–8d Rydberg states. Fluorescence from the 6p state is detected. (b) 6p fluorescence as a function of pump-probe delay time. Experiment (solid line). Theory (dots)

interval is the smallest temporal interval associated to a spectral perturbation (here the atomic resonance behaves as a Dirac function). We shall see in the next section how pulse shaping allows us to manipulate these coherent transients.

11.4.3 Coherent Control with Shaped Laser Pulses

11.4.3.1 Open-Loop Schemes. In simple systems, where a theoretical analysis of the interaction is available, it is possible to predict the effect of a given laser shape on the time evolution of the system. Several demonstrations have been achieved in atomic systems. The intrinsic nonlinearity of multiphoton transitions results in a strong sensitivity to shape effects.

The probability amplitude of a direct two-photon transition (with a nonresonant intermediate state $|i\rangle$) is given by

$$a_e = \frac{i}{2\pi} \int_{-\infty}^{+\infty} \tilde{E}(\omega_{eg} - \omega)\tilde{E}(\omega)\mathrm{P}\left(\frac{1}{\omega_{ig} - \omega}\right) d\omega, \qquad (11.32)$$

where $P(1/x)$ is the principal value of Cauchy. All the pairs of photons having a total energy equal to the two-photon transition contribute equally. If the electric field is real for all frequencies, all contributions to the integral in Eq. (11.32) add in phase. This produces the highest population transfer and corresponds to a Fourier transform limited pulse as predicted with an intuitive picture. Applying a phase step π of to the electric field changes its sign in a range of frequencies. Some contributions to Eq. (11.32) add now with opposite phase. The excitation probability decreases. A beautiful demonstration has been achieved in Cs atoms [11.93,94] and is displayed in Fig. 11.23. The position of the π-step is scanned over the spectral profile of the pulse. At both ends, the phase step does not affect the transition probability. When the step wavelength approaches the central wavelength, the increasing weight of frequency pairs having encountered a change of sign decreases the overall probability down to an almost negligible value. When the step position is exactly at the central wavelength, all the frequency pairs have changed their sign so that the probability is again at its maximum.

When an intermediate state is close to resonance, the two-photon transition probability has a second contribution corresponding to a sequential excitation involving the electric field amplitudes at the exact one-photon frequencies ω_{ei} and ω_{ig}. These two kinds of contributions have been encountered in the chirped pulse excitation of a ladder system (previous paragraph, Fig. 11.21 and 11.22). The excited state amplitude is now given by

$$a_e = \frac{1}{2}\tilde{E}(\omega_{fi}\tilde{E}(\omega_{ig}) + \frac{i}{2\pi} \int_{-\infty}^{+\infty} \tilde{E}(\omega_{eg} - \omega)\tilde{E}(\omega)\mathrm{P}\left(\frac{1}{\omega_{ig} - \omega}\right) d\omega. \quad (11.33)$$

These two contributions are in phase quadrature for a real electric field. As expected for a harmonically driven system, the first term (the on-resonance contribution) is shifted by $\pi/2$ compared with the second term (the off-resonance

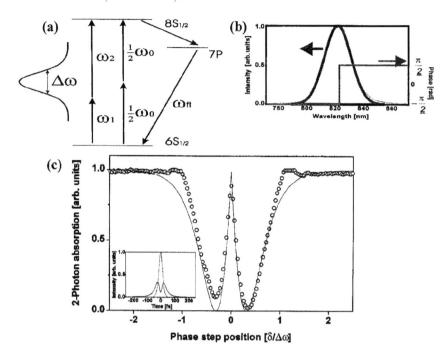

Fig. 11.23. Effect of a π-step in the spectral phase on a nonresonant two-photon transition in Cs [11.94]: (a) Levels involved (ground state 6s, excited states 8s, fluorescence observed through the 8s → 7p → 6s cascade); (b) spectral shape of the pulse and phase step applied; (c) excitation probability as a function of the position of the phase π-step

contributions). Also, the spectral components below and above the resonance excite the system in phase and π out of phase, respectively. This means that red-detuned photons and blue-detuned photons with respect to the resonance interfere destructively. These phase relations have been used to enhance the nonlinear response of Rb atoms [11.95]. Applying an adequate filter to the spectral phase can result in in-phase contributions. Figure 11.24 shows an example obtained with a $\pi/2$ phase window of 4 nm width (the splitting between the one-photon transitions). Scanning the window's central wavelength across the two-photon resonance produces a large variation of the transition probability. For a centered window, all contributions add constructively and an enhancement of a factor of 7 is observed.

In time-resolved experiments, shaped pulses can be used either to maximize the population transfer at a given time (within the laser pulse) or to change the temporal behavior. A first example takes advantage from the fact that above and below resonance, contributions (in a one-photon transition) are π out of phase [11.96]:

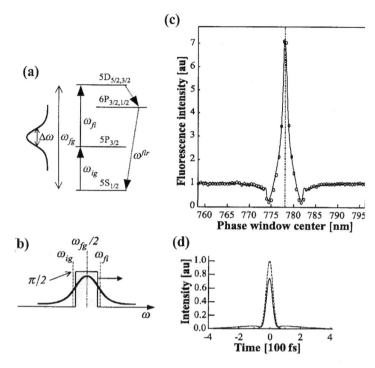

Fig. 11.24. Effect of a $\pi/2$ window in the spectral phase on a nearly resonant two-photon transition in Rb [11.95]; (a) Levels involved (ground state 5s, intermediate state 5p$_{3/2}$, excited state 5d, fluorescence observed through the 5d $\rightarrow$ 6p $\rightarrow$ 5s cascade); (b) spectral shape of the pulse and phase window; (d) temporal intensity (solid line) of the optimum shaped pulse, compared to the transform-limited pulse (dashed line) showing a 26% reduction of the peak intensity due to the broadening of the pulse.

$$a_e(t) = \frac{-\mu_{eg}}{\hbar} \int_{-\infty}^{+\infty} \frac{\tilde{E}(\omega)}{\omega_{eg} - \omega} \exp\left[i(\omega_{eg} - \omega)t\right] d\omega. \qquad (11.34)$$

Therefore these contributions add destructively and only the resonant part (due to the principal part in Eq. (11.34)) remains. Applying a π-step on the spectral phase at the resonance frequency can make these interferences constructive. This is demonstrated in Fig 11.25, which compares the result of the shaped pulse with a transform-limited pulse [11.96]. As can be seen, the transient population is increased by a factor of ca 1.8 at time 0. The population decreases then to a value lower than the asymptotic population reached with transform-limited pulses. This is due to the hole in amplitude that results from the phase step [11.98]. A third plot displays the result of an "optimized" pulse (step height and position), where a step of 0.7π is applied slightly after resonance.

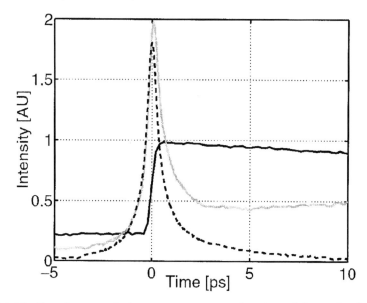

Fig. 11.25. Transient population measurement as function of the probe delay for the pulse shapes: transform limited pump pulse (solid black line), on resonant π-step shaped pulse (dashed line), shifted 0.7π step (solid grey line) [11.96]

The shape of the coherent transients described in the previous paragraph can be strongly modified by pulse-shaping techniques. The oscillations that appear after passage through resonance result from interferences between the atomic dipole and the instantaneous electric field. A change in the phase of the electric field after resonance can therefore be directly observable. Figure 11.26 displays an example of a π-phase step applied slightly after resonance [11.97]. The oscillations are π out of phase with respect to normal CT. When the position of the phase step is shifted away from resonance, a strong sensitivity down to the pixel resolution is observed [11.98]. This scheme can provide an efficient, sensitive tool for testing high-resolution pulse shapers.

The last example regards the use of pulse shaping to coherent anti-stokes Raman scattering (CARS). Since this is a second-order process, a behavior similar to two-photon transitions is expected. However, in general, several final states lie within the spectral range of the laser pulse. These can be, for instance, several vibrational modes. By applying phase steps or a sine phase function, it is possible to enhance some transitions while "killing" others. A resolution much higher than the spectral width can thus be achieved [11.99,100].

11.4.3.2 Closed-Loop Schemes. In molecules, a successful control is achieved when there is a cooperative interaction between a laser pulse shape and the precursor molecule's dynamics to alter the product distribution in

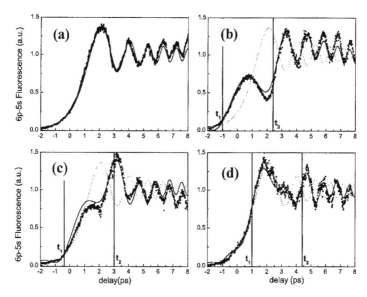

Fig. 11.26. Manipulation of coherent transients with a π-step shaped pulse applied at different frequencies after resonance, together with theoretical simulations [11.97,98]: (a) normal CT; (b–d) inverted CT with normal CT as a reference: (b) π-step 1 pixel from resonance; (c) π-step 2 pixels from resonance; (d) π-step 4 pixels from resonance.

a desirable and variable manner. Such optimal light fields might be computationally generated given the form of the electronic potential surfaces and the initial wave functions. However, when the complexity of dynamics is such that control cannot arise from trivial effects like intensity, pulse duration, chirp, pump-control sequences, as well as when the computational effort is too heavy to predict in advance the optimal pulse shape, it is interesting to implement experimentally a learning control scheme [11.64,65]. This control refers to the optimization of a desired task by means of suitable closed-loop laboratory procedure determining the optimal laser pulses. The achievement of the control task is monitored on an experimental response (fitness function f), for instance, a second harmonic generation (SHG) signal [11.101], the yield of a specified product of dissociation [11.102–104], ratio of ionization yields [11.105,106], fluorescence efficiency or fluorescence power [11.107], ratio of fluorescence signal from two different chromophore liquid samples [11.108], or stimulated Raman scattering signals [11.109]. The fitness f value depends on N laboratory variables defining the laser shape. These are, for instance, the voltage values applied to each individual liquid crystal pixel. The control task is to find an optimum of this f function. Essentially, the closed-loop laboratory procedure connects a shaper of the broadband laser pulse, an optimization algorithm, and the experimental response as can be seen in Fig. 11.27 [11.105]. In this learning control, the feedback is not built into the Hamiltonian of the

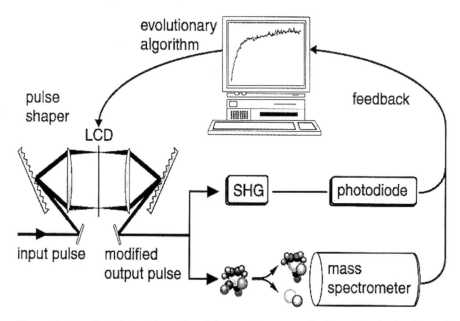

Fig. 11.27. Principle of a closed-loop scheme applied to the ionization of $C_6H_5Fe(CO)_2Cl$ [11.105]. The two products $C_6H_5FeCOCl^+$ and $FeCl^+$ are analyzed with a mass spectrometer. Each product can be independently maximized of minimized; the ratio of the two products can also be optimized. The chosen fitness function is used as feedback for the evolutionary algorithm. A second harmonic generation (SHG) signal is also monitered as an indication of the pulse shape.

system, but rather on the observables of the Hamiltonian. A typical example of the optimization of the ratio between two fragment ions ($FeCl^+$ versus $C_6H_5FeCOCl^+$) in the photoionization of $C_6H_5Fe(CO)_2Cl$ is displayed in Fig. 11.28 [11.105].

The optimization algorithm (or learning algorithm) must be adapted to the high number of degrees of freedom. Typically, with a 128-pixel SLM and 10 significant bits for the voltage values, the largest phase space corresponds to $2^{10*128} = 2^{1280}$ different pulse shapes. To reduce the phase space dimension, linear interpolation or phase function defined by few parameters to optimize [11.110] might be used across the pixel array, introducing smooth variations in phase and amplitude. Finally, when a temporal or frequency structure is identified, more subtle parameterization might be used in order to gain an insight in the most important control mechanisms [11.111]. Many different learning algorithms have been tested. They can be divided into deterministic and nondeterministic (random) algorithms. In the deterministic schemes, the main assumption is that the redirection of search is high enough to reach ergodicity so that a large fraction of the search space is explored. In the nondeterministic scheme, the search iteration includes a random stage [11.112]. The

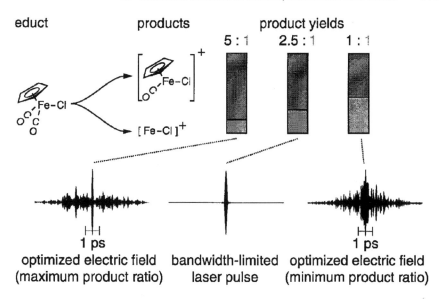

Fig. 11.28. Results of a closed-loop scheme (see Fig. 11.28) applied to the ionization of $C_6H_5Fe(CO)_2Cl$ [11.105]. The ratio of the two products $C_6H_5FeCOCl^+$ and $FeCl^+$ is either maximized (left) or minimized (right). The obtained results are compared with a Fourier transform limited pulse (middle).

main interest of the deterministic optimization algorithm and its derivatives [11.113] is the absence of algorithm parameters [11.113]. In general they are not suitable for learning control because they are appropriate for a smooth f function with one single broad minimum while in general the target f function has a lot of equally good local minima [11.114].

Thanks to the random facet of the nondeterministic algorithm, the search can get out of the small local minima and fall into the "good one." There are two main nondeterministic strategies commonly used in learning control: simulated annealing algorithms (SA) [11.115] and evolutionary algorithms (EA). The latter one includes genetic algorithms (GA, the most used) [11.116], evolutionary programming [11.117] and evolution strategy [11.118]. They are both robust against experimental noises and insensitive to the initial guess pulse shapes [11.119,120]. In the SA procedure, a new pulse shape is generated at each iteration via a random modification or an adept one (from a simplex scheme) of the N pulse shaper parameters. Then the new pulse shape is tested by recording the experimental signal, defining at the same time the new cost value C_i of this ith iteration. In the simplest procedure, the pulse modification is accepted when

$$x > e^{\frac{(C_i - C_{i-1})}{T}}, \tag{11.35}$$

where x is a random number generated at each iteration and uniformly distributed in the interval $[0,1]$ and T the "temperature" parameter. Progressively T is lowered, allowing fewer and fewer changes in the search. The speed of convergence is determined by the rate of temperature change: fast cooling might converge to a "bad" solution while slow cooling is computer-time-consuming. There is a large literature on the variations of this optimization algorithm [11.121]. The SA algorithms are well suited for problems with many equally good nearby minima and a few parameters to be adjusted [11.122]. However, for large differences between typical local minima and a global minimum, the EAs are the most appropriate. Indeed the most important is to pinpoint the basin defining the global minimum in the search space instead of improving every set of parameters as done in the SA procedure. To our knowledge, this principal limitation has restricted the application of SA to basic learning control investigated numerically on population inversion between electronic states of a diatomic molecule [11.123] or broadband pulse compression [11.124].

EAs are a class of global optimization search inspired by natural search processes in biological evolution. The EA procedure is defined by a "population" that is updated at each iteration, also called generation. This population is a set of "individuals" (for learning control, typical numbers are between 50 and 100), each corresponding to a laser pulse shape. An individual is a string of genes where each gene is a floating-point number specifying either the phase or the amplitude at the various frequency components of the laser pulse. Each individual of the population (pulse shape solution) interacts with the physical system such that the recorded signal is used as a fitness value of this solution. In each generation, a new population called the child population is built by the EA from the previous population called the parent population. EAs employ a trade-off between exploration by reproductive mechanism such as mutation or crossover operators and exploitation by cloning the fittest individuals of parent population. By iterating this process at each generation, the population of pulse solutions "evolves" toward near-optimal solutions. The genetic algorithms and the evolutionary programming (EP) vary in their reproductive mechanisms: GA models genotypic transformation while EP is based on phenotypic adaptation.[1] The algorithms differ by the fitness definition as well as by their operators. The latter are statistical as the selection rules and crossover mechanisms (average, elitism) for GAs and as mutation (gene creep) and score determinations for EP or physically motivated as time-domain crossover, smoothing and polynomial phase. Indeed, given the nature of the task and the measurements preformed to evaluate the success of one individual, it is not always clear how to assign a fitness that can be a function of different experimental data [11.108].

The influence of the replication process, the size of the parameter space, the influence of the noise level as well as the role of the parameterization on convergence speed have been investigated carefully in Motzkus' group in

---

[1] The information contained in human genes is the genotype and the human form is the corresponding phenotype.

[11.120]. More details can be encountered, for instance, in [11.125] and for the strategies tested on the learning control in the appendices of [11.119]. The optimization algorithm might be based on the automated competition between the various operators used in the GA, or even between the optimization algorithms as the GA, EP and the simulated annealing one. This is called an adaptive learning algorithm in which to each operator a fitness value is assigned. The interplay of the multiple operators in the selected adaptive strategy can yield further insight into the dynamic processes of the system [11.119].

The test of the algorithm parameters relative to the fitness function as well as the experimental conditions as the expected noise level or the pulse shaper characteristics, might be done on trivial control tasks such as:

- Pulse compression with as experimental fitness value, an SHG signal [11.101,119,120] or a two-photon absorption signal in a semiconductor [11.126]. Whatever the EAs used, a satisfactory convergence is reached after ~ 60 generations for a population of ~ 50 individuals. For a 1-kHz laser chain system, the pulse compression is achieved in less than 5 minutes.
- A two-photon transition in an atomic gas with a well-known energy scheme. The experimental fitness value might be a fluorescence signal, an ionization signal [11.127], or the control of the coherent transients (cf. Fig. 11.26) [11.97].

The learning control has been used in optics to compensate the phase distortions of the broadband pulse output of a noncollinear-type optical parametric amplifier [11.128] or to maximize the generation of a given harmonic generated in an atomic gas. The pulse shaping in this experiment was based on a deformable mirror [11.129]. In molecular liquids, the pioneering experiment realized in the group of Kent Wilson consisted of shifting the fluorescence of a dye molecule [11.107]. In the gas phase, the closed-loop learning process has converged to the optimal fragmentation of metal carbonyl by feedback checking the ion mass spectrum [11.102]. In transparent methanol or CCl4 liquids, using an intense phase-shaped pulse with a constant spectral intensity to induce a strong field molecular polarizability, it has been possible to control self-phase modulation (SPM) spectrum and even stimulated Raman scattering. In the latter the Stokes light is a part of the controlled coherent continuum induced by the shaped pulse [11.130]. Finally, it has been demonstrated on a 50% C_6H_6/C_6D_6 cell that the mode selection achieved in such a learning scheme results from an intramolecular coupling between the vibrational modes, rather than simple seeding one of the modes from the SPM continuum [11.119]. In polymers, learning control has allowed us to excite selectively ground state vibrational modes by optimizing the stimulated Raman processes via a fitness signal derived from the spectral distribution of the coherent antistokes Raman scattering [11.131].

In addition to the selective vibrational mode excitation, the decay times of
the different modes were extended. By controlling the phase shape of strong
field laser pulses, it has been possible to control the disturbance of the field-
free eigenstates of organic molecules (see next section) such that the cleavage
and rearrangement of organic functionality is achieved [11.106]. For learning
control in biologically interesting systems, we can outline the enhancement
of internal conversion dissipation against energy transfer dissipation in light
harvesting (the natural 50% branching ratio is increased up to 63%) using
a restricted sinusoidal phase optimization [11.110]. A similar sine spectral
phase function applied through the pulse shaper has also allowed to perform
a coherent anti-Stokes Raman microscope with a single pulse instead of two
or three different pulses [11.100]. The controlled modulation of the spectral
phase of the pulse exploits the quantum interferences between multiple paths
to selectively populate a given vibrational level. The fitness value is the CARS
signal while varying the periodicity of the sine function.

The learning control strategy finds an optimal pulse given the limitations of
the experimental apparatus. One key block in this approach is the pulse shaper
that is able to synthesize almost arbitrary waveforms [11.132]. The main advantage of the learning control is to reduce the search time and even to find
new innovative solutions unreachable in a sequential search. These learning
techniques are also used to enhance medical imaging and improve high-speed
communication architecture [11.133]. The system acts effectively as an analog
computer that surmounts the complexity of the problem and the attained optimal pulse might be rich of information on the underlying interatomic forces.
One outstanding dream is to join together in a laboratory the control and inversion of molecular dynamics problems [11.134] to improve the knowledge of
the molecular potentials. Although no general methods have been defined, in
a few examples, the dynamics induced by the optimized shaped pulses can be
understood and efficient comparison with a theoretical description has been
achieved with spectacular agreement [11.104,135].

11.5 Coherent Control in Strong Field

Laser radiation can be focused easily to a spot size of ca 100 μm or less.
Due to this temporal and spatial confinement, pulse energies higher than
10 μJ and shorter than 100 fs duration result in peak intensities higher than
10^{12} W/cm^2. The amplitude of the electric field E_L at this intensity level[2]
approaches 10^8 V/cm. This corresponds to the valence electron's binding energy. In general at this intensity, the lowest-order perturbation theory is no
longer adequate, and higher-order effects such as stimulated Raman processes
and ac Stark shifts become important.

---

[2] The connection between intensity and electric fields strength is $I[W/cm^2] =$
$1/(2Z_0)E^2[V/cm]$, where Z_0 is the vacuum impedance $Z_0 = \sqrt{\mu_0/\varepsilon_0} = 377V/A$.

Let us first introduce the basic equations and definitions for a two-level system $|1\rangle$ and $|2\rangle$. The $|1\rangle$ and $|2\rangle$ states are the diabatic states defined with an energy $\hbar\omega_1$ and $\hbar\omega_2$, respectively. These states are also called the bare states in contrast with the dressed states $|+\rangle$ and $|-\rangle$, namely the adiabatic states in which the light-coupling is included. The laser field is strong enough to deplete significantly the initial state $|1\rangle$. This depletion is quantitatively characterized by the Rabi frequency that is the rate at which the transition is coherently induced between the two levels:

$$\hbar\Omega(t) = -\mu_{21}.\mathbf{E}(t). \tag{11.36}$$

The time evolutions of the probability amplitudes are given by the differential equations

$$i\frac{d}{dt}\begin{bmatrix} c_1 \\ c_2 \end{bmatrix} = \frac{1}{2}\begin{bmatrix} -\Delta & \Omega \\ \Omega & \Delta \end{bmatrix}\begin{bmatrix} c_1 \\ c_2 \end{bmatrix}, \tag{11.37}$$

where the detuning Δ is zero for the resonance case and positive when the laser frequency is shorter than the transition frequency. We devote our attention to a class of interactions in which the rotating wave approximation remains valid.[3] The dressed states are obtained by diagonalizing the Schrödinger equation (11.37). The transformation from the diabatic basis to the adiabatic one is:

$$\begin{bmatrix} c_1(t) \\ c_2(t) \end{bmatrix} = \begin{bmatrix} \cos v(t) & \sin v(t) \\ -\sin v(t) & \cos v(t) \end{bmatrix}\begin{bmatrix} a_+(t) \\ a_-(t) \end{bmatrix} = \Re(v(t))\begin{bmatrix} a_+(t) \\ a_-(t) \end{bmatrix}, \tag{11.38}$$

where the angle $v(t)$ of the rotation matrix $\Re(v(t))$ depends on the Rabi frequency and the detuning:

$$v(t) = \text{artan}\left(\frac{\Omega(t)}{\Delta(t)}\right). \tag{11.39}$$

$a_+(t)$ and $a_-(t)$ are the probability amplitudes of respectively the $|+\rangle$ and $|-\rangle$ dressed states. Note that the time variation of the detuning can be introduced, for instance, by a linear chirp of the laser field (see section 11.4.1) as shown in Fig. 11.30.

The Hamiltonian in this adiabatic basis is defined as:

$$H_{adia}(t) = \Re(-v(t))H_{dia}\Re(v(t))i\hbar\Re(-v(t))\dot{\Re}(v(t)). \tag{11.40}$$

The first term is the diagonal part of the Hamiltonian and gives the eigen-energies of the adiabatic states while the second term is the off-diagonal part corresponding to nonadiabatic coupling. The population and energy vary as a function of time through the time-dependence of the laser field as

$$i\frac{d}{dt}\begin{bmatrix} a_1(t) \\ a_2(t) \end{bmatrix} = \begin{bmatrix} -\varepsilon(t) & -i\dot{v}(t) \\ i\dot{v}(t) & +\varepsilon(t) \end{bmatrix}\begin{bmatrix} a_1(t) \\ a_2(t) \end{bmatrix} \tag{11.41}$$

---

[3] $\Delta \ll \omega_L$ and $\Omega \ll \omega_L$.

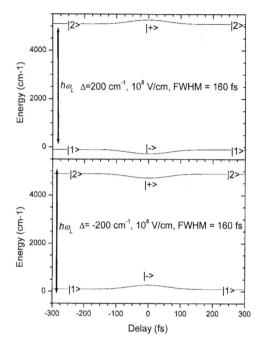

Fig. 11.29. AC-stark shift induced by a FWHM = 160 fs and 10^8 V/cm linearly polarized electric field on the two-level system $|1\rangle$ adn $|2\rangle$. In the case of a positive detuning (top), the energy difference of the adiabatic states $|+\rangle$ and $|-\rangle$ increases with the intensity, while it decreases for a negative detuning (bottom)

with the following expression for the eigenenergy $\varepsilon(t)$:

$$\varepsilon(t) = \frac{\hbar}{2}\sqrt{\Omega^2(t) + \Delta^2(t)}. \tag{11.42}$$

For instance, a one-photon resonant laser pulse with $E_L = 10^8$ V/cm, linearly polarized and parallel to a dipole moment μ_{21} of $\sim 1.10^{-30}$ C.m. induces a maximum light-coupling strength of $\hbar\Omega(t)$. The ac Stark shift of the two levels will reach a maximum of $\Omega/2 = 500\mathrm{cm}^{-1}$.

As shown in Fig. 11.29, the splitting of the adiabatic states increases or decreases for a positive (Fig. 11.29-top) or negative (Fig. 11-29 bottom) detuning. In the dressed state picture (Fig. 11.30), the coupling results always in a repulsion between the $|1, (n+1)\rangle$ and $|2, n\rangle$ states. Without any pulse (in Figs. 11.29 and 11.30 for $|t| > 250$ fs), the adiabatic states $|+\rangle$ and $|-\rangle$ correspond to the diabatic states $|1, (n+1)\rangle$ and $|2, n\rangle$. As shown in Figs. 11.29 and 11.30-top, for the off-resonance cases with no chirp of the laser pulse, the probability transition from $|1\rangle$ to $|2\rangle$ is zero: each adiabatic state $|+\rangle$ or $|-\rangle$ is correlated to the same diabatic state before and after the laser pulse interaction. Once the instantaneous pulse frequency is scanned as shown in Fig. 11.30 (middle and bottom), this probability transition becomes significant;

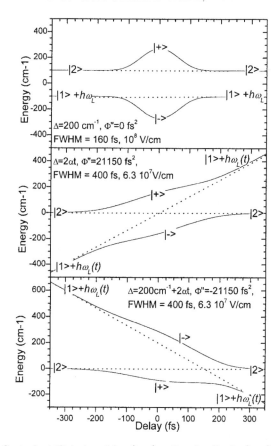

Fig. 11.30. AC-stark shift indeced by (top) a Fourier limited pulse in a two-levels system characterized by a $+200$-cm^{-1} detuning; the same laser pulse characteristics (160 fs, 10^8 V/cm) and the same energy scheme as Fig. 11.29 top; (middle) a pulse positively chirped from 160 fs to 400 fs in resonance and with the same energy/pulse and focalization length used in Fig. 11.29; (bottom) a pulse negatively chirped from 160 fs to 400 fs off-resonance by $+200$ cm^{-1} and with the same energy/pulse and focalization length used in Fig. 11.29.

each adiabatic state is correlated to different diabatic states before and after the laser pulse interaction. The dipole coupling lifts the degeneracy between the diabatic states at their crossing. Therefore this crossing becomes more or less avoided as a function of the laser intensity, guiding the population from the initial state $|1\rangle$ to the excited state $|2\rangle$. The efficiency of this transition process [11.136] might reach 100% once the adiabaticity criterion is fulfilled in the Schrödinger equation (11.41). Concretely, this means the population evolution follows the building of the adiabatic eigenstates with negligible nonadiabatic couplings (the off-diagonal terms of Eq. (11.41):

$$|\dot{v}(t)| \ll \varepsilon(t). \tag{11.43}$$

Taking into account Eqs. (11.39) and (11.42), the adiabaticity is encountered once:

$$|\dot{\Omega}(t)\Delta(t) - \Omega(t)\dot{\Delta}(t)(t)| \ll (\Omega^2(t) + \Delta^2(t))^{3/2}. \tag{11.44}$$

This latter inequality is achieved with a smooth pulse envelope and a slow frequency sweep to minimize $\dot{\Omega}(t)$ and $\dot{\Delta}(t)$, respectively. A large field intensity and a large detuning can also lead to adiabatic evolution. In the case of a partial adiabaticity, the system ends up in a coherent superposition of diabatic states. This interplay of the diabatic states has been observed as a function of the detuning as well as the laser field duration [11.137] on a two-photon ionization.

The main advantage of this adiabatic control is its generality and its robustness with respect to intensity or frequency fluctuations, once the laser field is high enough to achieve the adiabatic criterion (Eq. (11.44)). However, the main difficulty in this kind of control is to take into account all the energy levels involved, and more specifically the losses toward the dissociative and ionization continua. Indeed, there is always the possibility that other neighboring levels in the dressed state picture complicate the control scheme. A recent development is to implement a learning control strategy, controlling the intense pulse shape in order to achieve the control task [11.106,130] (see section 11.4.3.2).

After the first experiments realized at this intensity range, the idea of using the time-dependency of the laser intensity to shape the potentials has been introduced. For instance, a Rydberg radial wave packet has been tailored by depleting more or less differently the Rydberg levels composing this wave packet [11.138]. However, the state density is too high to get insight of the population redistribution as clearly as the freezing of a spin orbit wave packet in a three-level system [11.45].

In molecules, various ionization and dissociation multiphoton pathways coexist when exposed at this intensity range [11.139]. Their interplay, being at the same time a key point in the realization of new control schemes, have strongly impeded any conclusive identification of the phenomena. Experimentally, it is through the preparation of the initial state distribution [11.140] or via a differential detection [11.141–143] that unambiguous experimental evidences of these intense field effects have been obtained.

We illustrate the principles of wave packet tailoring via light-induced deformation of potential through the well-known work of Garraway and Suominen [11.144–147] on the mere sodium dimer. The three diabatic electronic states are represented in Fig. 11.31. In the dressed state picture, the optical resonances appear as curve crossings (Fig. 11.31(b)). The time evolution of the molecular system consists of a nontrivial balance between adiabatic following of the vibrational eigenstates and nonadiabatic transfers between them at the avoided crossings. The evolution of the three electronic state wave functions $\Psi_{i=X,A,\Pi}(R,t)$ is governed by the Schrödinger equation with the Hamiltonian

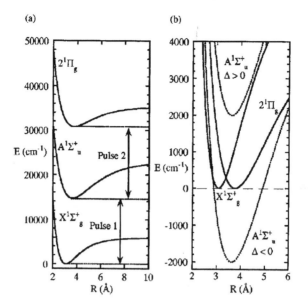

Fig. 11.31. (a) The three different diabatic electronic states used in the calculation of Garraway and Suominen [11.144–48] as well as the energy of the two laser pulses; (b) the shifted potentials by applying the rotating wave approximation, $\Delta = \Delta_1 = -\Delta_2 = 2000$ cm^{-1} is the detuning of the two pulses from the well bottoms of (a).

for the vibrational motion of the molecule written as

$$H = -\frac{\hbar^2}{2m}\frac{\partial^2}{\partial R^2}I + U(R,t), \tag{11.45}$$

where I is the identity matrix, R the internuclear distance and m the reduced mass. The matrix $U(R,t)$ includes the electronic potentials and light-coupling:

$$U(R,t) = \begin{bmatrix} U_x(R) & \hbar\Omega_1(t) & 0 \\ \hbar\Omega_1(t) & U_A(R) + \hbar\Delta_1 & \hbar\Omega_1(t) \\ 0 & \hbar\Omega_2(t) & U_\Pi(R) + \hbar(\Delta_2 + \Delta_1) \end{bmatrix}, \tag{11.46}$$

where $U_i(R)$ are the three electronic potentials; Δ_1 and Δ_2 are the constant detunings for A $\leftarrow$ X and Π $\leftarrow$ A bottom-bottom transitions, respectively. Eventually, $\Omega_1(t) = \hbar\mu_{A\leftarrow X}E_1(t)$ and $\Omega_2(t) = \hbar\mu_{\Pi\leftarrow A}E_2(t)$ are the two Rabi frequencies, independent of R for simplicity and function of time through the Gaussian pulse envelope such that

$$\Omega_{j=1,2}(t) = \Omega \exp(-[(t - t_j)/T]^2), \tag{11.47}$$

with t_j the arrival time of the laser pulse j. To simplify, the light-coupling strength Ω (= 3000 cm^{-1}) and the pulse duration T (= 5.42 ps) are identical for both pulses and the detunings are such that $\Delta = \Delta_1 = -\Delta_2$ (= −2200

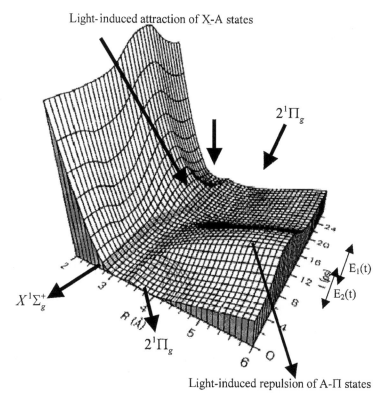

Light-induced attraction of X-A states

$2^1\Pi_g$

$E_1(t)$

$E_2(t)$

$X^1\Sigma_g^+$

$2^1\Pi_g$

Light-induced repulsion of A-Π states

Fig. 11.32. Space and time dependence of the light-induced potential responsible for the dynamics of the wave packet in Fig. 11.31. The active light-induced potential is the one with its minima always closest to the zero point energy in the dressed state picture [11.144].

cm^{-1}), ensuring two-photon resonance. The light-induced potentials are obtained by diagonalizing the matrix $\boldsymbol{U}(R,t)$.

When the laser pulses are applied in a counterintuitive way $(t_1 > t_2)$ as shown in Fig. 11.32, with a constant negative detuning Δ $(\Delta_2 > 0)$, the first light-induced shifting is the repulsion of the A-Π. When the electric field $E_2(t)$ increases (maximum at $t_2 = 10.8$ ps), the right-well of the double-well structure seen at $t = 0$ becomes progressively shallower and disappears. Once $E_1(t)$ is turned on $(t_1 = 16.3$ ps), the X and A states attract each other, broadening and displacing the bottom of the active light-induced potential to the right side, corresponding also to the minimum of the Π state. At this stage of the dynamics, there is a diabatic coupling from the X state onto the Π one. Indeed there is no crossing of the diabatic vibrational levels of the X and Π states, as observed in Fig. 11.30, so the population can never be transferred adiabatically from the X state onto the Π one. As shown in Fig. 11.33, the X and Π states exchange their population, while the population of

(a)

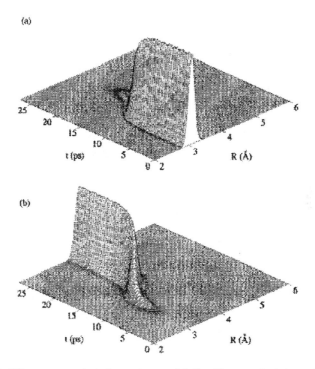

(b)

Fig. 11.33. The wave packet dynamics on (a) the X ground state and (b) the Π target state. The ground state population disappears when the two pulses arrive at $t_2 = 10.8$ ps and $t_1 = 16.3$ ps (a), while it appears slowly in the Π state (b). $T = 5.42$ ps. The adiabatic transition is from the X ($\nu = 0$) state onto the Π ($\nu = 0$) state [11.137]

the A state remains very low at any time since there is never a crossing of the A state with the active light-induced potential.

The signature of the diabatic coupling is the smooth positional shifting of the wave packet observed in Fig. 11.33. The not-perfect adiabatic following is the key point of this control and is achieved differently as a function of the laser intensity, detuning and chirp rate [11.147]. The overall control appears as a violation of the Franck-Condon principle, however the manipulation of the molecular states is achieved on a time scale longer than the vibrational period, allowing the displacement over a large distance. Control schemes manipulating excited vibrational levels [11.145], dissociative states [11.148] or efficient transfer to a dark molecular state [11.146] have also been proposed.

The goal of these previous examples was to control the population in the electronic excited states or the internal degrees of freedom of the molecules by playing on the laser-induced resonances and the electric dipole moment. Another main field of control is achieved on the external degrees of freedom. The rotation of molecules can be controlled and even stopped, thus creating

(static or dynamic) alignment or orientation. Beam acceleration, decelera-
tion or deflection has also been achieved by acting on the center of mass
motion [11.149]. A nonresonant intense laser pulse induces a dipole moment
in a molecular ground state. The interaction Hamiltonian involves thus the
permanent dipole moment μ and the dynamic polarizability tensor α:

$$H_{\text{ind}}(t) = -\frac{1}{2} \sum_{i,j=x,y,z} E_i(t)\alpha_{ij}E_j(t)\cos^2(\omega_L t) - \mu.\mathbf{E}(t)\cos(\omega_L t) \quad (11.48)$$

Here, (x, y, z) denotes the space-fixed Cartesian coordinates. For a linear or
symmetric top molecule with a quantization axis chosen along the body-fixed
Z-axis, the polarizability tensor in the molecular frame is reduced to two
components: $\alpha_{\parallel} = \alpha_Z$ and $\alpha_{\perp} = \alpha_X = \alpha_Y$. Therefore, for a linearly polarized
laser field along the Oz-axis, the laser-induced Hamiltonian of Eq. (11.48) is
dependent only on the polar Euler angle θ between the laser polarization Oz
and the molecular symmetry axis OZ with the following expression:

$$H_{\text{ind}}(t) = -\frac{1}{2}E^2(t)\cos^2(\omega_L t)\left[\alpha_{\parallel}\cos^2\theta + \alpha_{\perp}\sin^2\theta\right] - \mu E(t)\cos(\omega_L t)\cos\theta.$$
$$(11.49)$$

Taking into account that the laser period varies rapidly compared to the
envelope of the laser electric field $E(t)$ and that the laser is nonresonant,
the laser interaction with the permanent dipole moment is quenched. The
laser-induced dipole moment can be averaged over a laser period to give the
following expressions:

$$\mu_{\text{ind}}(t) = \tfrac{1}{4}E(t)\cos(\theta)\alpha_{eff}$$
$$(11.50)$$
$$H_{\text{ind}}(t) = -\mu_{\text{ind}}E(t)\cos(\theta)$$

with the effective polarization defined as $\alpha_{eff} = \alpha_{\parallel} - \alpha_{\perp}$. This laser-induced
Hamiltonian adds to the rotation Hamiltonian in the ground electronic state,
written for a rigid molecule as

$$H_{rot} = \sum_n \frac{J_n^2}{2I_n},$$

where n is the principle of axis inertia, I the inertia moment and J the angular
momentum operator. The orientation and alignment of the molecular axis are
given by the average values of M/J and $(M^2 - J(J+1))/(J(J+1))$, respectively
(M is the projection of J on the laser polarization axis). For a statistical M
distribution, no net alignment or orientation is observed. The intense laser field
induces Raman excitations in the ground electronic state ($\Delta J = \pm 2$) resulting
in rotational excitation corresponding to a rotational wave packet. Since M is
conserved with linear polarization, one obtains, on average, $J\hbar|M|$, leading to
a strong alignment. Thus the molecule might align parallel or perpendicular
to the polarization plane, depending on the sign of its effective polarizability.

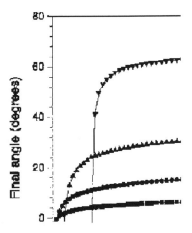

Fig. 11.34. Final angles reached by molecules with initial angles of 20° (squares), 40° (circles), 60° (up triangles) and 80° (down triangles) as a function of laser intensity [11.151].

Classical models have been derived to describe the angular motion of the molecule in the laser field. From the Lagrangian for a rigid molecule:

$$L = \frac{1}{2}I\dot{\theta}^2 - H_{\text{ind}} \tag{11.51}$$

(moment of inertia $I = mr^2(t)/2$) leads to the classical equation of motion:

$$\ddot{\theta} = \frac{\alpha_{\textit{eff}}}{2I}E^2(t)\sin(2\theta) - 2\frac{\dot{r}}{r}\dot{\theta}. \tag{11.52}$$

The first term of the angular acceleration causes the alignment, while the second term is a damping term that reduces the angular acceleration once the molecules dissociate. The mass of the molecule has two counteracting effects in this dynamics; a light molecule is easy to accelerate. However, at the same time, the internuclear distance is increasing quickly, leading to a larger moment of inertia, such that the damping term might become significant while the laser field is still on. As seen through Eq. (11.52), the angular acceleration depends on the time-dependency of the electric field as well as its intensity [11.150]. Figure 11.34 [11.151] shows the final angles of diatomic molecules with initial angles of 20°, 40°, 60° and 80° as a function of laser intensity. The molecules irradiated with the lowest-intensity laser pulses undergo the largest rotation. With the Gaussian laser pulses used in these calculations, molecules that experience higher intensities can end up less aligned. At fixed intensity, the width of the angular distribution decreases with increasing pulse duration, indicating that dynamic alignment occurs. For example, for a pulse energy of 3 mJ, the FWHM of the angular distribution of the I_2 molecules decreases from 64.4±2. to 25.2±0.8. when the duration is stretched from 500 fs to 3ps (with an intensity reduced from 1.6×1014 W/cm^2 to 2.65×10^{13} W/cm^2 [11.152].

This laser-induced alignment has been observed in many diatomic molecules [11.153–156] as well as on polyatomic molecules in which the laser polarization is chosen elliptic to force three molecular axes into the alignment [11.157]. The molecular alignment reached is truly significant [11.152] such that further perpendicular transitions are enhanced and parallel transitions suppressed (or vice versa). This macroscopic control of the orientation of the molecules allows, for instance, control of the branching ratio of further photodissociation processes [11.158].

Rotating the laser polarization during the alignment causes the aligned molecule to rotate. Two counterrotating circularly polarized fields, identical in intensity and spectrum, can be combined to produce this rotating polarization. This rotation of the polarization can be accelerated by chirping the two laser pulses with opposite signs, such that the rate of rotation increases from 0 to 6THz in 50 ps [11.159]. Molecular rotation is thus strongly accelerated up to values of several hundreds, where the molecular bond is broken by the centrifugal force [11.160]. This control of the dissociation is done in less than one ground state rotational period and very efficiently, since each of the hundred successive Raman transitions has nearly 100% probability.

In solid-state physics, the carrier densities are typically of 10^{18} to 10^{19} electrons/cm^3, involving large dipole moments between the band states. The threshold of strong field effects is lowered, typically around GW/cm^2 compared to the molecular physics case. In semiconductors, the dephasing increases with larger carrier density due to the elevated scattering processes. The main effect of the strong field interaction is the Rabi flopping of the optically induced excitation density. The first Rabi oscillations has been observed in the absorption continuum of a semiconductor off-resonance [11.161] or on-resonance [11.162]. Since then, many other works have been published on Rabi oscillations in semiconductors [11.163,164], as well as in microwave-irradiated Josephson-Junction circuit [11.165].

In semiconductor quantum wells, the dynamical Stark effect has recently been observed in quantum wells embedded in semiconductor microcavities [11.166]; the exciton states are dressed by the enhanced electromagnetic field within the cavity. Two sidebands appear in pump-probe experiments, whose spectral distance increases as the square root of the light intensity, as expected. These sidebands are reminiscent of the Mollow absorption spectrum occurring in two-level system [11.167,168] and observed in atomic systems as Stark effect excited by a CW laser [11.169]. More recently, the same effect could be observed in bare quantum wells [11.170]. Here, a strong interaction picture prevails against the perturbative description as soon as the exciton dipole interaction with the laser field becomes stronger than the exciton collision broadening.

For high pulse intensities, nearly full-state occupation occurs in an energy region substantially wider than the excitation bandwidth of the laser field of

120 fs. Hence scattering into these states is increasingly hindered by the Pauli principle resulting in lower dephasing rates [11.171]. Consequently, the quantum kinetic buildup of screening will tend to reduce carrier–carrier dephasing and allow for coherent dynamics in the later stages of the dynamics. Rabi oscillations have also been observed in semiconductor quantum dots [11.172]. In these artificial atoms, excitonic excitation from the ground state might be seen as a two-level system where the complete inversion induced by π-pulse is the deterministic creation of one electron-hole. A single quantum dot in a photodiode, once irradiated by a π-pulse, corresponds to a source of single electron delivered at the laser repetition rate [11.173].

Moreover, the interplay between Rabi oscillations and quantum interference in a temporal coherent control scheme has been observed [11.174,175]. The exciton dressed state in quantum dots have been evidenced, together with the state splitting under intense electromagnetic field [11.174], in analogy with the occurrence of Mollow triplet in atomic stimulated emission under intense light excitation [11.167,168].

11.6 Conclusion

The applications of coherent control or optimal control have reached wider and wider fields. As significant examples, we describe several cases of optimization of the generation of high-order harmonics, applications to information storage and the search for control of molecular chirality (several theoretical schemes based on frequency domain coherent control [11.176,177]).

The generation of coherent vacuum ultraviolet (VUV), extreme ultraviolet (XUV) or X-ray radiation is an emerging field that will open the route to a wealth of new applications and has attracted strong interest. Two examples are the possibility of carrying ultrafast X-ray diffraction of biological samples [11.178], or the generation of "attosecond" pulses [11.179–182]. The various approaches described in this chapter have been used to modify and control the efficiency of harmonic generation: chirped pulses [11.183], shaped pulses [11.129,182,184,185] (see Fig 11.35), or wave packet interferences in Ramsey-type spectroscopy [11.186].

Quantum information is also an exploding field in which coherent control schemes have been demonstrated in simple cases. A set of atomic Rydberg states have been used as quantum data register [11.187]. Wave packet interferences produced by a transform limited pulse and a shaped pulse are used to store and restore information [11.188]. Another possibility is to use half-cycle Terahertz pulses [11.189]. These schemes can be used to implement Grover's search algorithm [11.190,191]. Molecular Rydberg states have been used to achieve a multiple input AND gate [11.192]. Other quantum logic gates have been implemented in molecular vibrational levels [11.193,194].

Finally, an intense research activity started in solid-state systems, which aims to implement quantum logic gates based on the control of quantum states

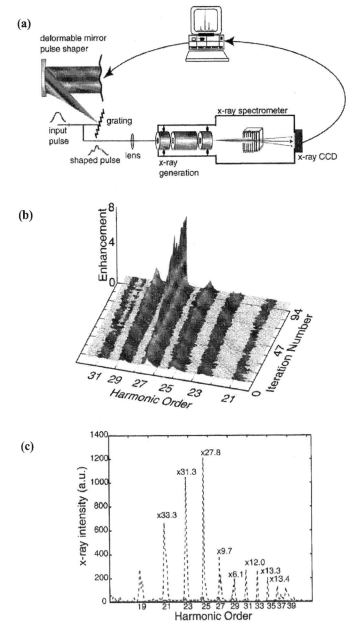

Fig. 11.35. Closed-loop scheme applied to enhance harmonic generation [11.182]: (a) Experimental setup: The intensity of the various harmonics numbers is recorded on the x-ray CCD in the x-ray spectrometer. Each harmonic number can be independently optimized; (b) result of the optimization of the 27th harmonic. The x-ray spectrum is presented as a function of the iteration number. An improvement by a factor of 8 is achieved; (c) maximal enhancement factor for each harmonic number.

by light pulses. (This effort is motivated by the integration and scalability potentialities provided by solid-state devices.) Two strategies are currently under investigation. The first consists of controlling *charge states* in quantum dots: exciton and exciton-molecule coherent dynamics within one single dot using shaped pulses [11.195–197], or charge manipulations in electronically coupled quantum dots [11.198,199]. All these proposals rely on electronic correlations in multicharged dots. The resulting coherent polarization dynamics allows us to build conditional operations. The second approach is based on coherent manipulations of localized *spin states* in separate quantum dots with light pulses. The interest of this approach relies on the fact that spin states are more robust against decoherence than charge states. In one proposed scheme [11.200], the long-range interaction between two spin eigenstates is obtained by Raman coupling via strong laser fields, mediated by the vacuum field of a high finesse microcavity. In another scheme, spin interaction is generated by an optical pulse via virtual excitation of delocalized excitons of the semiconductor matrix [11.201]. This effect is analogous to the Rudemann-Kittel-Kasuya-Yoshida interaction between two magnetic impurities, which is mediated by delocalized conduction electrons or excitons in magnetic materials, except that the intermediate electron-hole pair is produced here by external light. The quantum manipulation can be achieved in the adiabatic limit and is robust against decoherence by spontaneous emission. This leads to some effective light-induced exchange interaction, allowing us to generate and control the entanglement of spin states in two neighboring quantum dots.

References

[11.1]   E.A. Manykin, A.M. Alfanasev: *Sov. Phys. JETP* **25**, 828 (1967)
[11.2]   M. Shapiro, P. Brumer: *Int. Rev. Phys. Chem.* **13**, 187 (1994)
[11.3]   R.J. Gordon, S.A. Rice: *Ann. Rev. Phys. Chem.* **48**, 601 (1997)
[11.4]   P. Brumer, M. Shapiro: Structure and Dynamics of Electronic Excited States, Springer, Berlin, (1999)
[11.5]   W. Pötz, W.A. Schroeder: *Coherent Control in Atoms, Molecules and Semiconductors* Kluwer Academic (1999)
[11.6]   S.A. Rice, M.S. Zhao: Optical Control of Molecular Dynamics, Wiley, New York (2000)
[11.7]   T. Brixner, N.H. Damrauer, G. Gerber: Advances in Atomic, Molecular, and Optical Physics, 46, (2001)
[11.8]   D.J. Tannor, S.A. Rice: *J. Chem. Phys.* **83**, 5013 (1985)
[11.9]   R. Kosloff, S.A. Rice, P. Gaspard, S. Tersigni, D.J. Tannor: *Chem. Phys.* **139**, 201 (1989)
[11.10]   E.D. Potter, J.L. Herek, S. Pedersen, Q. Liu, A.H. Zewail, *Nature* **355**, 66 (1992)
[11.11]   J.L. Herek, A. Materny, A.H. Zewail: *Chem. Phys. Lett.* **228**, 15 (1994)
[11.12]   U. Gaubatz, P. Rudecki, S. Schiemann, K. Bergmann: *J. Chem. Phys.* **92**, 5363 (1990)

[11.13]   N.V. Vitanov, T. Halfmann, B.W. Shore, K. Bergmann: Ann. Rev. Phys. Chem. **52**, 763 (2001)

[11.14]   N.V. Vitanov, M. Fleischhauer, B.W. Shore, K. Bergmann: Advances in Atomic, Molecular, and Optical Physics, **46** (2001)

[11.15]   M. Shapiro, J.W. Hepburn, P. Brumer, *Chem. Phys. Lett.* **149**, 451 (1988).

[11.16]   C. Chen, Y.-Y. Yin, D.S. Elliott: *Phys. Rev. Lett.* **64**, 507 (1990)

[11.17]   S.M. Park, S.P. Lu, R.J. Gordon: *J. Chem. Phys.* **94**, 8622 (1991)

[11.18]   S.P. Lu, S.M. Park, Y. Xie, R.J. Gordon: *J. Chem. Phys.* **96**, 6613 (1992)

[11.19]   L. Zhu, V.D. Kleiman, X. Li, S.P. Lu, K. Trentelman, R.J. Gordon, *Science* **270**, 77 (1995)

[11.20]   L. Zhu, K. Suto, J.A. Fiss, R. Wada, T. Seideman, R.J. Gordon: *Phys. Rev. Lett.* **79**, 4108 (1997)

[11.21]   J.A. Fiss, L. Zhu, R.J. Gordon, T. Seideman: *Phys. Rev. Lett.* **82**, 65 (1999)

[11.22]   V.D. Kleiman, L. Zhu, J. Allen, R.J. Gordon: *J. Chem. Phys.* **103**, 10800 (1995)

[11.23]   G. Xing, X. Wang, X. Huang, R. Bersohn, B. Katz: *J. Chem. Phys.* **104**, 826 (1996)

[11.24]   X. Wang, R. Bersohn, K. Takahashi, M. Kawasaki, H.L. Kim: *J. Chem. Phys.* **105**, 2992 (1996)

[11.25]   H. Lefebvre Brion, T. Seideman, R.J. Gordon: *J. Chem. Phys.* **114**, 9402 (2001)

[11.26]   H.G. Muller, P.H. Bucksbaum, D.W. Schumacher, A. Zavriyev: *J. Phys. B* **23**, 2761 (1990)

[11.27]   Y.-Y. Yin, C. Chen, D.S. Elliott, A.V. Smith: *Phys. Rev. Lett.* **69**, 2353 (1992)

[11.28]   N.B. Baranova, A.N. Chudinov, B.Y. Zel'dovich: *Opt. Commun.* **79**, 116 (1990)

[11.29]   E. Charron, A. Giusti-Suzor, F.H. Mies: *Phys. Rev. Lett.* **75**, 2815 (1995)

[11.30]   B. Sheehy, B. Walker, L.F. DiMauro: *Phys. Rev. Lett.* **74**, 4799 (1995)

[11.31]   E. Dupont, P.B. Corkum, H.C. Liu, M. Buchanan, Z.R. Wasilewski: *Phys. Rev. Lett.* **74**, 3596 (1995)

[11.32]   R. Atanasov, A. Hache, J.L.P. Hughes, H.M. van Driel, J.E. Sipe: *Phys. Rev. Lett.* **76**, 1703 (1996)

[11.33]   E. Charron, A. Giusti-Suzor, F.H. Mies: *Phys. Rev. Lett.* **71**, 692 (1993)

[11.34]   T. Zuo, A.D. Bandrauk: *Phys. Rev. A* **54**, 3254 (1996)

[11.35]   G. Kurizki, M. Shapiro, P. Brumer: *Phys. Rev. B* **39**, 3435 (1989)

[11.36]   A. Hache, Y. Kostoulas, R. Atanasov, J.L.P. Hughes, J.E. Sipe, H.M. van Driel: *Phys. Rev. Lett.* **78**, 306 (1997)

[11.37]   F. Wang, C. Chen, D.S. Elliott: *Phys. Rev. Lett.* **77**, 2416 (1996)

[11.38]   E. Luc Koenig, M. Aymar, M. Millet, J.M. Lecomte, A. Lyras: *Eur. Phys. J.D* **10**, 205 (2000)

[11.39]   M.A. Bouchene, V. Blanchet, C. Nicole, N. Melikechi, B. Girard, H. Ruppe, S. Rutz, E. Schreiber, L. Wöste: *Eur. Phys. J.D* **2**, 131 (1998)

[11.40]   N.F. Scherer, A.J. Ruggiero, M. Du, G.R. Fleming: *J. Chem. Phys.* **93**, 856 (1990)

[11.41]   N.F. Scherer, R.J. Carlson, A. Matro, M. Du, A.J. Ruggiero, V. Romero-rochin, J.A. Cina, G.R. Fleming, S.A. Rice, *J. Chem. Phys.* **95**, 1487 (1991)

[11.42]   V. Blanchet, M.A. Bouchene, B. Girard: *J. Chem. Phys.* **108**, 4862 (1998)

[11.43]   V. Blanchet, C. Nicole, M.A. Bouchene, B. Girard: *Phys. Rev. Lett.* **78**, 2716 (1997)

[11.44]   C. Nicole, M.A. Bouchene, S. Zamith, N. Melikechi, B. Girard, *Phys. Rev. A* **60**, R1755 (1999)
[11.45]   C. Nicole, M.A. Bouchene, B. Girard: *J. Mod. Optics* **49**, 183 (2002)
[11.46]   J.F. Christian, B. Broers, J.H. Hoogenraad, W.J. Van der Zande, L.D. Noordam: *Opt. Commun.* **103**, 79 (1993)
[11.47]   O. Kinrot, I.S. Averbukh, Y. Prior: *Phys. Rev. Lett.* **75**, 3822 (1995)
[11.48]   M.A. Bouchene, C. Nicole, B. Girard: *Opt. Commun.* **181**, 327 (2000)
[11.49]   A.P. Heberle, J.J. Baumberg, K. Köhler: *Phys. Rev. Lett.* **75**, 2598 (1995)
[11.50]   M. Woerner, J. Shah: *Phys. Rev. Lett.* **81**, 4208 (1998)
[11.51]   S. Haacke, R.A. Taylor, R. Zimmermann, I. Bar-Joseph, B. Deveaud: *Phys. Rev. Lett.* **78**, 2228 (1997)
[11.52]   N. Garro, S.P. Kennedy, A.P. Heberle, R.T. Phillips: *Phys. Stat. Sol. B* **221**, 385 (2000)
[11.53]   A.P. Heberle, J.J. Baumberg, E. Binder, T. Kuhn, K. Köhler, K.H. Ploog: *IEEE J. Quantum Elect.* **2**, 769 (1996)
[11.54]   X. Marie, P. Le Jeune, T. Amand, M. Brousseau, J. Barrau, M. Paillard, R. Planel: *Phys. Rev. Lett.* **78**, 3222 (1997)
[11.55]   X. Marie, P. Renucci, S. Dubourg, T. Amand, P. Le Jeune, J. Barrau, J. Bloch, R. Planel: *Phys. Rev. B* **59**, R2494 (1999)
[11.56]   P.C.M. Planken, I. Brener, M.C. Nuss, M.S.C. Luo, S.L. Chuang: *Phys. Rev. B* **48**, 4903 (1993)
[11.57]   M.U. Wehner, M.H. Ulm, D.S. Chemla, M. Wegener: *Phys. Stat. Sol. B* **206**, 281 (1998)
[11.58]   M.U. Wehner, M.H. Ulm, D.S. Chemla, M. Wegener: *Phys. Rev. Lett.* **80**, 1992 (1998)
[11.59]   C.-K. Sun, Y.-K. Huang, J.-C. Liang, A. Abare, S.P. DenBaars: *Appl. Phys. Lett.* **78**, 1201 (2001)
[11.60]   U. Özgür, L. Chang Won, H.O. Everitt: *Phys. Rev. Lett.* **86**, 5604 (2001)
[11.61]   S.P. Kennedy, N. Garro, R.T. Phillips; *Phys. Rev. Lett.* **86**, 4148 (2001)
[11.62]   S. Ogawa, H. Nagano, H. Petek, A.P. Heberle: *Phys. Rev. Lett.* **78**, 1339 (1997)
[11.63]   H. Petek, A.P. Heberle, W. Nessler, H. Nagano, S. Kubota, S. Matsunami, N. Moriya, S. Ogawa: *Phys. Rev. Lett.* **79**, 4649 (1997)
[11.64]   R.S. Judson, H. Rabitz: *Phys. Rev. Lett.* **68**, 1500 (1992)
[11.65]   W.S. Warren, H. Rabitz, M. Dahleh: *Science* **259**, 1581 (1993)
[11.66]   A.M. Weiner: *Rev. Sci. Instr.* **71**, 1929 (2000)
[11.67]   T. Brixner, G. Gerber: *Opt. Lett.* **26**, 557 (2001)
[11.68]   M.A. Dugan, J.X. Tull, W.S. Warren: *J. Opt. Soc. Am. B* **14**, 2348 (1997)
[11.69]   E. Zeek, K. Maginnis, S. Backus, U. Russek, M. Murnane, G. Mourou, H. Kapteyn, G. Vdovin: *Opt. Lett.* **24**, 493 (1999)
[11.70]   F. Verluise, V. Laude, Z. Cheng, C. Spielmann, P. Tournois: *Opt. Lett.* **25**, 575 (2000)
[11.71]   G. Stobrawa, M. Hacker, T. Feurer, D. Zeidler, M. Motzkus, F. Reichel, *Appl. Phys. B* **72**, 627 (2001)
[11.72]   R. Trebino, K.W. DeLong, D.N. Fittinghoff, J.N. Sweetser, M.A. Krumbügel, B.A. Richman, D.J. Kane: *Rev. Sci. Instr.* **68**, 3227 (1997)
[11.73]   C. Dorrer, B. de Beauvoir, C. Le Blanc, J.-P. Rousseau, S. Ranc, P. Rousseau, J.-P. Chambaret, F. Salin: *Appl. Phys. B* **70**, S77 (2000)
[11.74]   L. Lepetit, G. Cheriaux, M. Joffre: *J. Opt. Soc. Am. B* **12**, 2467 (1995)

[11.75]   M. Wefers, K. Nelson: *IEEE J. Quantum Elect.* **32**, 161 (1996)

[11.76]   J.C. Vaughan, T. Feurer, K.A. Nelson: *J. Opt. Soc. Amer. B* **19**, 2489 (2002)

[11.77]   R.M. Koehl, S. Adachi, K.A. Nelson: *J. Chem. Phys.* **110**, 1317 (1999)

[11.78]   R.M. Koehl, K.A. Nelson: *J. Chem. Phys.* **114**, 1443 (2001)

[11.79]   N.S. Stoyanov, D.W. Ward, T. Feurer, K.A. Nelson: *J. Chem. Phys.* **117**, 2897 (2002)

[11.80]   B. Kohler, V.V. Yakovlev, J. Che, J.L. Krause, M. Messina, K.R. Wilson, N. Schwentner, R.M. Whitnell, Y. Yan: *Phys. Rev. Lett.* **74**, 3360 (1995)

[11.81]   C.J. Bardeen, Q. Wang, C.V. Shank: *Phys. Rev. Lett.* **75**, 3410 (1995)

[11.82]   A. Assion, T. Baumert, J. Helbing, V. Seyfried, G. Gerber: *Chem. Phys. Lett.* **259**, 488 (1996)

[11.83]   J.S. Melinger, A. Hariharan, S.R. Gandhi, W.S. Warren: *J. Chem. Phys.* **95**, 2210 (1991)

[11.84]   D.J. Maas, C.W. Rella, P. Antoine, E.S. Toma, L.D. Noordam: *Phys. Rev. A* **59**, 1374 (1999)

[11.85]   G. Cerullo, C.J. Bardeen, Q. Wang, C.V. Shank: *Chem. Phys. Lett.* **262**, 362 (1996)

[11.86]   C.J. Bardeen, V.V. Yakovlev, J.A. Squier, K.R. Wilson: *J. Am. Chem. Soc.* **120**, 13023 (1998)

[11.87]   C.J. Bardeen, J.S. Cao, F.L.H. Brown, K.R. Wilson: *Chem. Phys. Lett.* **302**, 405 (1999)

[11.88]   A.H. Buist, M. Muller, R.I. Ghauharali, G.J. Brakenhoff, J.A. Squier, C.J. Bardeen, V.V. Yakovlev, K.R. Wilson: *Opt. Lett.* **24**, 244 (1999)

[11.89]   B. Broers, H.B. van Linden van den Heuvell, L.D. Noordam: *Phys. Rev. Lett.* **69**, 2062 (1992)

[11.90]   P. Balling, D.J. Maas, L.D. Noordam: *Phys. Rev. A* **50**, 4276 (1994)

[11.91]   B. Chatel, J. Degert, S. Stock, B. Girard: *Phys. Rev. A,* **68**, 041402(R) (2003)

[11.92]   S. Zamith, J. Degert, S. Stock, B. De Beauvoir, V. Blanchet, M.A. Bouchene, B. Girard: *Phys. Rev. Lett.* **87**, 033001 (2001)

[11.93]   D. Meshulach, Y. Silberberg: *Nature* **396**, 239 (1998)

[11.94]   D. Meshulach, Y. Silberberg: *Phys. Rev. A* **60**, 1287 (1999)

[11.95]   N. Dudovich, B. Dayan, S.H.G. Faeder, Y. Silberberg, *Phys. Rev. Lett.* **86**, 47 (2001)

[11.96]   N. Dudovich, D. Oron, Y. Silberberg: *Phys. Rev. Lett.* **88**, 123004 (2002)

[11.97]   J. Degert, W. Wohlleben, B. Chatel, M. Motzkus, B. Girard: *Phys. Rev. Lett.* **89**, 203003 (2002)

[11.98]   W. Wohlleben, J. Degert, A. Monmayrant, B. Chatel, M. Motzkus, B. Girard: *Appl. Phys. B* accepted (2004)

[11.99]   D. Oron, N. Dudovich, D. Yelin, Y. Silberberg, *Phys. Rev. Lett.* **88**, 063004 (2002)

[11.100] N. Dudovich, D. Oron, Y. Silberberg: *Nature* **418**, 512 (2002)

[11.101] T. Baumert, T. Brixner, V. Seyfried, M. Strehle, G. Gerber: *Appl. Phy. B* **65**, 779 (1997)

[11.102] A. Assion, T. Baumert, M. Bergt, T. Brixner, B. Kiefer, V. Seyfried, M. Strehle, G. Gerber: *Science* **282**, 919 (1998)

[11.103] S. Vajda, A. Bartelt, E.-C. Kaposta, T. Leisner, C. Lupulescu, S. Minemoto, P. Rosendo-Francisco, L. Wöste: *Chem. Phys.* **267**, 231 (2001)

[11.104] C. Daniel, J. Full, L. González et al., *Chem. Phys.* **267**, 247 (2001)
[11.105] T. Brixner, B. Kiefer, G. Gerber: *Chem. Phys.* **267**, 241 (2001)
[11.106] R.J. Levis, G.M. Menkir, H. Rabitz: *Science* **292**, 709 (2001)
[11.107] C.J. Bardeen, V.V. Yakovlev, K.R. Wilson, S.D. Carpenter, P.M. Weber, W.S. Warren, *Chem. Phys. Lett.* **280**, 151 (1997)
[11.108] T. Brixner, N.H. Damrauer, P. Niklaus, G. Gerber: *Nature* **414**, 57 (2001)
[11.109] R.A. Bartels, T.C. Weinacht, S.R. Leone, C. Kapteyn, M.M. Murnane: *Phys. Rev. Lett.* **88**, 033001 (2002)
[11.110] J.L. Herek, W. Wohlleben, R.J. Cogdell, D. Zeidler, M. Motzkus: *Nature* **417**, 533 (2002)
[11.111] T. Hornung, R. Meier, M. Motzkus: *Chem. Phys. Lett.* **326**, 445 (2000)
[11.112] J.A. Nelder, R. Mead: *Comp. J.* **7**, 308 (1965)
[11.113] N.A. Gershenfeld: The Nature of Mathematical Modeling, Cambridge University Press (1999)
[11.114] M. Hacker, G. Strobrawa, T. Feurer: *Opt. Exp.* **9**, 191 (2001)
[11.115] S. Kirkpatrick, C.D. Gelatt, M.P. Vecchi: *Science* **220**, 671 (1983)
[11.116] J.H. Holland: Adaptation in Natural and Artificial Systems, Ann Arbor (1975)
[11.117] D.B. Fogel, A.J. Owens, M.J. Walsh: *Artificial Intelligence through Simulated Evolution*, New York (1966)
[11.118] T. Back and H.-P. Schwefel: *Evol. Comp.* **1**, 1 (1993)
[11.119] B.J. Pearson, J.L. White, T.C. Weinacht, P.H. Bucksbaum: *Phys. Rev. A* **63**, 063412 (2001)
[11.120] D. Zeidler, S. Frey, K.-L. Kompa, M. Motzkus: *Phys. Rev. A* **64**, 023420 (2001)
[11.121] D.A. Stariolo, C. Tsallis, *Annu. Rev. Comp. Physic.* (1995).
[11.122] B. Amstrup, J.D. Doll, R.A. Sauerbrey, G. Szabó, A. Lorincz: *Phys. Rev. A* **48**, 3830 (1993)
[11.123] A. Glass, T. Rozgonyi, T. Feurer, R. Sauerbrey, G. Szabo: *Appl. Phys. B* **71**, 262 (2000)
[11.124] D. Meshulach, D. Yelin, Y. Silberberg: *Opt. Commun.* **138**, 345 (1997)
[11.125] L. Chambers: Practical Handbook of Genetic Algorithm: New Frontiers, CRC Press, New York (1995)
[11.126] U. Siegner, M. Haiml, J. Kunde, U. Keller: *Opt. Lett.* **27**, 315 (2002)
[11.127] T. Hornung, R. Meier, D. Zeidler, K.-L. Kompa, D. Proch, M. Motzkus: *Appl. Phys. B* **71**, 277 (2000)
[11.128] D. Zeidler, T. Hornung, D. Proch, M. Motzkus: *Appl. Phys. B* **70**, S125 (2001)
[11.129] R. Bartels, S. Backus, E. Zeek, L. Misoguti, G. Vdovin, I.P. Christov, M.M. Murnane, C. Kapteyn: *Nature* **406**, 164 (2000)
[11.130] T.C. Weinacht, J.L. White, P.H. Bucksbaum: *J. Phys. Chem. A* **103**, 10166 (1999)
[11.131] D. Zeidler, S. Frey, W. Wohlleben, M. Motzkus, F. Busch, T. Chen, W. Kiefer, A. Materny: *J. Chem. Phys.* **116**, 5231 (2002)
[11.132] H. Rabitz, R. de Vivie-Riedle, M. Motzkus, K. Kompa: *Science* **288**, 824 (2000)
[11.133] D. Goswami, M.R. Fetterman, W.S. Warren, W. Yang: *J. Opt. Comm.* **21**, 694 (2001)
[11.134] Z.M. Lu, H. Rabitz: *J. Phys. Chem.* **99**, 13731 (1995)

[11.135] C. Daniel, J. Full, L. Gonzalez, C. Lupulescu, J. Manz, A. Merli, S. Vajda,
          L. Woste: *Science* **299**, 536 (2003)
[11.136] N.V. Vitanov: *Phys. Rev. A* **59**, 988 (1999)
[11.137] J.G. Story, D.I. Duncan, T.F. Gallagher: *Phys. Rev. Lett.* **70**, 3012 (1993)
[11.138] R.R. Jones, C.S. Raman, D.W. Schumacher, P.H. Bucksbaum: *Phys. Rev.
          Lett.* **71**, 2575 (1993)
[11.139] P. Dietrich, P. Corkum: *J. Chem. Phys.* **97**, 3187 (1992)
[11.140] C. Wunderlich, H. Figger, T.W. Hansch: *Chem. Phys. Lett.* **256**, 43 (1996)
[11.141] P.H. Bucksbaum, A. Zavriyev, H.G. Muller, D.W. Schumacher: *Phys. Rev.
          Lett.* **64**, 1883 (1990)
[11.142] V. Schyja, T. Lang, H. Helm: *Phys. Rev. A* **57** (1998)
[11.143] R.B. Vrijen, J.H. Hoogenraad, H.G. Muller, L.D. Noordam: *Phys. Rev.
          Lett.* **70**, 3016 (1993)
[11.144] B.M. Garraway, K.-A. Suominen: *Phys. Rev. Lett.* **80**, 932 (1998)
[11.145] M. Rodriguez, K.-A. Suominen, B.M. Garraway: *Phys. Rev. A* **62**, 053413
          (2000)
[11.146] V.S. Malinovsky, J.L. Krause: *Chem. Phys.* **267**, 47 (2001)
[11.147] V.S. Malinovsky, J.L. Krause: *Phys. Rev. A* **63**, 043415 (2001)
[11.148] K. Suominen, B.M. Garraway, S. Stenholm: *Phys. Rev. A* **45**, 3060 (1992)
[11.149] H. Stapelfeldt, H. Sakai, E. Constant, P.B. Corkum: *Phys. Rev. Lett.* **79**,
          2787 (1997)
[11.150] S. Banerjee, D. Mathur, G. Ravindra Kumar: *Phys. Rev. A* **63**, 045401
          (2001)
[11.151] E. Springate, F. Rosca-Pruna, H.L. Offerhaus, M. Krishnamurthy, M. J.J.
          Vrakking: *J. Phys. B* **34**, 4939 (2001)
[11.152] F. Rosca-Pruna, M.J.J. Vrakking: *Phys. Rev. Lett.* **87**, 153902 (2001)
[11.153] D. Normand, L.A. Lompre, C. Cornaggia: *J. Phys. B* **25**, L497 (1992)
[11.154] B. Friedrich, D. Herschbach: *Phys. Rev. Lett.* **74**, 4623 (1995)
[11.155] J.H. Posthumus, J. Plumridge, M.K. Thomas, K. Codling, L.J. Frasinski,
          A.J. Langley, P.F. Taday: *J. Phys. B* **31**, L553 (1998)
[11.156] C. Ellert, P.B. Corkum: *Phys. Rev. A* **59**, R3170 (1999)
[11.157] J.J. Larsen, K. Hald, N. Bjerre, H. Stapelfeldt: *Phys. Rev. Lett.* **85**, 2470
          (2000)
[11.158] J.J. Larsen, I. Wendt-Larsen, H. Stapelfeldt: *Phys. Rev. Lett.* **83**, 1123
          (1999)
[11.159] D.M. Villeneuve, S.A. Aseyev, P. Dietrich, M. Spanner, M.Y. Ivanov, P.B.
          Corkum: *Phys. Rev. Lett.* **85**, 542 (2000)
[11.160] J. Karczmarek, J. Wright, P. Corkum, M. Ivanov: *Phys. Rev. Lett.* **82**,
          3420 (1999)
[11.161] S.T. Cundiff, A. Knorr, J. Feldmann, S.W. Koch, E.O. Göbel, H. Nickel:
          Phys. Rev. Lett. **73**, 1178 (1994)
[11.162] C. Fürst, A. Leitenstorfer, A. Nutsch, G. Tränkle, A. Zrenner: *Physica
          status solidi (b)* **204**, 20 (1997)
[11.163] O.D. Mücke, T. Tritschler, M. Wegener, U. Morgner, F.X. Kärtner, *Phys.
          Rev. Lett.* **87**, 057401 (2001)
[11.164] A. Schülzgen, R. Binder, M.E. Donovan, M. Lindberg, K. Wundke, H.M.
          Gibbs, G. Khitrova, N. Peyghambarian: *Phys. Rev. Lett.* **82**, 2346 (1999)
[11.165] Y. Nakamura, Y.A. Pashkin, J.S. Tsai: *Phys. Rev. Lett.* **87**, 246601 (2001)
[11.166] F. Quochi, G. Bongiovanni, A. Mura, J.L. Staehli, B. Deveaud, R.P. Stan-
          ley, U. Oesterle, R. Houdre: *Phys. Rev. Lett.* **80**, 4733 (1998)

[11.167] B.R. Mollow: *Phys. Rev.* **188**, 1969 (1969)
[11.168] B.R. Mollow: *Phys. Rev. A* **5**, 2217 (1972)
[11.169] F.Y. Wu, S. Ezekiel, M. Ducloy, B.R. Mollow: *Phys. Rev. Lett.* **38**, 1077 (1977)
[11.170] M. Saba, F. Quochi, C. Ciuti, D. Martin, J.L. Staehli, B. Deveaud, A. Mura, G. Bongiovanni: *Phys. Rev. B* **62**, R16322 (2000)
[11.171] L. Bányai, Q.T. Vu, B. Mieck, H. Haug: *Phys. Rev. Lett.* **81**, 882 (1998)
[11.172] T.H. Stievater, X. Li, D.G. Steel, D. Gammon, D.S. Katzer, D. Park, C. Piermarocchi, L.J. Sham: *Phys. Rev. Lett.* **87**, 133603 (2001)
[11.173] A. Zrenner, E. Beham, S. Stufler, F. Findeis, M. Bichler, G. Abstreiter: *Nature* **418**, 612 (2002)
[11.174] H. Kamada, H. Gotoh, J. Temmyo, T. Takagahara, H. Ando: *Phys. Rev. Lett.* **87**, 246401 (2001)
[11.175] H. Htoon, T. Takagahara, D. Kulik, O. Baklenov, A.L. Holmes, Jr., C.K. Shih: *Phys. Rev. Lett.* **88**, 1 (2002)
[11.176] M. Shapiro, E. Frishman, P. Brumer: *Phys. Rev. Lett.* **84**, 1669 (2000)
[11.177] P. Kral, M. Shapiro: *Phys. Rev. Lett.* **87**, 1 (2001)
[11.178] C. Zenghu, A. Rundquist, W. Haiwen, M.M. Murnane, H.C. Kapteyn: *Phys. Rev. Lett.* **79**, 2967 (1997)
[11.179] M. Hentschel, R. Kienberger, C. Spielmann, G. Reider, A.N. Milosevic, T. Brabec, P. Corkum, U. Heinzmann, M. Drescher, F. Krausz: *Nature 414*, 509 (2001)
[11.180] I.P. Christov, R. Bartels, H.C. Kapteyn, M.M. Murnane: *Phys. Rev. Lett.* **86**, 5458 (2001)
[11.181] I.P. Christov: *J. Opt. Soc. Am. B* **18**, 1877 (2001)
[11.182] R. Bartels, S. Backus, I. Christov, H. Kapteyn, M. Murnane: *Chem. Phys.* **267**, 277 (2001)
[11.183] D.G. Lee, J.-H. Kim, K.-H. Hong, C.H. Nam: P*hys. Rev. Lett.* **87**, 243902 (2001)
[11.184] C. Xi, I.C. Shih: *Phys. Rev. A* **64**, 021403 (2001)
[11.185] P. Balcou, R. Haroutunian, S. Sebban et al., *Appl. Phys. B* **6**, 509 (2002)
[11.186] S. Cavalieri, R. Eramo, M. Materazzi, C. Corsi, M. Bellini: *Phys. Rev. Lett.* **89**, 133002 (2002)
[11.187] T.C. Weinacht, J. Ahn, P.H. Bucksbaum, *Phys. Rev. Lett.* **80**, 5508 (1998)
[11.188] J. Ahn, T.C. Weinacht, P.H. Bucksbaum: *Science* **287**, 463 (2000)
[11.189] J. Ahn, D.N. Hutchinson, C. Rangan, P.H. Bucksbaum: *Phys. Rev. Lett.* **86**, 1179 (2001)
[11.190] L.K. Grover: *Phys. Rev. Lett.* **79**, 325 (1997)
[11.191] L.K. Grover: *Phys. Rev. Lett.* **79**, 4709 (1997)
[11.192] Z. Amitay, R. Kosloff, S.R. Leone: *Chem. Phys. Lett.* **359**, 8 (2002)
[11.193] V.V. Lozovoy, M. Dantus: *Chem. Phys. Lett.* **351**, 213 (2002)
[11.194] C.M. Tesch, R.de Vivie-Riedle: *Phys. Rev. Lett.* **89**, 157901 (2002)
[11.195] F. Troiani, U. Hohenester, E. Molinari: *Phys. Rev. B* **62**, R2263 (2000)
[11.196] P.-C., C. Piermarocchi, L.J. Sham: *Phys. Rev. Lett.* **87**, 067401 (2001)
[11.197] G. Chen, T.H. Stievater, E.T. Batteh, Xiaoqin-Li, D.G. Steel, D. Gammon, D.S. Katzer, D. Park, L.J. Sham: *Phys. Rev. Lett.* **88**, 117901 (2002)
[11.198] E. Biolatti, R.C. Iotti, P. Zanardi, F. Rossi: *Phys. Rev. Lett.* **85**, 5647 (2000)
[11.199] M. Bayer, P. Hawrylak, K. Hinzer, S. Fafard, M. Korkusinski, Z.R. Wasilewski, O. Stern, A. Forchel: *Science* **291**, 451 (2001)

[11.200] A. Imamoglu, D.D. Awschalom, G. Burkard, D.P. DiVincenzo, D. Loss, M. Sherwin, A. Small: *Phys. Rev. Lett.* **83**, 4204 (1999)
[11.201] C. Piermarocchi, C. Pochung, L.J. Sham, D.G. Steel: *Phys. Rev. Lett.* **89**, 167402 (2002)

Attosecond Pulses

E. Constant and E. Mével

With 17 figures

12.1 Introduction

It is now possible to generate optical pulses with a duration corresponding to only a few optical cycles of the electromagnetic radiation in a wide spectral range. The generation of such short pulses is usually performed in two ways.

Low-energy pulses can be obtained directly from femtosecond oscillator and the minimum pulse duration supported by broadband gain media, such as the titanium-doped sapphire (Ti:Sa), is currently around 10 fs. Careful intracavity gain shaping allows ultimate pulse duration below 5 fs in Ti:Sa oscillators [12.1,2]. However amplification of these pulses results in a spectral narrowing, and the shortest pulses obtained until now above the mJ level are 17.5 fs long [12.3].

Those amplified pulses can then be further shortened by postcompression techniques and the shortest high-energy visible pulses obtained by these techniques have a duration of ~ 5 fs [12.4,5], which corresponds to only 2 optical cycles of the central wavelength and could not be significantly shortened in this wavelength range.

The possibility to generate and use such short pulses allowed many fascinating applications in the time domain through pump-probe techniques. However, many events occur on a subfemtosecond time scale, and their study requires shorter pulses. This includes, for instance, most of the electronic motion, the evolution of molecules in highly excited states or even the evolution of molecular hydrogen, which is one of the best-known systems in theory but still cannot be experimentally studied in the time domain. Obtaining pulses shorter than 1 fs and entering the attosecond domain (1 as $= 10^{-18}$ s) will therefore open new applications in physics.

As for femtosecond pulses generated in mode-locked oscillators, attosecond pulses require phase-locked frequencies over an even larger bandwidth. There are two known processes that provide a coherent spectra over a bandwidth broad enough to support attosecond emission. A first one is based on stimulated Raman scattering [12.6–8]. A subsequent spectral broadening of a UV pulse is obtained through phase modulation by impulsively exciting coherent vibration of molecules [12.9,10]. This method promises subfemtosecond sub-cycle pulse generation. So far, experimental tests have demonstrated emission of a periodic train of 6 fs pulses [12.11] and generation of a single 4 fs pulse [12.12], both at 400 nm. Here, we will focus on the second process, which allows attosecond emission in the XUV spectral range. This process relies on high-order harmonic generation (HHG), which was discovered in 1987 [12.13,14]. The light emitted through high-order harmonic generation consists of a comb of many VUV lines (equally spaced and of approximately equal amplitudes) in a spectral range that can be broader than 100 eV. Shortly after the discovery of this process, it was understood that the emitted spectrum was large enough to support trains of attosecond pulses [12.15–17] The theoretical understanding of high-order harmonic generation [12.18–21] even quickly revealed that these harmonics can be phase-locked [12.22] and that an isolated attosecond pulse can be created through high-order harmonic generation [12.23,24].

To illustrate how this can be done, we first present the process of high-order harmonic generation and its main characteristics. Although this process is now very well simulated with quantum models, we will present a simple semiclassical model [12.18,19] for understanding how the response of an atom excited by a strong laser field can lead to high-order harmonic generation. Historically, this model allowed the understanding of this process. Although this model may appear oversimplified, it contains the essence of high-order harmonic generation.

Being a coherent process, high-order harmonic generation not only relies on the single-atom response but also on collective effects such as phase matching (for instance, see [12.25] and references therein), which will only be briefly mentioned in this chapter.

After this presentation, we describe how high-order harmonic generation can be considered as a source of train of attosecond pulses and how a single attosecond pulse can be extracted from this train. Then we present some techniques for measuring harmonic pulses or attosecond pulses, and finally we present possible applications of attosecond pulses.

12.2 High-Order Harmonic Generation: A Coherent, Short-Pulse XUV Source

High-order harmonic generation occurs when an ultrashort intense laser field is focused onto an atomic gas target. The typical experimental setup, depicted in Fig. 12.1 consists of a short-pulse laser focused by a lens (or mirror) onto a gas medium located in a vacuum chamber where the emitted VUV light can propagate.

High-order harmonics are produced by focusing short laser pulses in a gas medium at intensities where substantial ionization of the gas can occur.

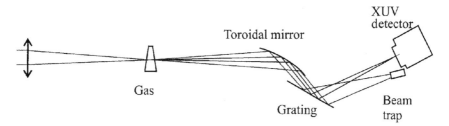

Fig. 12.1. Schematic setup for high-order harmonic generation. An intense laser is focused onto a gas target and nonlinearly interacts with it. The light emitted in the forward direction is then analyzed by a VUV spectrometer

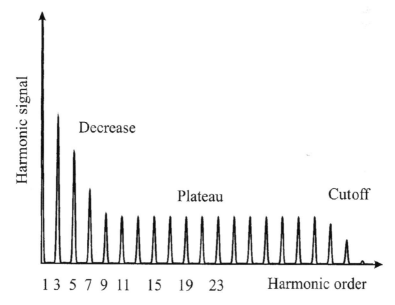

Fig. 12.2. Typical spectrum obtained through HHG as a function of the harmonic order ($q = \omega_q/\omega_0$). It shows a fast decrease in efficiency ("perturbative like") followed by a region where the efficiency is approximately constant (plateau) and then falls down very rapidly (cutoff)

Typically, when rare gas atoms are irradiated by sub-100-fs pulses, suitable intensities for high-order harmonic generation range from 10^{13} to 10^{15} W/cm^2. Inside the gas medium, the intense laser non-linearly interacts with the ionizing gas. This nonlinear interaction leads to the creation of new frequencies and, after the gas medium, one can observe odd high-order harmonics of the fundamental frequency copropagating with the fundamental laser beam.

A schematic spectrum obtained through high-order harmonic generation is presented in Fig. 12.2, which illustrates some important features of high-order harmonics generation:

- The spectrum consists of lines separated by twice the frequency of the fundamental laser field (ω_0) and only odd harmonics are present ($\omega_q = q\omega_0 = (2n + 1)\omega_0$).
- After a fast decrease in efficiency for the low-order harmonics (the efficiency being defined as the ratio of the energy contained in a given harmonic peak to the initial energy of the fundamental pulse), the high-order harmonics all have roughly the same amplitude (presence of a plateau) and then abruptly disappear for frequencies higher than the cutoff frequency, ω_c. Although the initial decrease is typical of perturbative interactions, the presence of such a plateau is the signature of a nonperturbative interaction regime.

The efficiency of high-order harmonic generation and its spectral extension depends a lot on the gas (mainly through its ionization potential I_p) and on the fundamental laser characteristics (mainly through its frequency ω_0, peak intensity $I_{\max}$, pulse duration τ, and polarization state). Generally, one can observe that:

- The higher I_p, the higher the cutoff frequency.
- The smaller ω_0, the higher the cutoff frequency.
- The shorter τ, the higher the cutoff frequency.
- For a given VUV wavelength of the plateau and a given atomic species, the smaller ω_0, the lower the efficiency.
- For high harmonics, the efficiency is maximal when the fundamental field is linearly polarized and quickly decreases when the ellipticity is increasing.
- When the intensity of the fundamental pulse is increased, the high-order harmonic generation efficiency first increases quickly and then saturates or remains roughly constant when the peak intensity exceeds the so-called saturation intensity (at which all atoms are ionized).

As will be seen in the next paragraph, all these characteristics arise from the single-atom response to the strong field excitation.

HHG is also a coherent process and many of the VUV beam properties depend on collective effects. For instance the efficiency depends nonlinearly on the gas pressure [12.26], which implies that the radiation emitted by different atoms can interfere constructively. The emitted harmonic beam is a coherent beam that can be very well collimated with a divergence similar to the one of the fundamental beam [12.27–29].

12.3 Semiclassical Picture of HHG

High-order harmonic generation is now very well understood theoretically and quantum calculation (resolution of the time-dependent Schrödinger equation,

TDSE) [12.30,31] or even approximate quantum calculations (Strong field approximation, SFA) [12.20,21] associated with propagation codes can simulate very nicely most of the experimental results.

In this chapter we will present a much simpler theory, so called semiclassical theory (or the simple man model), which also reproduces reasonably well the experimental findings and provides a clear understanding of the high-order harmonic generation process. Strictly speaking, this semiclassical model is valid only in the so-called tunnel regime defined by

$$\hbar\omega_0 \ll I_p \ll U_p, \tag{12.1}$$

where I_p is the ionization potential of the atom and U_p is the so-called ponderomotive energy that corresponds to the average kinetic energy of a free electron oscillating in the laser field (of frequency ω_0 and intensity I). It also represents the typical energy that the intense laser can provide to an electron and scales as I/ω_0^2. For a short pulse centered at a wavelength of 800 nm (Ti:Sa), U_p (in eV) is equal to $6I$ (with I in units of 10^{14} W/cm^2). The description of this semiclassical model for high-order harmonic generation requires some prerequisite knowledge about atomic ionization and about intense laser-electron interaction.

12.3.1 Atomic Ionization in the Tunnel Domain

In this tunnel domain, the condition $\hbar\omega_0 \ll I_p$ implies that the absorption of many photons is necessary to ionize an atom. An accurate description of the ionization mechanisms requires therefore many simulations, and nonperturbative methods are necessary. However, in this domain the photonic character of the laser field can be partly neglected and quasistatic approximate models have been successfully developed to simulate strong field-atom ionization via tunnel ionization.

According to these models, an atom is described as an ionic core and an electron. The ionic core creates a Coulomb potential in which the electron evolves. The intense laser field is simply considered as a slowly oscillating electric field $E(t) = E_0(t)\cos(\omega_0 t)$. When an atom is irradiated by the strong laser field, the total potential is the sum of the ionic core potential and the electric interaction potential $q\boldsymbol{E}.\boldsymbol{r}$. It results in a potential barrier in the direction of the laser electric field. Therefore, when this field is very strong the electron has some probability to escape the ionic core by tunneling through the potential barrier. This is illustrated in Fig. 12.3, where one can see the potential created by the ionic core only (left) and the total potential in presence of the field (right).

The probability for an electron to tunnel through a potential barrier is well known and depends on the initial energy level (I_p for an atom initially in its ground state) and on the width of the barrier (defined by the field strength, $E = (2I/\varepsilon_0 c)^{1/2}$). A general formula to describe tunnel ionization of the atom

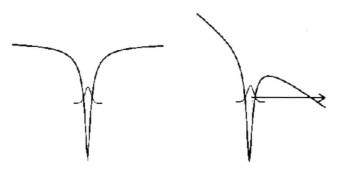

Fig. 12.3. Schematic of the potential created by the ionic core only (left) and by the strong field + ionic core (right). When the field is strong enough, the electron can escape the core attraction and reach the continuum by tunneling through the potential barrier

was derived by Ammossov, Delone and Krainov [12.32]. This instantaneous ionization rate W_{dc} (in s^{-1}) is given by

$$W_{dc} = \omega_s |C_{n^*l^*}|^2 G_{lm} (4\omega_s/\omega_t)^{2n^*-m-1} \exp(-4\omega_s/3\omega_t), \tag{12.2}$$

where $\omega_s = I_h/\hbar$, (with I_h being the ionization potential of hydrogen), $\omega_t = qE/(2m_e I_p)^{1/2}$, (with I_p the ionization potential of the atom, m_e and q being, respectively, the mass and charge of the electron, Z, the charge of the final ion, and E the laser electric field), $n^* = Z(I_h/I_p)^{1/2}$, $G_{lm} = (2l + 1)(l+|m|)!(2^{-|m|})/|m|!(l-|m|)!$ (l and m being the azimuthal and magnetic quantum numbers, respectively), $|C_{n^*l^*}|^2 = 2^{2n^*}[n^*\Gamma(n^*+l^*+1)\Gamma(n^*-l^*)]^{-1}$ with $l^* = 0$ for $l \ll n$ or $l^* = n^* - 1$ otherwise.

For a slowly varying pulse envelope, this rate can be averaged over one optical period of the laser field and leads to

$$W_{ac}(\text{in } s^{-1}) = (3\omega_t/2\pi\omega_s)^{1/2}W_{dc}. \tag{12.3}$$

This rate agrees well with experimentally observed ionization in the tunnel domain [12.33], and this simple formula is therefore widely used. Moreover, the relative crudity of the model (classical field, one discrete state, a continuum) is validated by the structureless photoelectron spectra detected in this regime [12.34].

12.3.2 Electronic Motion in an Electric Field

When an atom is ionized, an electron is released in a strong electric field, and the corresponding electric force will impose its motion. Even by neglecting the core attraction and considering a free electron, the electron motion is strongly influenced by the time of ionization, t_i. Its classical motion is defined by:

$$m\frac{\partial^2 \mathbf{r}}{\partial t^2} = q\mathbf{E}\cos(\omega_0 t). \tag{12.4}$$

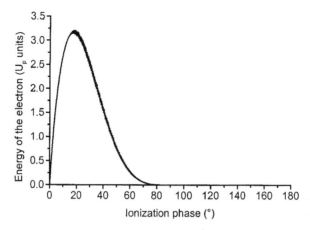

Fig. 12.4. Kinetic energy of the electrons, released with no kinetic energy at their first return at their initial position ($x = 0$) as a function of the phase at ionization

For a linearly polarized field (along x), we have

$$m\frac{\partial^2 x}{\partial t^2} = qE\cos(\omega_0 t).$$ (12.5)

from which we can find the electron velocity:

$$V(t) = [qE/(m\omega_0)](\sin\omega_0 t - \sin\omega_0 t_i) + V_0$$ (12.6)

(V_0 being the initial velocity) in the slowly varying envelope approximation ($E(t) \approx cte$ during one optical cycle). For a zero initial kinetic energy, its average kinetic energy (ponderomotive energy) is then:

$$U_p = q^2 E^2/(4m\omega_0^2).$$ (12.7)

Again, this ponderomotive energy is a very important parameter in the tunnel regime in which we can consider that an electron mainly interacts with the electric field of the laser. It can provide to the electron an energy that is on the order of U_p.

For the specific case where the electron is released in the continuum without any kinetic energy ($V_0 = 0$), the kinetic energy of the electron when it comes back at its initial position ($x = 0$) is simply defined by the ionization time, t_i, as shown in Fig. 12.4. One can see, in this figure, that the maximum kinetic energy of the electrons coming back at their initial position, $x = 0$, is $3.17U_p$ and is achieved for an ionization phase $\omega_0 t_i = 18°$.

High-order harmonic generation usually occurs at the limit of this tunnel regime ($\hbar\omega_0 \leq I_p \approx U_p$), but it was quickly discovered that U_p is a key parameter for this process. Indeed, the cutoff frequency, ω_c, is linked to U_p by:

$$\hbar\omega_c = I_p + 3.17U_p$$ (12.8)

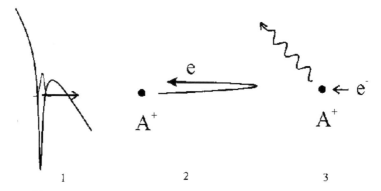

Fig. 12.5. Schematic of the three steps of HHG: (1) Ionization, (2) acceleration by
the laser electric field, (3) radiative recombination

12.3.3 Semiclassical View of HHG

In the "simple man model," high-order harmonic generation consists of a
three-step phenomena; the steps are illustrated in Fig. 12.5.

1. An atom irradiated by a strong laser field can be ionized at a time t_i,
   creating an ion (the parent ion) and a free electron with no initial kinetic
   energy.
2. The electron is accelerated by the laser electric field. When the electric
   field changes its sign, it may (or not) drive back the electron in the vicinity
   of the parent ion.
3. If the electron collides with the parent ion, the ion and electron can then
   recombine and the total extra energy (i.e., kinetic energy, E_c, of the electron plus the binding energy of the atom (i.e., I_p)) is released by emitting
   a photon of energy $\hbar\omega = E_c + I_p$.

This model allows us to understand the cutoff law $\hbar\omega_c = I_p + 3.17U_p$.
Indeed, an electron released (with zero initial velocity) at a given time, accelerated and driven back to its initial position, has a maximum kinetic energy
of $3.17U_p$ (Fig. 12.4) when the collision occurs. Therefore the emitted photon energy cannot exceed $I_p + 3.17U_p$. This cutoff law was discovered through
numerical simulations [12.35], interpreted with the semiclassical model as presented earlier [12.18,36] and confirmed experimentally [12.37].

It should then be understood that this process is probabilistic and one
cannot say whether the atom is ionized or not but rather that at some time the
atom has a probability to get ionized. These three processes can therefore lead
to the emission of a VUV photon (or more precisely, to the emission of VUV
radiation) every time that the atom has a strong probability to be ionized
(first step). This occurs when the electric field is maximum and therefore
with a periodicity of half an optical cycle of the fundamental. A single atom
therefore has some probability to emit XUV photons periodically.

Furthermore, one should not consider the electron as a classical particle but rather as a quantum wave packet that oscillates with its own phase. In terms of waves, the three steps correspond to a transition from the initial wave function (ground state of the atom) to the continuum, then an evolution of the wave packet in the continuum and finally a transition from the continuum back to the initial wave function associated with the emission of a radiation [12.20]. The phase of this radiation is therefore imposed by the evolution of the wave packet phase in the continuum (which is imposed by the fundamental field) and linked to the phase of the fundamental. As will be stressed later, one finds that this phase accumulated by the wave packet evolution in the continuum is proportional to the laser intensity $\varphi_q = \alpha_q I$ [12.39,40]. The important point is that the phases of the XUV radiation are linked to the fundamental laser phase. These harmonics are therefore phase locked with the fundamental laser [12.41].

The fact that the XUV phase is linked to the laser phase makes the process coherent. This implies that not only the XUV energy bursts are emitted periodically but also that the associated XUV field is emitted periodically in phase.

With this model, one can see high-order harmonic generation as a periodic emission of coherent radiation (note that this picture is valid, provided the field strength is constant and the atomic ground state is not significantly depleted by ionization). The symmetry of the ion potential ensures that the periodicity of the XUV emission is $T_0/2$ and that the phase of the emitted XUV light is locked to the phase of the fundamental in the same way at each half optical period. However, the sign of the fundamental field changes at each half optical period, and therefore does the XUV field. The real periodicity of the emission is therefore T_0 with an antiperiodicity of $T_0/2$ (periodicity of $T_0/2$ in intensity). The total XUV emitted field, at a frequency ω_q, is therefore:

$$E_{\omega q} = \int A_{\omega q} \exp -i(\omega_q t + \varphi_q). \tag{12.9}$$

Or by considering the periodicity of the process and summing over many half optical period (with index n):

$$E_{\omega q} = \sum A_{\omega q}(-1)^n \exp -i(\omega_q(t + nT_0/2) + \varphi_q) \tag{12.10a}$$

$$E_{\omega q} = A_{\omega q} \exp -i(\omega_q t + \varphi_q) \sum \exp -i(\omega_q nT_0/2 - n\pi). \tag{12.10b}$$

The first term $(A_{\omega q} \exp -i(\omega_q t + \varphi_q))$ corresponds to a wave oscillating at the frequency ω_q, and the second term $(\sum \exp -i(\omega_q nT_0/2 - n\pi))$ defines its amplitude. For a large number of half optical periods $(n \gg 1)$, this amplitude is

$$\sum \exp -i(\omega_q nT_0/2 - n\pi) = (1 - \exp -ni(\omega_q T_0/2 - \pi))/(1 - \exp -i(\omega_q T_0/2 - \pi)), \tag{12.11}$$

which is large (and independent on n) only for $(1 - \exp -i(\omega_q T_0/2 - \pi)) = 0$ which implies $\omega_q = (2n + 1)\omega_0$. This periodicity and the coherence of the process ensure therefore that the emitted spectrum will be composed only of peaks at frequencies ω_q that are odd multiples of the fundamental frequency, $\omega_q = q\omega_0 = (2n + 1)\omega_0$, as observed experimentally.

The fact that the atomic response is imposed (in amplitude and phase) by the fundamental laser field also has some important consequences in the spatial domain since the spatial coherence of the fundamental laser beam is partly transferred to the harmonic beam. Indeed, all atoms irradiated by the same field will emit the same radiation. This makes the process of high-order harmonic generation a (spatially) coherent process and to compute the total field emitted by a macroscopic medium, one therefore needs to sum the amplitudes of the fields emitted by the different atoms. The total emitted intensity is then obtained by squaring the total amplitude, and this coherence implies that the net efficiency of high-order harmonic generation is imposed not only by the single-atom response but also by collective effects such as phase matching [12.25] and reabsorption of the harmonics [12.42,43]. These collective effects are only briefly mentioned here for the sake of simplicity.

The coherence of high-order harmonic generation implies that in order to maximize the total number of XUV photons emitted in a given direction, phase matching should be obtained. In other words, the radiation emitted by several atoms must interfere constructively (Fig. 12.6). This is well known in standard nonlinear optics where the phase-matching condition states that all radiation emitted in different part of the medium must exit the medium with the same phase to interfere constructively. This implies that

$$\mathbf{\Delta K} = 0 \qquad (12.12)$$

for perfect phase matching, or

$$\mathbf{\Delta K L_{med}} < \pi \qquad (12.13)$$

for approximate phase matching in a medium of length L_{med}, where $\mathbf{\Delta K} = \mathbf{k_q} - q\mathbf{k_0}$ being the wave-vector mismatch between the fundamental wave (of wave-vector $\mathbf{k_0}$) and the harmonic wave (wave-vector $\mathbf{k_q}$).

Similar notation can be extrapolated for HHG, provided that one considers that the phase of the emitted radiation also depends on the intensity of the fundamental beam at the place of emission via $\varphi_q = \alpha_q I$. One therefore needs to include an additional wave-vector $\mathbf{k_I} = -\nabla \alpha I$ to obtain the generalized phase-matching condition [12.25]:

$$\mathbf{\Delta K} = \mathbf{k_q} + \mathbf{k_I} - q\mathbf{k_0} = \mathbf{0}. \qquad (12.14)$$

Once phase matching is achieved, the total emitted field should scale as the number of irradiated atoms (proportional to the gas pressure) and the number of XUV photons emitted in a given direction should scale with the square

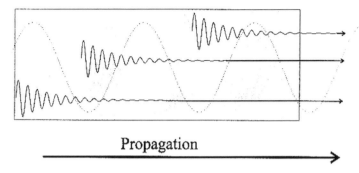

Propagation

Fig. 12.6. Phase matching and absorption effect in HHG

of the pressure. This implies that optimizing high-order harmonic generation requires use of high pressures. However, high-order harmonics can also be reabsorbed by the emitting medium and this absorption increases with the gas pressure. When phase matching is achieved and absorption becomes important, the number of emitted XUV photons gets roughly independent of the gas pressure and optimization of the single-atom response, and increasing the emitting volume allows the optimization of the high-order harmonic generation efficiency [12.42].

12.4 High-Order Harmonic Generation as an Attosecond Pulse Source

The typical shape of harmonic spectra (constituted on peaks of constant amplitude and spaced by $2\hbar\omega_0$) associated with the coherence of this XUV radiation indicates that this emission is the result of a periodic light emission with a periodicity of $T_0/2$. Furthermore, for an event to be periodic, its natural time scale needs to be smaller than the period. This implies that the time of emission of these XUV photons occurs at well-defined times within each half optical cycle of the fundamental. For the typical case of Ti: Sapphire fundamental laser beams (centered at 800 nm, $T_0 = 2.66$ fs) the typical time scale of the XUV emission is therefore well defined compared to 1.33 fs and is in the attosecond range.

As will be seen, the high-order harmonic generation can indeed be seen as the emission of a train of attosecond pulses (after spectral filtering). In the spectral domain, this implies a well-defined phase relation all the frequencies, which is not always clearly visible in calculations (this phase relation will be discussed at the end of this chapter). In the following, we will mainly consider this problem in the time domain which turns out to be much simpler to use to figure out how high-order harmonic generation can be seen as an emission of a train of attosecond pulses and how this emission can be confined to generate a single isolated attosecond pulse.

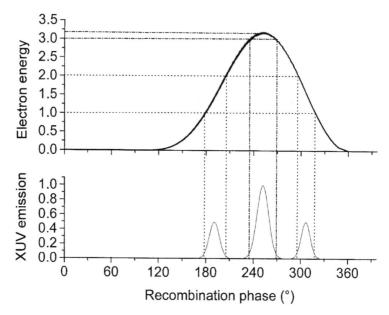

Fig. 12.7. (top) Energy (in U_p units) of the recolliding electron as a function of the recollision time, t_r; (bottom) schematic of the corresponding XUV emission (see text) after selection of the cutoff harmonics (peak centered at around $t_r = 0.7T_0$, i.e., $\omega t_r = 252°$) or of some plateau harmonics (peaks around $t_r = 0.53T_0$, i.e., $\omega t_r = 192°$ and $t_r = 0.85T_0$, i.e., $\omega t_r = 306°$). (In both cases, only electron released for $0 < t_i < T_0/2$ are considered)

According to the quasiclassical model of HHG, the energy of the XUV photon emitted during recollision is simply defined by the fundamental field intensity and by the phase, $\omega_0 t_i$ (or the "birth time," t_i), at which the electron is released in the continuum. Equivalently, one can draw as a function of the recombination time, t_r, the energy of the emitted XUV photons as shown in Fig. 12.7.

In this figure, one can clearly see that the photons with the highest energy (cutoff photons with energy close to $I_p + 3.2U_p$) are emitted only around the time $t_r = 0.7T_0$. Using a spectral filter that transmit-only frequencies higher than $I_p + 3U_p$, only the photons emitted around $\omega t_r = 0.7T_0$ (modulo $T_0/2$) would be transmitted. The transmitted light would therefore consist of very short bursts of light (attosecond pulses) separated by $T_0/2$.

In Fig. 12.7, one can also see that for the photons with energy smaller than the cutoff frequency (for instance, $I_p + U_p < \hbar\omega_q < I_p + 2U_p$), there are two possible times of emission within each half optical cycle of the fundamental. These two times of emission correspond to two different electron trajectories (often referred to as the two quantum paths), which end at the time of recollision with the same kinetic energy. In the temporal domain, these two trajectories still give rise to light emission at well-defined times and after

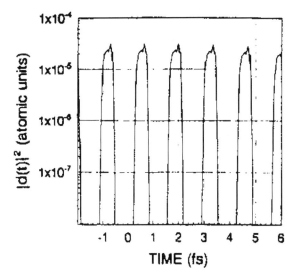

Fig. 12.8. High-order harmonic generation seen in the temporal domain after selection of the cutoff harmonics by spectral filtering. (from [12.23])

spectral filtering (for instance, transmission of light having frequencies between $I_p + U_p$ and $I_p + 2U_p$), each of them should lead to a subfemtosecond emission of XUV light. The total emission should therefore consist of a train of two sets of attosecond peaks with a periodicity of $T_0/2$ with two attosecond pulses every half optical cycle of the fundamental [12.22].

Figure 12.8 shows the results of semiclassical calculations performed under realistic conditions (harmonics generated in neon with a 25-fs pulse at a peak intensity of 6×10^{14} W/cm^2 resulting in a cutoff energy of 140 eV) and where the spectral filtering is performed by a 200 nm thick silver filter that transmits only photons with energy larger than 100 eV.

This basic approach for attosecond pulse generation is valid only for the single-atom response; it gets more complex when propagation effects or even distribution of peak intensities (due to the spatial profile of the fundamental laser beam in the gas medium) are taken into account. Surprisingly enough, it was found that under certain conditions phase matching may help to clean up the pulse train. In this case, the simple picture of the attosecond pulse train emission can still be valid after propagation in a macroscopic gas medium, as was simulated in [12.22].

When the plateau harmonics are considered in the spectral domain, the phase of the harmonics changes a lot from one to the next [12.22]. This could imply that no phase relation exists between neighbor harmonics. However, this arises only because in quantum single-atom response simulations, the two quantum paths are considered simultaneously and the interference between them leads to an almost chaotic phase [12.40,44]. When these two quantum paths are considered separately, a clear phase relation between neighbor harmonics appears as requested to generate a train of attosecond pulses [12.22]. The artificial separation of these two paths is justified [12.45] because of their very different emission characteristics, which lead, after propagation in a macroscopic medium, to very different spatial characteristics (generally a well-centered beam for the short quantum path and a much larger one, or even an annular beam, for the long quantum path) [12.39]. Indeed, phase matching itself can fully destroy one of the quantum path emission while optimizing the emission arising from the other. This was clearly observed in [12.22], where the temporal shape of the harmonic emission was obtained by quantum calculations associated with a propagation code. Therefore, for plateau harmonics, a proper spectral filtering associated with specific interaction geometry in a macroscopic medium can lead to emission of a train of attosecond pulses.

12.4.1 Emission of an Isolated Attosecond Pulse

The basic idea to extract a single attosecond pulse out of this pulse train is simply to prevent the harmonic emission during most of the fundamental pulse and to confine the emission to a time shorter than half an optical cycle of the fundamental. Preventing the harmonic emission is relatively easy; several techniques have been proposed.

The simplest one is to use an extremely short fundamental pulse (5–7 fs) to generate the harmonics. Simply because U_p changes with time, the highest harmonics with energy $I_p + 3.2U_p^{\mathrm{max}}$ (where U_p^{max} is the maximum ponderomotive energy) can only be emitted at a very well-defined time [12.46]. This could lead to the emission of a single attosecond pulse, as was recently claimed [12.47], although the harmonic selection should be very careful. Indeed, even without shot-to-shot intensity fluctuations, the cutoff position can change a lot because of changes in the absolute phase of the fundamental pulse (even with the same envelope, two pulses oscillating as either a cosine or a sine would lead to very different harmonic spectra with such short pulses) [12.48,49]. The price to pay for the simplicity of the technique is therefore that it can only be used with intense extremely short pulses, only for the cutoff harmonics and only after the rejection of most of the laser shots.

The second technique [12.23,24] relies on the extreme sensitivity of high-order harmonic generation on the ellipticity of the fundamental pulse [12.50–53]. Indeed, as soon as this pulse is slightly elliptically polarized, high-order harmonic generation is strongly suppressed. As illustrated in Fig. 12.9, modulating temporally the ellipticity of the fundamental pulse is a way to confine high-order harmonic generation in a well-defined temporal gate.

The reason of the strong influence of the fundamental polarization on the high-order harmonic generation efficiency can be understood with the semiclassical model and is illustrated in Fig. 12.10. A slight ellipticity of the fundamental does not strongly affect the ionization process (step 1 of HHG), but when the free electron is accelerated by an elliptic field, it evolves in the

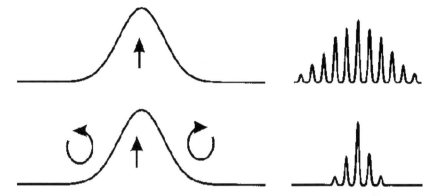

Fig. 12.9. Temporal profile and polarization state of the fundamental pulse (left) and the corresponding harmonic emission with (top) a linearly polarized fundamental pulse (bottom) a fundamental pulse where polarization evolves temporally and is linear around the maximum of the pulse

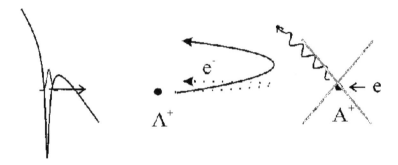

Fig. 12.10. The three-steps process of high-order harmonic generation in an elliptically polarized fundamental field

continuum on a curve that will always drive it away from the parent ion. Therefore, the probability of recollision is strongly reduced, as is the emission of XUV photons.

This sensitivity of high-order harmonic generation on the ellipticity of the fundamental pulse was studied in [12.50–53] and for very high-order harmonics the emission efficiency is typically reduced by a factor of 2 for an ellipticity of 10% (this ellipticity being defined as the ratio between the fields Ex and Ey oscillating with a $\pi/2$ dephasing). Modulating the ellipticity of the fundamental field in such a way that it remains smaller than 0.1 for less than one optical period of the fundamental field would therefore confine the XUV photon emission to less than an optical cycle of the fundamental. After proper spectral filtering, this implies the emission of a single XUV attosecond pulse as was theoretically observed in the first paper proposing this technique [12.23] and recently corroborated by calculations [12.54].

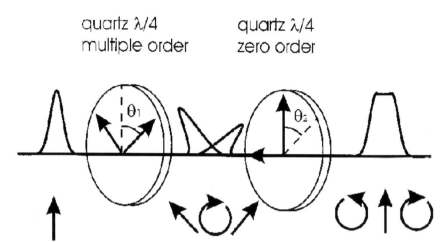

quartz λ/4
multiple order

quartz λ/4
zero order

Fig. 12.11. Experimental setup for modulating the polarization state of a short pulse (top) and the corresponding polarization of the pulse (bottom)

Several methods have been proposed to modulate temporally the ellipticity of the fundamental pulse [12.23,24,54,55]. A first technique derived from [12.23,24] was experimentally investigated in [12.56]. It demonstrated that a temporal gate for harmonic generation could be obtained. Nevertheless, the method was found impractical to vary the width and position of the gate relative to the pulse envelope. Essentially, the experiment resulted in a spectral narrowing of harmonics compatible with a reduction of the frequency chirp induced by the intensity-dependent intrinsic atomic phase [12.38,39]. The two techniques proposed in [12.54] and [12.55], respectively, were recently tested experimentally for high-order harmonic generation. Here, the width and position of the gate relative to the pulse envelope could be varied. In contrast to [12.56], the contribution of the frequency chirp was overcome and a spectral broadening of harmonics consistent with an effective confinement of the harmonic emission was observed [12.57,58].

This technique of polarization control [12.55] (Fig. 12.11) relies on the use of birefringent plates, which naturally have two different axes along two perpendicular directions. When a linearly polarized pulse is sent through this plate with an initial polarization at 45° of the plate axis, two delayed and perpendicularly polarized pulses exit the plate. Choosing the thickness of the plate allows us to control the delay between the two pulses and the phase difference between these pulses. For instance, a 1-mm-thick quartz plate induces a delay of 30 fs when a pulse centered at 800 nm crosses it. When a 30-fs pulse crosses this plate, the output pulse is first linearly polarized along the fast axis and at the end of the pulse the polarization is also linear along the slow axis. Between these two extremes, there is a time interval when the two

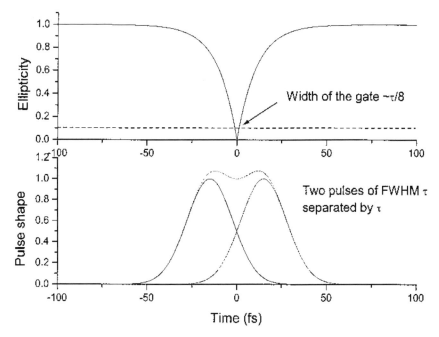

Fig. 12.12. Time-dependent ellipticity and resulting pulse shape after sending a short pulse through a multiple-order quarter wave plate (with thickness inducing a delay equals s to the initial pulse duration) and a zero order quarter wave plate

pulses have the same amplitude and the dephasing is chosen to be $\pi/2$: the total field is then circularly polarized.

Simply using a 1-mm-thick quartz quarter wave plate to transmit a 30-fs pulse makes it possible therefore to create a pulse having a polarization that evolves from linear to circular and back to linear. An additional zero order quarter wave plate can then change a circular polarization in linear and a linear polarization in circular. Using a multiple-order wave plate associated with a zero-order wave plate allows us therefore to change a linearly polarized pulse in a pulse with a temporally modulated polarization suitable for confining high-order harmonic generation.

The evolution of the ellipticity of the pulse (shown in Fig. 12.12) depends only on the initial pulse duration and the time during which the ellipticity is smaller than 10% is set to $\tau/8$ where τ is the initial pulse duration. If one considers that the efficiency of high-order harmonic generation is reduced by 50% for an ellipticity of 10% [12.50], this polarization modulation should create a temporal gate of width $\tau/8$ where high-order harmonic generation is confined. Using 15-fs pulses would therefore makes it possible to confine the high-order harmonic generation to less than one optical cycle and therefore to obtain an isolated attosecond pulse.

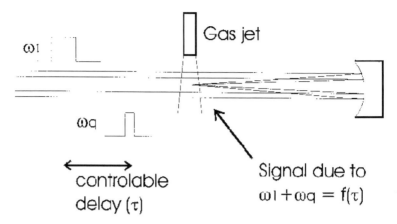

Fig. 12.13. Typical setup for cross correlation of a harmonic pulse with an infrared pulse

Furthermore, with this simple setup, one can tune the width of the gate simply by rotating the zero-order wave plate. This rotation does not affect the central part of the pulse (a circular polarization is changed in linear for all angles, θ_2, of the zero-order wave plate) but changes the maximum ellipticity. Indeed, when the orientation of the second plate axis is identical to the one of the first plate ($\theta_2 = 0$), the output field is always linearly polarized (gate of infinite width). Such an experiment was performed with a 30-fs input pulse and, although no temporal measurement was performed, the spectral broadening due to the temporal gating was consistent with a harmonic emission confined to ~ 5 fs [12.57].

12.5 Techniques for Measurement of Attosecond Pulses

Although several techniques exist to generate isolated subfemtosecond pulses, the main problem lies in the measurement of these pulses. Three possible techniques will be outlined in this chapter: the cross correlation, laser streaking and autocorrelation. Other techniques are now appearing [12.59–61] and some have even already led to the observation of "attosecond localization of light" [12.59], but the complexity of these techniques or of the full analysis of the results puts their presentation beyond the scope of this chapter.

12.5.1 Cross Correlation

This technique relies on the observation of photo-electrons created by ionization of atoms in the combined field of an XUV pulse and an IR pulse (Fig. 12.13).

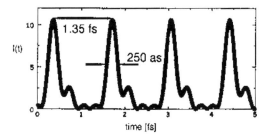

Fig. 12.14. Attosecond pulse train reconstructed by measuring the relative phase of several harmonics (from [12.64])

When atoms are ionized by an XUV harmonic pulse, the energies of the photo-electrons, E_c, are given by the energy of the XUV photons, $\hbar\omega_q$, minus the ionization potential I_p ($E_c = \hbar\omega_q - I_p$). The photo-electrons spectra consist therefore in peaks separated by $2\hbar\omega_0$. When this ionization takes place in presence of a weak IR field (with photon energy $\hbar\omega_0$), the atom can absorb (or emit) extra IR photons during ionization, and this extra energy is given to the photo-electron. The photo-electron spectra then consist in peaks separated by $2\hbar\omega_0$ and some extra peaks in between called side bands [12.62]. This extra absorption (or emission) of an IR photon can only occur if the XUV and IR fields are temporally overlapped (if the IR photon arrives after the XUV ionization, the photo-electron energy is not modified because a free electron cannot absorb photons). The amplitude of the side bands changes with the temporal overlap of the XUV pulse and the dressing IR pulse. Recording the amplitude of the side bands as a function of the delay between the IR and XUV pulses provides the cross correlation between the two pulses. The resolution of the cross correlation is limited by the accuracy with which one knows the IR pulse shape and can be on the order of a few fs.

This technique can be extended to multiphoton IR absorption (which results in a shift of the ionization potential [12.63]). It can also be used with interferometric stability (as compared to the fundamental wavelength) to observe the amplitude modulation of side bands as a function of the delay between the XUV pulse and the IR pulse. This experiment was recently performed and led to the observation of a modulation of the side band amplitude when the delay was changed by a fraction of the fundamental wavelength. This shows that the XUV pulse positions are well locked to the phase of the fundamental laser, as theoretically expected. Recording this evolution of the side band amplitude for several harmonics also allowed the authors to estimate the relative dephasing between neighbor harmonics. Under the assumption that all the frequencies of a given harmonic have the same phase, the temporal shape of the XUV pulse was estimated (Fig. 12.14) and the authors confirm that high-order harmonic generation consists trains of attosecond pulses [12.64].

With these cross-correlation techniques, one needs to resolve the side bands and therefore to observe well-defined harmonics. This prevents its use for isolated attosecond pulses for which the XUV spectra should be continuous.

12.5.2 Laser Streaking

A second technique for characterizing the XUV pulse relies on a similar principle except that many IR photons are absorbed instead of one. The basic principle is then that the XUV photon triggers the ionization in the presence of a strong IR field (the deflecting field). According to classical calculations, the effect of the IR field on the photoelectron depends on the exact time at which the electron is released (t_i) and can affect it through its energy or the direction at which the electron is detected. When a far infrared intense field (for instance, the laser field from a CO_2 laser) is used as the deflecting field, it can provide such a large kinetic energy to the photo-electrons that the initial kinetic energy is negligible and the photo-electron spectra is only determined by the time of ionization [12.65]. This technique can then have a subfemtosecond resolution. However, it is very difficult to synchronize an intense far IR laser with an XUV pulse, preventing to use this technique as a routine.

An alternative laser streaking technique was recently implemented in a clever way that allowed the authors to use directly the fundamental IR field to deflect photo-electrons released by a short XUV pulse. The initial kinetic energy of the photons released by the XUV field was then comparable to the energy provided by the deflecting field and the final energy of the photo-electron depends on both the XUV energy and the dressing IR field. A careful choice of the studied system still allowed the authors to get a very good temporal resolution.

The basic principle of the technique, outlined in Fig. 12.15, is that an XUV pulse ionizes atoms in the presence of an intense IR pulse (both linearly polarized) and photo-electrons are detected in a well-defined direction perpendicular to the laser polarization.

In this experiment, the electron time of flight is measured and provides the electron velocity and its kinetic energy. When ionization is triggered by the XUV pulse only, the photo-electron energy distribution corresponds to the XUV spectrum. When ionization takes place in the presence of the IR pulse that deflects the electrons perpendicularly to the time-of-flight axis (V_2 in Fig. 12.15), the net velocity of the electrons along the time-of-flight axis appears smaller than without the IR field. According to the simple man model, this extra transverse velocity

$$V_2 = qE(t_i)\sin(\omega_0 t_i)/m\omega_0 \qquad (12.15)$$

depends on the exact time of ionization and changes within one quarter of the optical cycle of the fundamental. When the delay is averaged over all phases, only the infra-red pulse envelope, $E(t)$, will determine the shift and spread in

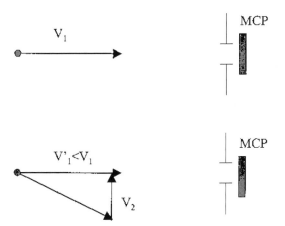

Fig. 12.15. Schematic setup for subcycle laser streaking

the flight velocity of the photo-electrons. Changes in this energy as a function of the XUV and IR pulse delays allows us to estimate the XUV pulse duration and an XUV duration smaller than 2.66 fs was obtained [12.66] by generating the harmonics and probing them with a 7-fs IR pulse. The accuracy of the method is then limited by the measure of the probe pulse duration.

When the IR-XUV delay is controlled with an interferometric accuracy [12.47], the velocity V_2 changes within half an optical cycle of the fundamental pulse. It is therefore very well suited to measure XUV pulses shorter than half of the optical cycle of the deflecting pulse. However, the measurement cannot discriminate between the measurement of a single isolated pulse or two (or more) pulses separated by half an optical cycle of the fundamental pulse and can only be used once it is proven that a single attosecond pulse is generated. It is therefore clear that the method offers a subfemtosecond accuracy and enables the measure of ultrashort XUV pulse duration shorter than half an optical cycle of the fundamental.

12.5.3 Autocorrelation

In the visible domain, the standard technique to measure ultrashort pulses is autocorrelation. It relies on splitting a pulse in two identical pulses, recombining them after a control of their relative delay and measuring the efficiency of a nonlinear process as a function of this delay (see Chapter 7).

While an XUV autocorrelation was already performed with low-order harmonics [12.67,68], several problems remain to extend this technique to very high-order harmonics. These problems are the observation of an XUV-induced nonlinear process and the splitting/recombination and delay control between XUV pulses.

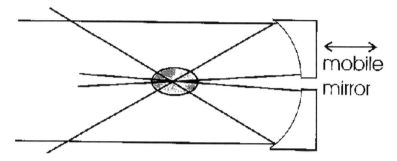

Fig. 12.16. A nondispersive quasi-interferometric autocorrelator

12.5.4 XUV-induced Nonlinear Processes

According to simulations and measurement, the achievable intensities in the XUV domain are very high with high-order harmonic generation. Indeed 10^9 photons per pulse, focused on a spot of few microns [12.69] and associated with a duration of 10 fs would lead to intensities higher than 10^{10} W/cm^2, which are sufficient to observe a nonlinear process such as two photon ionization of atoms [12.67,68,70,71].

12.5.5 Splitting, Delay Control and Recombination of Attosecond Pulses

In order to perform an autocorrelation, one also needs to spatially recombine two identical pulses while controlling the delay between them. So far, two alternatives have been pursued:

- generation of two identical XUV pulses using two identical fundamental beams and controlling their relative delay [12.67,68].
- generation of a single XUV pulse and subsequent splitting and recombination [12.72,73]

Generating two independent XUV sources has the great advantage of simplicity for the delay control because it only requires to control the relative delay between two identical infrared pulses. The main drawback is that the two harmonic sources need to be identical, which is quite difficult to achieve. This technique already gave promising results and allowed several teams to study the intercoherence [12.41,74,75] of two XUV sources or even to perform the autocorrelation of low-order harmonic pulses [12.67,68].

Splitting a single harmonic beam in two and recombining them after controlling their relative delay remains difficult mainly because the XUV light is highly absorbed by the optics, and very few optics, such as beam splitters, have been developed so far. This limits the number of optics to the minimum and prevents the use of a standard Michelson geometry. However, alternative techniques are under development and it was recently shown that quasi-interferometric autocorrelation could be obtained (in the infrared) by using an interferometer relying on the wave front division rather than on the amplitude division [12.73].

Indeed, focusing an XUV beam with a spherical mirror gets a single focused beam, but simply by cutting the mirror in two one can get a splitting of the beam and recombine them at focus (Fig. 12.16). Moving one of the mirrors then allows the control of the relative delay between the two pulses for small delays [12.73]; it does not change the spatial overlap of the two beams. This nondispersive technique has the advantage of using only a single XUV source and a single optic thereby enhancing the throughput, but it has not yet led to any autocorrelation in the XUV domain. This technique is also usable to perform pump probe experiments with attosecond pulses with an attosecond resolution.

An alternative technique was recently developed for a dispersionless XUV autocorrelator and offers an attosecond resolution. Although it requires several reflections, this technique has the great advantage to allow a control of the spectral content of the studied harmonic pulse [12.72].

12.6 Applications of Attosecond Pulses

Because attophysics is a new domain, the applications of attosecond pulses are only starting to emerge, and they mainly concern the ultrafast evolution of electronic motion [12.43,47,76,77]. For instance, XUV attosecond pulses seem particularly suitable for time-resolved inner-shell spectroscopy. Typically, when an inner-shell electron is kicked out, the resulting vacancy is refilled by an electron from an outer shell within a time scale ranging from a few femtosecond to hundreds of attoseconds, depending on the binding energy of the released electron. The energy released by the inner shell transition can liberate a second, so-called Auger electron. A method derived from subcycle laser streaking has been proposed to measure the duration of such an Auger transition [12.43,47,77]. It was also suggested that attosecond pulses may time resolve the process of electron valency, for instance, in H_2^+ and in benzene structures [12.78].

The motion of heavier particles such as atoms in molecules is usually accessible with femtosecond pulses. However, highly excited states are still poorly known because of their short lifetimes and fast evolution. Also, even if the hydrogen molecule is perfectly known on a theoretical basis, the temporal evolution is still inaccessible experimentally because of its very fast evolution [12.79]. This is also the case for most of the hydrogenated molecules in which the proton motions are very fast.

Here, we present a possible application of attosecond pulses for studying the vibrational dynamic of the H_2^+ molecule (Fig. 12.17), which has a vibrational period on the order of 14 fs. Although this molecular ion is perfectly well

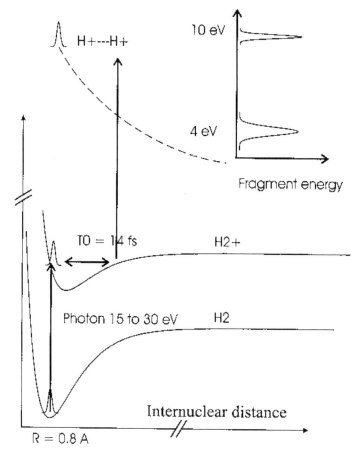

Fig. 12.17. Schematic of a pump probe experiment using attosecond pulses to image vibrational wave functions in H_2^+

known in theory, its quick evolution has so far put it beyond our experimental capabilities for time-resolved studies.

For this experiment, two XUV pulses need to be synchronized and focused onto the same spot; this can be performed with the dispersionless autocorrelator. The first pump pulse would induce a transition from the H_2 ground state to the H_2^+ ground state by ionizing the hydrogen molecule. In order to selectively excite this transition, the XUV pulse spectra can range from 15.5 to 30 eV, which can be achieved by generating harmonics with a Ti:Sapphire laser in Xenon or Krypton (cutoff close to the 19th harmonic, which has an energy of 29.5 eV) and spectrally filtering with an aluminium filter (transmission above ~ 15 eV). This attosecond pump pulse would create a vibrational wave packet in H_2^+, which can evolve from the initial internuclear distance of ~ 0.8 to 2–3 Å in approximately 7 fs. A second pulse (probe) could then ionize H_2^+ and thereby create two protons. These two protons then repel each other via Coulomb repulsion, and this Coulomb energy is transferred to the protons as kinetic energy (Coulomb explosion). A measurement of the kinetic energy of the protons then provides the internuclear separation of H_2^+ at the time of second ionization and the distribution of kinetic energies provides an image of the distribution of internuclear distances at that time, i.e., an image of the vibrational wave packet of H_2^+ [12.79]. Because the kinetic energy of the fragments change with the internuclear distance (the energy of each fragment being Ec (eV) $= 7.7/R$ where R is the internuclear distance in Å) and should peak around 10 eV at $t = 0$ ($R_0 \sim 0.77$ Å) and 3.8 eV at $t = 7$ fs, following the kinetic energy distribution of the fragments should allow us to follow the evolution of one of the fastest molecular vibrational wave packets. To get the full image of the wave packet, the probe pulse should have a spectrum extending from 35 to 15 eV and should therefore be slightly broader than the first pump pulse but even with two identical pulses (ranging from 15.5 to 30 eV), one should be able to image the wave packet between 1 Å and large internuclear distances.

12.7 Conclusion

This paper presents an overview of the main basic aspects of high-order harmonic generation and how this process can be used to generate attosecond pulse trains or an isolated attosecond pulse. It also introduces few basic techniques for the measurement of XUV subfemtosecond pulses and some possible applications of these ultrashort XUV pulses.

Acknowledgments

Part of the work presented here was supported by the EC in the frame of the ATTO network (contract n°HPRN-2000-00133). We also acknowledge the "Région Aquitaine" for its partial support.

References

[12.1]   D.H. Sutter, G. Steinmeyer, L. Gallmann, N. Matuschek, F. Morier-Genoud, U. Keller, V. Scheuer, G. Angelow, T. Tschudi: *Opt. Lett.* **24**, 631 (1999)

[12.2]   U. Morgner, F.X. Kärtner, S.H. Cho, Y. Chen, H.A. Haus, J.G. Fujimoto, E.P. Ippen, V. Scheurer, G. Angelow, T. Tschudi: *Opt. Lett.* **23**, 411 (1999)

[12.3]   Z. Cheng, F. Krausz, C. Spielmann: *Opt. Comm.* **201**, 145 (2002)

[12.4]   M. Nisoli, S. De Silvestri, O. Svelto: *Appl. Phys. Lett.* **68**, 2793 (1996)

[12.5]   S. Sartania, Z. Cheng, M. Lenzner, G. Tempea, C. Spielmann, F. Krausz: *Opt. Lett.* **22** 1562 (1997)

[12.6]   A.E. Kaplan: *Phys. Rev. Lett.* **73**, 1243 (1994)

[12.7]   S.E. Harris, A.V. Sokolov: *Phys. Rev. Lett.* **81**, 2894 (1998) '

[12.8]   A. Nazarkin, G. Korn, T. Elsaesser: *Opt. Comm.* **203**, 403 (2002)

[12.9]   A.V. Sokolov, D.R. Walker, D.D. Yavuz, G.Y. Yin, S.E. Harris: *Phys. Rev. Lett.* **85**, 562 (2000)

[12.10] A. Nazarkin, G. Korn, M. Wittmann, T. Elsaesser: *Phys. Rev. Lett.* **83**, 2560 (1999)

[12.11] M. Wittmann, A. Nazarkin, G. Korn: *Opt. Lett.* **26**, 298 (2001)

[12.12] N. Zhavoronkov, G. Korn: *Phys. Rev. Lett.* **88**, 203901 (2002)

[12.13] A. McPherson, G. Gibson, H. Jara, U. Johann, T.S. Luk, I. McIntyre, K. Boyer, C.K. Rhodes: *J. Opt. Soc. Am. B* **4**, 595 (1987)

[12.14] M. Ferray, A. L'Huillier, K.F. Li, L.A. Lompré, G. Mainfray, C. Manus: *J. Phys. B* **21**, L31 (1988)

[12.15] S. Gladkov, N. Koroteev: *Sov. Phys. Usp.* **33**, 554 (1990)

[12.16] G. Farkas, C. Toth: *Phys. Lett. A* **168**, 447 (1992)

[12.17] S.E. Harris, J.J. Macklin, T.W. Hansch: *Opt. Comm.* **100**, 487 (1993)

[12.18] P.B. Corkum: *Phys. Rev. Lett.* **71**, 1994 (1993)

[12.19] K.J. Schafer, B. Yang, L.F. DiMauro, K.C. Kulander: *Phys. Rev. Lett.* **70**, 1599 (1993)

[12.20] M. Lewenstein, P. Balcou, M. Yu. Ivanov, A. L'Huillier, P.B. Corkum, *Phys. Rev. A* **49**, 2117 (1994)

[12.21] W. Becker, S. Long, J.K. McIver: *Phys. Rev. A* **50**, 1540 (1994)

[12.22] P. Antoine, A. L'Huillier, M. Lewenstein: *Phys. Rev. Lett.* **77**, 1234 (1996)

[12.23] P.B. Corkum, N.H. Burnett, M.Y. Ivanov: *Opt. Lett.* **19**, 1870 (1994)

[12.24] M.Y. Ivanov, P.B. Corkum, T. Zuo, A. Bandrauk: *Phys. Rev. Lett.* **74**, 2933 (1995)

[12.25] P. Balcou, P. Salières, A. L'Huillier, M. Lewenstein: *Phys. Rev. A* **55**, 3204 (1995)

[12.26] C. Altucci, T. Starczewski, E. Mével, C.G. Wahlström, B. Carré, A. L'Huillier: *J. Opt. Soc. Am. B* **13**, 148 (1996)

[12.27] J.W.G. Tisch, R.A. Smith, J.E. Muffet, M. Ciarroca, J.P. Marangos, M. H.R. Hutchinson: *Phys. Rev A* **49**, R28 (1994)

[12.28] P. Saliéres, T. Ditmire, K.S. Budil, M.D. Perry, A. L'Huillier: *J. Phys. B* **27**, L217 (1994)

[12.29] J. Peatross, D.D. Meyerhofer: *Phys. Rev. A* **51**, R906 (1995)

[12.30] K.C. Kulander, K.J. Schafer, J.L. Krause: *Atoms in Intense Radiation Fields*, Academic Press, New York (1992).

[12.31] J.H. Eberly, Q. Su, J. Javanainen: *Phys. Rev. Lett.* **62**, 881 (1989)

[12.32] M.V. Ammosov, M.B. Delone, V.P. Kraïnov: *Sov. Phys. JETP* **64**, 1191 (1986)

[12.33] T. Auguste, P. Monot, L.A. Lompré, G. Mainfray, C. Manus: *J. Phys. B* **25**, 4181 (1992)

[12.34] E. Mével, P. Breger, R. Trainham, G. Petite, P. Agostini, A. Migus, J.P. Chambarret, A. Antonetti: *Phys. Rev. Lett.* **70**, 406 (1993)

[12.35] J. L. Krause, K. Schafer, K.C. Kulander: *Phys. Rev. Lett.* **68**, 3535 (1992)

[12.36] K.C. Kulander, K.J. Schafer, J.L. Krause: *Super Intense Laser-Atom Physics* **316** of NATO Advanced Study Institute, Series B: Physics, Plenum, New York, (1993).

[12.37] A.L'Huillier, M. Lewenstein, P. Salières, P. Balcou, M. Yu. Ivanov, J. Larsson, C.G. Wahlström: *Phys. Rev. A* **48**, 3433 (1993).

[12.38] C.G. Wahlström, J. Larsson, A. Persson, T. Starczewski, S. Svanberg, P. Salières, P. Balcou, A.L'Huiller: *Phys. Rev. A* **48**, 4709 (1993)

[12.39] P. Salières, A. L'Huillier, and M. Lewenstein: Phys. Rev. Lett. 74, 3776 (1995)

[12.40] M. Lewenstein, P. Salères, A. L'Huillier: *Phys. Rev. A* **52**, 4747 (1995)

[12.41] R. Zerne, C. Altucci, M. Bellini, M. B. Gaarde, T. W. Hänsch, A. L'Huillier, C. Lynga, C.G. Wahlström, *Phys. Rev. Lett* **79**, 1006 (1997)

[12.42] E. Constant, D. Garzella, E. Mével, P. Breger, C. Dorrer, C. Le Blanc, F. Salin, P. Agostini: *Phys. Rev. Lett.* **82**, 1668 (1999)

[12.43] M. Schnürer, Z. Cheng, M. Hentschel, G. Tempea, P. Kálmán, T. Brabec, F. Krausz: *Phys. Rev. Lett.* **83**, 722 (1999)

[12.44] C. Kan, C.E. Capjack, R. Rankin, N.H. Burnett: *Phys. Rev. A* **52**, R4336 (1995)

[12.45] M.B. Gaarde, F. Salin, E. Constant, P. Balcou, K.J. Schafer, K.C. Kulander, A.L'Huillier: *Phys. Rev. A* **59**, 1367 (1999)

[12.46] I.P. Christov, M.M. Murnane, H.C. Kapteyn: *Phys. Rev. Lett.* **78**, 1251 (1997)

[12.47] M. Hentschel, R. Klenberger, C. Spielmann, G.A. Reider, N. Milosevic, T. Brabec, P. Corkum, U. Heinzmann, M. Drescher, F. Krausz, *Nature* **414**, 509 (2001)

[12.48] A.de Bohan, P. Antoine, D.B. Milošević, B. Piraux: *Phys. Rev. Lett.* **81**, 1837 (1998)

[12.49] G. Tempea, M. Geissler, T. Brabec: *J. Opt. Soc. Am. B* **16**, 669 (1999)

[12.50] K.S. Budil, P. Salières, A. L'Huillier, T. Ditmire, M.D. Perry: *Phys. Rev. A* **48**, R3437 (1993)

[12.51] P. Dietrich, N.H. Burnett, M. Ivanov, P.B. Corkum: *Phys. Rev. A* **50**, R3585 (1994)

[12.52] N.H. Burnett, C. Kan, P.B. Corkum: *Phys. Rev. A* **51**, R3418 (1995)

[12.53] P. Antoine, A. L'Huillier, M. Lewenstein, P. Salières, B. Carré: *Phys. Rev. A* **53**, 1725 (1996)

[12.54] V.T. Platonenko, V.V. Strelkov: *J. Opt. Soc Am. B* **16**, 435 (1999)

[12.55] E. Constant, Ph.D. thesis

[12.56] C. Altucci, C. Delfin, L. Ross, M. B. Gaarde, A. L'Huillier, I. Mercer, T. Starczewski, C.G. Wahlström: *Phys. Rev. A* **58**, 3934 (1998)

[12.57] O. Tcherbakoff, E. Mével, D. Descamps, J. Plumridge, E. Constant: *Phys. Rev. A* **68**, 043804 (2003)

[12.58] M. Kovacev, Y. Mairesse, E. Priori, H. Merdji, O. Tcherbakoff, P. Monchicourt, P. Breger, E. Mével, E. Constant, P. Salières, B. Carré, and P. Agostini: *European Physics Journal* **D26**, 79 (2003)

[12.59] N.A. Papadogiannis, B. Witzel, C. Kalpouzos, D. Charalambidis: *Phys. Rev. Lett.* **83**, 4289 (1999)

[12.60] C. Dorrer, E. Cormier, I.A. Walmsley, L.F. DiMauro: Private communication

[12.61] F. Quéré, J. Itatani, G.L. Yudin, P.B. Corkum: Private communication

[12.62] V. Véniard, R. Taieb, A. Maquet: *Phys. Rev. Lett.* **74**, 4161 (1995)

[12.63] E.S. Toma, H.G. Muller, P.M. Paul, P. Breger, M. Cheret, P. Agostini, C.le Blanc, G. Mulot, G. Cheriaux: *Phys. Rev. A* **62**, 061801 (R) (2000)

[12.64] P.M. Paul, E.S. Toma, P. Berger, G. Mullot, F. Augé, P. Balcou, H.G. Muller, P. Agostini: *Science* **292**, 1689 (2001)

[12.65] E. Constant, V.D. Taranukhin, A. Stolow, P.B. Corkum: *Phys. Rev. A* **56**, 3870 (1997)

[12.66] M. Drescher, M. Hentschel, R. Kienberger, G. Tempea, C. Spielmann, G. A. Reider, P.B. Corkum: *Science* **291**, 1923 (2001)

[12.67] Y. Kobayashi, T. Sekikawa, Y. Nabekawa, S. Watanabe: *Opt. Lett.* **23**, 64 (1998)

[12.68] D. Descamps, L. Roos, C. Delfin, A. L'Huillier, C.-G. Wahlström, *Phys. Rev. A* **64**, 031404 (R) (2001)

[12.69] L.Le Déroff, P. Salières, B. Carré: *Opt. Lett. 23*, 1544 (1998)

[12.70] D. Xenakis, O. Faucher, D. Charalambidis, C. Fotakis: *J. Phys. B* **29**, L457 (1996)

[12.71] E.J. McGuire: *Phys. Rev. A* **24**, 835 (1981)

[12.72] E. Goulielmakis, G. Nersisyan, N.A. Papadogiannis, D. Charalambidis, G.D. Tsakiris, K. Witte: *Appl. Phys. B* **74**, 197 (2002)

[12.73] E. Constant, E. Mével, V. Bagnoud, F. Salin: *J. Phys. IV France* **11**, Pr2-537 (2001)

[12.74] M. Bellini, C. Lynga, M.B. Gaarde, T.W. Hänsch, A. L'Huillier, C.G. Wahlström: *Phys. Rev. Lett.* **81**, 297 (1998)

[12.75] C. Lynga, M.B. Gaarde, C. Delfin, M. Bellini, T.W. Hänsch, A.L'Huillier, C.-G. Wahlström: *Phys. Rev. A* **60**, 4823 (1999)

[12.76] A.H. Zewail: *J. Phys. Chem. A* **104**, 5660 (2000)

[12.77] F. Krausz: *Optics & Photonics News* **13**, 62 (2002)

[12.78] A.H. Zewail: *Faraday Discuss. Chem. Soc.* **91**, 207 (1991)

[12.79] A.D. Bandrauk, S. Chelkowski: *Phys. Rev. Lett.* **87**, 273004-1 (2001)

Index

426   Index

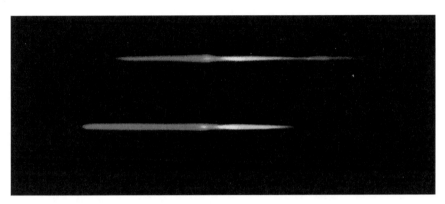

Fig. 8.38. Photograph of probe continuum as it appears on a white screen. *Top*: without excitation beam on sample. *Bottom*: with excitation beam on sample. (For experimental conditions, see text)

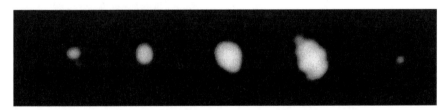

Fig. 8.42. Diffraction pattern observed on the screen in the setup shown in Fig. 8.41. *Left* to *right*: second order, first order, pulse 1, pulse 2, first order

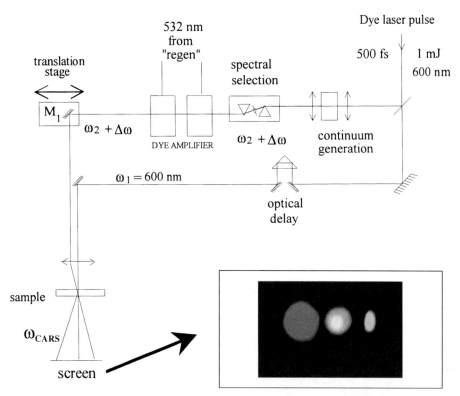

Fig. 8.39. Experimental setup used for demonstration of CARS signal generation. "regen": Nd³⁺:YAG regenerative amplifier